DNA Virus Replication

EDITED BY

Alan J. Cann

University of Leicester,
Leicester

OXFORD

UNIVERSITY PRESS

OXFORD

UNIVERSITY PRESS

Great Clarendon Street, Oxford OX2 6DP

Oxford University Press is a department of the University of Oxford
and furthers the University's aim of excellence in research, scholarship,
and education by publishing worldwide in

Oxford New York

Athens Auckland Bangkok Bogotá Buenos Aires Calcutta
Cape Town Chennai Dar es Salaam Delhi Florence Hong Kong Istanbul
Karachi Kuala Lumpur Madrid Melbourne Mexico City Mumbai
Nairobi Paris São Paulo Singapore Taipei Tokyo Toronto Warsaw

with associated companies in Berlin Ibadan

Oxford is a registered trade mark of Oxford University Press
in the UK and in certain other countries

Published in the United States
by Oxford University Press Inc., New York

A catalogue record for this book is available from the British Library

Library of Congress Cataloging in Publication Data
DNA virus replication / edited by Alan J. Cann.
(Frontiers in molecular biology ; 26)
Includes bibliographical references and index.
1. DNA viruses. 2. Viruses – Reproduction. 3. Viral proteins. I. Cann, Alan. II. Series.
QR394.5 .D625 2000   579.2′4–dc21   99-046601

ISBN 0 19 963713–X (Hbk)
ISBN 0 19 963712–1 (Pbk)

Typeset by Footnote Graphics, Warminster, Wilts

Printed by the Bath Press, Avon

Preface

DNA viruses, i.e. viruses with DNA genomes, have always been the most important model systems for eukaryotic DNA replication. Add to this the clinical significance of these human pathogens – 99% of the population of the world is infected with at least one of the viruses discussed in this volume (hepatitis B virus, Epstein–Barr virus, or herpes simplex virus) – and it is difficult to overstate the importance of this group. What is clearly not possible is to summarize the enormous research effort involving these diverse viruses in a single volume. I have attempted to circumvent this problem by concentrating on the theme of protein–protein interactions in DNA virus replication.

In the opening chapter, Michael Nassal discusses the replication of hepatitis B virus and amply illustrates some of the difficulties which have hampered development of experimental systems for this important pathogen. Although sometimes not regarded as a DNA virus in the same way as adenoviruses or herpesviruses, hepatitis B virus qualifies since the virus particles contain DNA and this is the infectious material. In Chapter two, Dennis McCance describes protein–protein interactions in the context of human papillomavirus replication and pathogenesis. Nigel Stow describes recent advances in understanding of herpes simplex virus DNA replication and Martin Rowe the Epstein–Barr virus proteins involved in cell transformation. In Chapters 5 and 6, Patrick Moore and his colleagues and John Sinclair describe pathogenic strategies used by two human herpesviruses, Kaposi's sarcoma-associated herpes virus and cytomegalovirus. The last two chapters by Alex Zantema and Alex van der Eb and Bill Wold and G. Chinnadurai discuss aspects of infection by adenoviruses—control of transcription by E1A proteins and regulation of apoptosis. Although this list may not do complete justice to the vitality of research in this area, I hope it serves to whet the appetite, and point the way through the complex literature in this field. For more details, readers may care to consult the 1232 references cited by the contributors to this volume!

My thanks go to David Hames (joint series editor) for his guidance in shaping this volume, and to the editorial staff at Oxford University Press. Most importantly, thanks must go to the contributors who were prepared to share their combined knowledge with the wider research community.

Leicester University   A. J. C
October 1999

Contents

3 Molecular interactions in herpes simplex virus DNA replication 66

NIGEL D. STOW

8. Adenovirus proteins that regulate apoptosis 200

WILLIAM S. M. WOLD AND G. CHINNADURAI

Contributors

YUAN CHANG
Department of Pathology, College of Physicians and Surgeons, Columbia University, 630 West 168 Street, New York, NY 10032, USA.

G. CHINNADURAI
Institute for Molecular Virology, St. Louis University School of Medicine, 3681 Park Avenue, St. Louis, MO 63104, USA.

J. EIKE FLOETTMANN
Department of Medicine, Tenovus Building, University of Wales College of Medicine, Cardiff CF4 4XX, UK.

SUKHANYA JAYACHANDRA
Department of Pathology, College of Physicians and Surgeons, Columbia University, 630 West 168 Street, New York, NY 10032, USA.

DENNIS J. MCCANCE
Department of Microbiology and Immunology, University of Rochester, Rochester, NY 14642, USA.

PATRICK S. MOORE
Department of Pathology, College of Physicians and Surgeons, Columbia University, 630 West 168 Street, New York, NY 10032, USA.

MICHAEL NASSAL
University Hospital, Department of Internal Medicine II/Molecular Biology, Hugstetter Strasse 55, D-79106, Freiburg, Germany.

MARTIN ROWE
Department of Medicine, Tenovus Building, University of Wales College of Medicine, Cardiff CF4 4XX, UK.

JOHN SINCLAIR
Department of Medicine, Level 5, University of Cambridge, Addenbrook's Hospital, Hills Road, Cambridge CB2 2QQ, UK.

NIGEL D. STOW
Medical Research Council Virology Unit, Church Street, Glasgow G11 5JR, UK.

ALEX J. VAN DER EB
Molecular Carcinogenesis Laboratory, Leiden University Medical Centre, Wassenaarseweg 72, PO Box 9503, 2333 AL Leiden, Netherlands.

WILLIAM S. M. WOLD
Department of Molecular Microbiology and Immunology, St. Louis University School of Medicine, 1402 S. Grand Boulevard, St. Louis, MO 63104, USA.

ALT ZANTEMA
Molecular Carcinogenesis Laboratory, Leiden University Medical Centre, Wassenaarseweg 72, PO Box 9503, 2333 AL Leiden, Netherlands.

Abbreviations

AcMNPV	*Autographa californica* nuclear polyhedrosis virus
Ad	adenovirus
AIDS	acquired immune deficiency syndrome
CaMV	cauliflower mosaic viruses
CBP	creb-binding protein
CMV	cytomegalovirus
CTL	cytotoxic T lymphocyte
DHBV	duck hepatitis B virus
DHFR	dihydrofolate reductase
EBNA	EBV nuclear antigen
EBNA-LP	EBV leader protein
EBV	Epstein–Barr virus
E. coli	*Escherichia coli*
EGF	epidermal growth factor
EGFR	epidermal growth factor receptor
GCR	G protein-coupled receptor
GFP	green fluorescent protein
GTFs	general transcription factors
HAT	histone acetyl transferase
HBV	hepatitis B virus
HCMV	human cytomegalovirus
HIV	human immunodeficiency virus
HPV	human papillomavirus
HSV	herpes simplex virus
IE	immediate early
KS	Kaposi's sarcoma
KSHV	Kaposi's sarcoma-associated herpes virus (HHV-8)
LATs	HSV latency associated transcripts
LMP	EBV latent membrane protein
MIEP	major immediate early promoter
mRNA	messenger RNA
NK	natural killer cell
ORF	open reading frame
PCR	polymerase chain reaction
PMNL	polymorphonuclear leucocytes
Rb	retinoblastoma
RT	reverse transcriptase
SDS	sodium dodecyl sulfate

SDS–PAGE	SDS–polyacrylamide gel electrophoresis
SV40	simian virus 40
TPA	12-*O*-tetradecanoyl phorbol-13-acetate
TS	thymidylate synthase
UV	ultraviolet
WHV	woodchuck hepatitis virus

1 | Macromolecular interactions in hepatitis B virus replication and particle assembly

MICHAEL NASSAL

1. Introduction

Hepatitis B virus (HBV) is the causative agent of acute and chronic hepatitis B in humans. Chronic infection in particular poses a serious problem to public health. According to the World Health Organization (WHO), about 5% of the world population are HBV carriers, with a greatly increased risk for developing liver cirrhosis and, eventually, primary liver cell carcinoma. It is estimated that more than one million deaths per year are due to liver disease associated with chronic HBV infection. While an effective prophylactic vaccination based on a recombinant virus envelope protein is available, there is as yet no generally effective therapy (1, 2). The only currently approved treatment is the high-dose systemic administration of interferon-α but virus elimination is achieved in less than one-third of the patients. Hence the search for novel targets to combat chronic HBV infection continues to be one driving force for efforts to understand in detail the molecular mechanisms underlying the viral infectious cycle.

Because of its many unique properties, this virus has turned out to be a similarly rewarding, though difficult, system to study fundamental aspects of virology and molecular biology. Two of the central problems are that HBV cannot feasibly be propagated in cell culture, and that as yet no system exists to biochemically study the authentic activity of its key replication enzyme, the P protein. None the less, research over the past years, mainly employing genetic approaches and the few available animal models, has revealed many unique features of how the virus replicates its tiny 3 kb genome (Fig. 1). Most (but not all) of the functions of the few viral gene products during the virus life cycle have been disclosed. Aided by powerful new methods to detect protein–protein and protein–nucleic acid interactions, for instance a variety of one-, two-, or three-hybrid systems (3, 4), emphasis is currently shifting towards

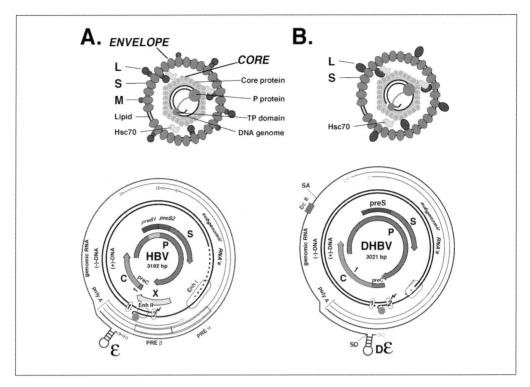

Fig. 1 Structural and genetic organization of the hepadnaviruses. (a) HBV. The envelope contains the surface proteins L, M, and S. The preS domain of L occurs in an inward and an outward topology. The irregularly shaped objects symbolize the chaperone Hsc70 that co-purifies with virions and S particles. The HBV genome inside the capsid is present as partially double-stranded circular DNA. The 5′end of the complete (–)DNA strand is covalently linked to the TP domain of P protein. In the lower panel, the heavy black lines represent the DNA genome with the four major open reading frames (ORFs) indicated in the centre. Boxes marked '1' and '2' symbolize the direct repeats DR1 and DR2, Enh I and Enh II indicate the two enhancer elements. The outer lines depict the genomic and subgenomic RNAs. Arrowheads indicate transcriptional start sites, the symbolic hairpin the encapsidation signal ε, and PREα and β the location of the post-transcriptional regulatory element (PRE) and on the preS/S mRNAs. (B) DHBV. The principal architecture of the virion and the genome organization are similar as in HBV. However, DHBV does not encode an M protein, and it has no homologue of the X gene. DεII denotes a region that in addition to the Dε element is required for packaging of the pgRNA. SD, splice donor; SA, splice acceptor.

elucidating the interactions between the virus components and cellular factors (5). Prominent examples are the dependence of P protein activity on chaperones and the nucleocytoplasmic trafficking of viral nucleic acids by as yet unidentified components of the host machinery. The new technologies have already substantially widened our knowledge but they have also produced some controversial and, for the time being, confusing results. After a brief summary over the complete infectious cycle, this review will survey the experimental systems that are currently available to study HBV, and focus on their individual contributions to our understanding of hepadnavirus genome replication and particle assembly. As will be evident there is as yet no single system that allows to address all aspects of HBV infection.

2. The HBV infectious cycle

In contrast to the other viruses described in this volume, HBV is not a DNA virus *stricto sensu*. The term hepadnaviruses, of which HBV is the prototype, is an acronym for *hepa*totropic *DNA* viruses, merely indicating that infectious extracellular virions contain DNA. This DNA, however, is produced by reverse transcription from an RNA template (6). On this fundamental level, hepadnaviruses are relatives of retroviruses. However, retrovirus virions contain two molecules of ssRNA, and their replication involves integration into the host genome as a provirus which is not the case for hepatitis B viruses. Hepadnaviruses, together with a number of plant viruses (e.g. cauliflower mosaic viruses, CaMV), have therefore been classified as pararetroviruses. A few years ago it appeared that additional fundamental differences existed between the two groups (7, 8). More recently, the borders became fuzzier again. Notable examples are the existence of splicing in pararetroviruses (9), e.g. in CaMV and duck hepatitis B virus (DHBV), but not HBV. On the retrovirus side especially the foamy viruses have features thought to be atypical of retroviruses, e.g. an additional internal rather than solely LTR-based promoters, and the generation of their reverse transcriptase (RT) as a separate entity rather than as a gag-pol fusion protein; their virions even appear to carry mostly DNA rather than RNA (10, 11). This diversity probably reflects that viruses, liberated from the necessity to faithfully replicate million base pair genomes like their hosts, exploit diverse replication mechanisms.

The next paragraphs will present a short summary of the hepadnavirus infectious cycle. Using this as a conceptual framework it should be easier to appreciate the possibilities, and the limitations, of the currently available experimental systems.

The structural and genetic organization of the type members of mammalian and avian hepadnaviruses, HBV and DHBV, is shown in Fig. 1. Major features of extracellular, infectious virions (12) are an outer envelope containing three, or two, respectively, differently sized but sequence-related surface proteins (L/large, M/middle, and S/small for HBV; L and S for DHBV); an icosahedrally symmetric inner capsid, or core particle, built from a single species of core protein; and a partially double-stranded, circular DNA molecule of about 3 kb with unusual properties: none of the strands is covalently closed, the (+)strand is incomplete, and the 5′ end of the complete (−)strand is covalently linked to the terminal protein (TP) domain of the P protein which is also an integral component of the nucleocapsid. There is evidence that also a cellular chaperone of the Hsp70 family, and possibly the small co-chaperone p23, complexed with P protein, are present in the particles (see Sections 4.1 and 3.3.1).

A characteristic of hepadnavirus genomes is their small size and compact organization. All nucleotides have coding function, about half of them in two different reading frames, and consequently all regulatory elements are overlappingly arranged with coding regions (Fig. 1). The various subgenomic and genomic transcripts are initiated at several internal promoters, and end after a single common polyadenylation signal. The principal translation products are the capsid and the P

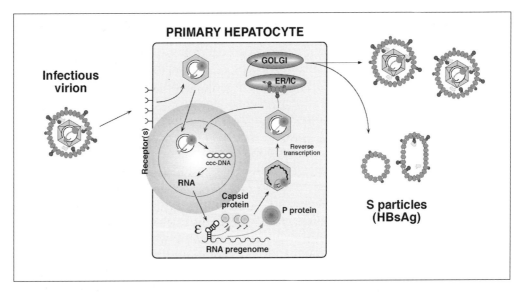

Fig. 2 The HBV infectious cycle. Virions infecting primary hepatocytes release nucleocapsids into the cytoplasm. These deliver the DNA genome to nucleus where it is converted into cccDNA to serve as transcriptional template. After nuclear export, all transcripts are translated; the RNA pregenome, in addition, interacts with its two gene products to form immature, RNA-containing nucleocapsids. Inside the capsids, the RNA is reverse transcribed into DNA by P protein. The DNA genome can either be re-delivered to the nucleus, or the matured nucleocapsids, via interaction with the surface proteins, bud into a pre-Golgi compartment of the secretory pathway and are exported as enveloped virions. Budding of the surface proteins without capsids yields empty spherical and filamentous S particles.

protein, encoded by the C and the P genes, the L and S surface proteins, specified by the preS/S and S genes, and the X protein, or HBx, encoded by the X ORF which has no equivalent in DHBV. Some of the transcripts have staggered 5′ ends which either include or exclude upstream initiator codons. This allows for the synthesis of additional, N terminally extended gene products like the precore protein, a secretory non-assembling variant of the core protein, and, in the mammalian viruses, of the M protein. In this way, the same protein modules are repeatedly used to generate products with new properties.

Figure 2 shows a simplified view of the hepadnavirus infectious cycle which incorporates the combined data obtained with the different experimental systems described below. Since further details may be found in previous reviews (7, 13, 14) only the major steps are highlighted here, except for a short summary on new results concerning the role of host cell factors during the uptake of DHBV. The current lack of knowledge on the cellular receptor(s) for human HBV is one of the central unresolved issues in hepatitis B research.

Infection is initiated by binding of the enveloped virion to a suitable cell, i.e. a primary hepatocyte from a susceptible host, a process mediated by the preS region of the large envelope protein. The narrow host-range and strict liver-tropism of the hepadnaviruses suggest that specific structures on the cell surface are recognized, but except for the DHBV data discussed below very little is known about this and the immediately following steps. Once released into the cytoplasm the capsid transports the partially double-stranded, P protein-linked genome to the nucleus, probably by an importin α/β-dependent mechanism (15). There it is converted into cccDNA which acts as template for the transcription of various subgenomic and genomic RNAs. RNA export from the nucleus involves a *cis*-acting transport signal called 'post-transcriptional regulatory element' or PRE (16–19) which also appears to contain binding sites for proteins mediating RNA stability (20). In the cytoplasm, all transcripts are used as mRNAs. One of them, the pregenomic or pgRNA, encodes both the capsid protein and the reverse transcriptase, the P protein. Specifically this transcript, together with P protein, is then packaged into new, immature capsids and inside the capsid reverse transcribed into the partially double-stranded DNA. Such mature capsids can either be redirected to the nucleus to amplify the pool of cccDNA (usually between 10 and 100 copies per cell), or they interact with the envelope proteins in the ER, or a later compartment of the secretory pathway, into which they bud as enveloped virions and are secreted by normal vesicular transport. Which of these routes is followed is regulated by the L protein (21) and again it is the preS domain of this protein that is required for capsid envelopment (22). The apparent contradiction of preS having essential functions on both sides of the membrane has been solved by the observation that the preS domains occur in two opposite topologies (23), both in HBV and in DHBV (24). During passage through the cellular secretory pathway about half of them remain on the cytosolic side, topologically equivalent to the virus interior, while the other half is translocated through the membrane of the corresponding compartment and hence ends up on the surface of the virion.

For HBV, little is known about what these preS domains recognize on the surface of susceptible cells. A variety of candidate proteins have been identified based on their ability to bind to the envelope proteins (25) but none of them has been shown to be a bona fide receptor allowing productive entry of the virus into the cell. The most comprehensive data so far have recently been collected about a duck protein that appears to play a critical role in DHBV infection, emphasizing the value of this animal model that allows to directly study infection in cell culture. This large glycoprotein protein, first termed gp180 (26) or p170 (27), is a membrane protein of the carboxypeptidase D family (28, 29). However, since only a small fraction of it was found at the cell surface its putative role in DHBV infection was met with skepticism. However, the enzyme can shuttle between the Golgi apparatus and the plasma membrane (30). Also, the ability of separately expressed wild-type and mutant preS domains to compete with DHBV infection in primary duck liver cells is strictly correlated with their ability to bind to gp180 *in vitro* (31), and a soluble form of gp180 as well as antibodies against gp180 can effectively interfere with DHBV infection (32). These data provide strong evidence for a critical role of gp180 in DHBV infection, and plausible models as to how the shuttling gp180 might be involved in internalization and/or intracellular trafficking of the virus can be invoked. However, a number of questions remain. One is that the protein displays a broad tissue distribution rather than liver specificity. Another is that expression of duck gp180 in the chicken hepatoma cell line LMH, which supports DHBV production once the viral genome has been introduced into it by artificial means, does not render the cells susceptible to DHBV infection; this would have been the classical *experimentum crucis* to prove that the molecule is really the receptor for the virus. The most likely explanation is that productive infection requires more than one cellular protein. Prominent examples for such a mechanism are the recently discovered HIV-1 co-receptors which, in addition to the classical CD4 molecule, are essential for infection (33). LMH cells may lack such a second (or more) essential molecule(s). Thus even if gp180 is only part of the story, its characterization should greatly improve the chances for identifying the missing component(s), e.g. by screening of duck liver cDNA libraries in LMH cells that are permanently transfected with gp180. Unfortunately, it is as yet not clear to what extent these DHBV data bear on the human virus system.

3. Experimental systems to study HBV replication

Understanding the complete life cycle of a pathogenic virus obviously requires an experimental infection system that should consist of both infectable cultured cells and an animal model for *in vivo* studies, including pathogenic aspects of infection. Ideally, this would be complemented by *in vitro* systems allowing individual steps of the life cycle to be analysed by biochemical means. Unfortunately, such a perfect model system is not currently available for HBV.

3.1 Animal models

One of the key properties of hepadnaviruses is their restricted host range. Only humans and chimpanzees are definitely susceptible to infection with HBV. For ethical reasons experimental infections are prohibited, or very much restricted, respectively, for these host species. A second severe limitation is that only primary hepatocytes but none of the known established cell lines can be efficiently infected by HBV. Several important aspects of the hepadnavirus infectious cycle have therefore been worked out using the few available animal systems (34, 35).

3.1.1 Established animal models: possibilities and limitations

In non-primate mammals, hepatitis B viruses have been found to be endemic in the North American woodchuck (*Marmota monax*; WHV), Beechey ground squirrels (*Spermophilus beecheyi*; GSAHV), and arctic ground squirrels (*Spermophilus parryii*; ASHV). Why these species, but apparently not others, are hosts for the correspond-ing hepadnaviruses remains enigmatic. The size and organization of their genomes is rather similar to that of HBV. Since ground squirrels are difficult to keep and do not reproduce well in captivity, woodchucks have become the prime mammalian animal for hepatitis B infection (36). However, woodchucks are quite large, hibernating and relatively slowly reproducing animals, so only few laboratories with appropriate animal facilities have access to them as experimental animals. On the other hand, chronic WHV infections are common in woodchucks and often lead to primary liver cell carcinoma. That this happens more frequently than with HBV infections in humans is probably due to a particular duplication of the N-myc gene in woodchucks that presents a favourite target for integration of the virus genome with subsequent up-regulation of N-Myc expression (37). None the less, the woodchuck/ WHV model is still the best available system for analysing the pathogenesis of chronic hepadnavirus infections, and it is increasingly used for studying antiviral therapies (38). The major obstacles towards an analysis on the molecular level are the lack of inbred animals, and until recently, of sequence information about immunologically relevant host molecules. With the recent cloning of several cytokine and other important genes (39), many pathogenic aspects of hepadnavirus infections should now be addressable. Notably, however, there are as yet no cell lines available that allow the efficient generation of virions from cloned transfected genomes.

The other well established animal system is the Pekin duck (*Anas domesticus*) with DHBV, the type member of the *avihepadna*viruses (Fig. 1). Further avian hepatitis B viruses have been found in the grey heron (HHBV), and recently in storks, but the experimental advantages of the duck system are obvious. The animals are easy to keep, they can efficiently be infected *in vivo*, and they provide a convenient source for primary hepatocytes for *in vitro* infection studies. Furthermore, the P protein of DHBV is, for unknown reasons, still the only hepadnavirus reverse transcriptase whose authentic enzymatic activity can be studied by biochemical means (Section 3.3.1).

Since infectious DHBV particles can also efficiently be produced in avian hepatoma cell lines by transfection of cloned virus DNA, this is the only hepadnavirus system for which the full range of experimental approaches is accessible; however, there are also restrictions. The most obvious is the rather large evolutionary distance of DHBV and its host from HBV and humans. Although many principal features of the mammalian hepadnaviruses are preserved, there are some distinct differences (Fig. 1). On the genetic level, for instance, DHBV lacks an X ORF, it has no M protein, but it produces substantial amounts of a spliced message (40). Many ducks are endogenously infected with DHBV without showing any symptoms or increased frequency of liver cancer, suggesting that ducks are not the model of choice for investigating the long-term pathogenic effects of hepadnavirus infection; probably, the virus and its host have undergone a long co-evolution towards peaceful coexistence. On the other hand, ducks can clear the infection, and DHBV replication appears to be similarly affected by cytokine-based host defence mechanisms as that of HBV (41). The recent cloning of several duck cytokine genes should therefore provide an opportunity to disclose the underlying mechanisms under true *in vivo* conditions.

3.1.2 New hepadnaviruses and alternative hosts

These considerations underscore the importance of developing new animal models that close the evolutionary gap between chimpanzees, and woodchucks and ducks as experimental hosts for hepadnaviruses. Two novel mammalian systems hold some promise in this direction. One is the woolly monkey HBV (WMHBV), a hepadnavirus recently discovered in woolly monkeys (*Lagothrix lagotricha*), a New World primate (42). This virus is apparently closely related to HBV but is more distant in sequence from all known HBV isolates than all HBV isolates from humans are among each other. This has recently been confirmed by the sequence of a second isolate of WMHBV which differs from the first at only 14 nucleotides (M. Nassal and A. Wahl-Feuerstein, in preparation). Woolly monkeys are an endangered species and therefore cannot be used as experimental animals. While chimpanzees and other Old World primates appear poorly, if at all, susceptible to WMHBV infection (42), the virus has been successfully transmitted to the closely related and more common spider monkey (*Ateles geoffroyi*). If transmission to other New World primates is also possible, e.g. to marmosets, capuchins, or squirrel monkeys which have long been used as experimental animals, this may provide a route toward a new infection model that should much more directly bear on HBV infection in humans.

Another potentially useful animal is the tree shrew, or tupaia. Tupaias have for a long time been considered as the most primitive primates; now they form an order of their own (*Scandentia*). They are common in Asia where chronic hepatitis B is most widespread, and several reports from China described the successful infection of tupaias with human HBV (e.g. 43). The animals are rather small and can be easily kept in captivity when they reproduce efficiently. These are important advantages over the other mammalian hepadnavirus hosts. However, thus far the date on their susceptibility to HBV infection is not entirely convincing. HBsAg production upon inoculation with human HBV positive serum *in vivo* has been detected (44), and *de novo* synthesis of viral RNAs upon incubation of primary tupaia hepatocyte cultures with such sera has been demonstrated (J. Köck, F. V. Weizsäcker, S. McNelly, and H. E. Blum, unpublished data). However, direct evidence for the generation of complete infectious HBV particles from such cells is scarce, and if it does occur, the efficiency is very low. One limitation is probably the initial entry of HBV into the tupaia cells. Possibly, this barrier can be overcome by the use of replication-defective adenovirus vectors that carry a complete hepadnavirus genome. Such vectors should be able to efficiently transduce the hepadnavirus genome into tupaia hepatocytes, both *in vitro* and *in vivo*. The enhanced levels of hepadnavirus gene products should make it possible, in the best case, to unequivocally establish a productive infection or, at least, to clearly define potential host-specific intracellular blocks in the hepadnavirus infectious cycle. In the long run this system might provide a chance to obtain a tupaia-adapted hepadnavirus strain which can then be used for further studies.

The ultimate experimental animal would probably be the mouse, not only for logistic reasons but also because of the vast amount of knowledge about this animal and the multitude of mouse strains with defined genetic backgrounds, including many knock-out mice that would allow the roles of individual host genes in infection to be defined. Possibly, transduction of hepadnavirus genomes using adenovirus or other vectors may help to overcome the species barrier which might, however, be higher for the mouse (see below) than for tupaias.

3.1.3 Chimerical animals

Another way of obtaining a non-host animal that is susceptible to hepadnavirus infection is the generation of chimeric animals in which part of the liver stems from a susceptible host. Some important progress in this direction has come from the availability of immunodeficient mice with strongly decreased xenograft rejection. One system takes advantage of transgenic mice carrying the urokinase type plasminogen activator transgene (uPA) and lacking mature B and T lymphocytes due to a knockout of the recombination activation gene 2 (RAG2). The livers of such uPA/RAG2 mice can be destroyed and be repopulated with hepatocytes from, in this case, woodchucks. Such chimeric mice are susceptible to WHV infection, and they produce virus at relatively high titres (45). Work in progress focuses on adapting this system to mice with human hepatocytes. The other system (46) relies on grafting of small pieces of human liver under the kidney capsule of mice. The transplanted human cells are susceptible to infection by HBV, and by hepatitis C virus. Major limitations are that producing such chimeric mice is technically demanding and, by necessity, only immunosuppressed animals can be used. Hence questions regarding the immune response to hepadnavirus infection can only be addressed to a limited extent.

3.1.4 HBV transgenic mice

An intensely used system for such immunological questions are HBV transgenic mice (47, 48). Several mouse lineages have been established which, from integrated overlength HBV genomes, produce all viral gene products and even infectious virions. An exciting finding made with these mice was that adoptively transferred cytotoxic T cells can not only kill HBV-infected hepatocytes but efficiently suppress virus production by a non-cytolytic mechanism mediated, in particular, by interferons and tumour necrosis factor α (49, 50). This provides a very plausible model for the clearance of acute hepadnavirus infections, and very recently, experimental evidence supporting its validity has been obtained in two experimentally infected chimpanzees (51). If clearance were to occur by killing of infected cells, the entire liver would be destroyed. Fortunately, fulminant hepatitis caused by such an overshooting cytotoxic T cell response is a relatively rare event. The major shortcoming of the transgenic mouse model is that HBV production results from the integrated virus genome. Hence, suppression of HBV replication can only be transient, and the generated virions do not re-infect the mouse liver. Moreover, no evidence for the presence in the transgenic mouse hepatocytes of viral cccDNA, the central intracellular molecule during infection, has been obtained. Hence mouse cells seem to lack factors that are required to transport the viral capsid to the nuclear pore and/or to deliver the genome into the nucleus. This will be an additional host-range determinant to be overcome when establishing a mouse model of hepadnavirus infection. This restriction probably also applies to a rat model in which transient viraemia was induced by *in vivo* transfection of an HBV expression construct with cationic liposomes (52).

3.2   Transfection of cloned hepadnavirus DNA

The most generally valuable system to study the principal molecular mechanisms of hepadnavirus replication has been the transfection of cloned viral DNA into suitable cell lines. Most of the information concerning the viral life cycle has been derived by analysing the phenotypic effects of specifically mutated virus genes and genomes, i.e. by reverse genetics. The limited transfection efficiencies can significantly be improved by using baculovirus (53) or adenovirus vectors (S. Ren and M. Nassal, unpublished data). For HBV, the commonly used cells are the highly differentiated human hepatoma lines HepG2 and HuH7, for DHBV it is the chicken hepatoma cell line LMH. While these lines cannot be efficiently infected, they are competent for the production of all viral gene products and infectious virions. Hence all steps of the viral life cycle downstream of the transcription of the genomic and subgenomic RNAs can be studied. Since many of these steps apparently depend on host factors and can as yet not be analysed with purified components (see below), the transfection system is still the most important tool for producing relevant information (5). The following paragraph gives a short outline of the overall process of hepadnavirus genome replication; its mechanistic details derived from *in vitro* studies are discussed in Section 3.3

3.2.1   Hepadnavirus reverse transcription—genetic studies

As summarized in Fig. 2, the major steps of hepadnavirus replication downstream of transcription are the export of the viral RNAs from the nucleus, translation of the viral proteins, formation of replication-competent nucleocapsids containing pgRNA and P protein, reverse transcription of the RNA, and export of the matured DNA-containing capsids as enveloped virions via interaction with the surface proteins and budding into the cell's secretory compartments. All host factors involved in these events are obviously provided by the cell. On the other hand, whole cells cannot easily be manipulated from outside. None the less, a gross picture of the mechanism of hepadnavirus genome replication has emerged from such studies. One line of research focused on the question of how the selective encapsidation of just the pgRNA but not other viral or cellular transcript is achieved, and how P protein is co-packaged with this RNA. It turns out that the pgRNAs of all hepadnaviruses contain a structured, *cis*-acting RNA element close to their 5′ end that is absolutely required for the packaging process; it was therefore termed encapsidation signal, or ε (54). Subsequently, it was shown that the stem–loop structure is recognized by P protein (55)—in contrast to retroviruses where genome packaging is mediated by the gag proteins (56). Hence hepadnavirus RNA and P protein encapsidation are directly coupled. The predicted stem–loop structure for HBV ε (Fig. 3) was experimentally confirmed using short *in vitro* transcribed ε RNAs (57, 58). The functional consequences of disturbing that structure could be monitored by encapsidation assays in which the ε signal was fused to an unrelated reporter RNA, e.g. derived from the *E. coli* lacZ gene (Fig. 4). If a corresponding reporter plasmid is co-transfected with a helper construct containing a complete HBV genome with a deletion in its ε

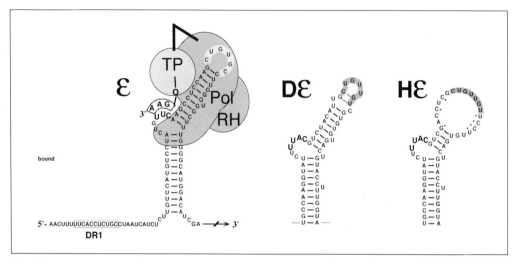

Fig. 3 Secondary structures of hepadnavirus RNA encapsidation signals. All ε elements consist of a lower and an upper stem, separated by a bulge, and an apical loop. For HBV, the 5′ end of the pgRNA with the ε stem–loop is shown schematically in a complex with P protein, depicted with its three genetically defined domains terminal protein (TP), polymerase (Pol), and RNase H (RH). The 3′ part of the bulge serves as template for a short DNA primer that becomes covalently linked to a Tyr residue in the TP domain. DR1, 5′ copy of direct repeat 1. The template regions in the bulge of the DHBV and heron HBV (HHBV) encapsidation signals Dε and Hε are highlighted in bold face, the regions corresponding to the HBV ε loop are shaded in grey. Note that in Hε the upper stem carries several nucleotide exchanges versus Dε (asterisks) that prevent extensive base pairing.

sequence, the truncated pgRNA can still act as mRNA for core and P protein but is not encapsidated. Instead, the artificial chimeric RNA will end up in a capsid where it can be detected by RNase protection assays or similar techniques. It should be noted, however, that in the authentic situation P protein preferentially packages the same RNA molecule from which it has been translated, i.e. the pgRNA (55). Fortunately, this *cis*-preference is not absolute and the reporter RNA encapsidation assay has been successfully used to show that gross alterations of the ε structure, such as deletions of the single-stranded bulge or loop regions, inhibited packaging and, by inference, interaction with P protein. Similar experiments with DHBV revealed that here the 5′ proximal Dε element is not sufficient for encapsidation but that a second, as yet poorly characterized downstream element ('region II') is required (59).

Because ε consists of about 60 nt, a more detailed structure–function analysis would have required transfection of hundreds of individual point mutants. A modified approach uses pools of reporter RNAs instead, in which segments with ε carry mixtures of nucleotides rather than one defined sequence (Fig. 4). From such pools of several thousand different molecules functional RNAs are selected by their ability to be encapsidated, and their sequences are subsequently revealed by RT-PCR amplification of the capsid-borne RNA (60). Applied to RNAs with random sequence in the 6 nt bulge region, a surprising variety of sequence combinations in the 3′ part of

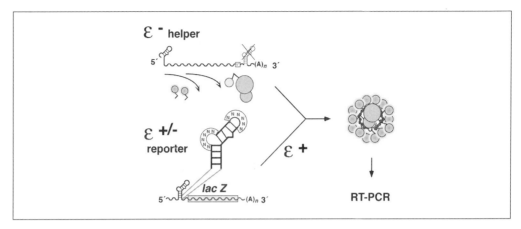

Fig 4 Identification of functional HBV ε signals by reporter RNA encapsidation. This is the central assay used to define functionally important determinants within ε in transfected hepatoma cell lines. In the format shown here, two plasmids are co-transfected: the helper plasmid encodes a 5′ terminally truncated pgRNA which because of the deletion in its 5′ ε is not packaged but produces core and P protein (ε^-); the 3′ copy of ε is functionally silent. The reporter plasmid codes for an artificial RNA containing a lacZ fragment preceded by an ε element which consists of either a defined sequence, or a pool of sequences with randomized regions ($\varepsilon^\pm$), e.g. the bulge and the loop. Functional ε signals (ε^+) lead to encapsidation of the reporter RNA. The packaged RNA is isolated from the capsids, and amplified by RT-PCR for sequence analysis.

the bulge was compatible with encapsidation, suggesting that the sequence in this region is not important for recognition by P protein.

By contrast, a selection toward maintaining the two 5′ proximal nt (C and U) was clearly detectable. This led us to propose that the base of the bulge might contain an element that is recognized by Protein (Fig. 3). This suggestion has very recently been substantiated by *in vitro* experiments with the DHBV system (Section 3.3.1). However, many of the encapsidation-competent bulge mutants, after their integration as 5′ε into complete HBV genomes, did not support reverse transcription of the encapsidated RNA. This was surprising since at that time replication was assumed to initiate at the other end of the pgRNA wit the 3′ DR1* element; hence encapsidation and initiation of reverse transcription appeared to be completely unrelated events (Fig. 5A). More direct clues against this classical replication model came from experiments with *in vitro* translated DHBV P protein (see below) which suggested that the first few nucleotides of (–)DNA are copied from the ε bulge and not by *de novo* initiation with DR1* (61, 62). Evidence that this revised replication model holds also for HBV within a living cell was again obtained in the transfection system (63). First, most of the replication-defective ε bulge mutants described above could be rescued by introducing the same mutations into DR1*—thus restoring the natural identity between the sequence in the 3′ part of the ε bulge and the 5′ proximal region of DR1* (Fig. 5A). Final proof for the synthesis of the first few nucleotides of the new DNA from the ε bulge as template and a subsequent translocation of the covalent P protein primer complex to 3′ DR1* was obtained with one of the rare bulge mutants which despite having non-identical bulge and DR1* sequences yielded clearly detectable

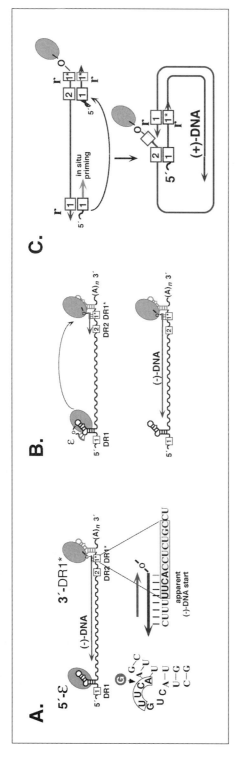

Fig. 5 Genetically derived models of hepadnavirus reverse transcription. (A) The classical view. pgRNA encapsidation and initiation of its reverse transcription were regarded as P protein dependent but unrelated events. Binding of P to 5′ ε mediates packaging, interaction with 3′ DR1* initiates replication. pgRNA is shown as wavy line, (−)DNA as leftward pointing arrow. Primer extensions (rightward pointing arrow) mapped the apparent start site of (−)DNA to opposite the UUCA motif within DR1*. Note that the identical motif occurs within the bulge of ε (boxed residues). (B) The revised model. Binding of P to 5′ ε mediates both RNA packaging and replication initiation. The first nucleotides of (−)DNA are copied from the UUC, or possibly UUCA, motif in the ε bulge, followed by translocation of the covalent complex to the same motif in 3′ DR1*. Genetic proof for the revised model with HBV was obtained with a mutant carrying G >C mutation in ε (encircled G in A) and a wild-type DR1*. (C) (+)DNA formation. Concomitantly with (−)DNA synthesis the RNase H activity of P degrades the pgRNA template, except for an oligoribonucleotide at the 5′ end including the DR1 sequence. This RNA primer translocates to DR2 from where it is extended to the 5′ end of the (−)DNA template within DR1*. This part of DR1* and the immediately preceding nucleotides comprise a small terminal redundancy ('r'). In a final template switch, the r sequence at the 3′ end of the (−)DNA substitutes for its 5′ terminal copy, leading to circularization and formation of relaxed circular (RC) DNA. RNA primers that fail to translocate can be extended 'in situ', yielding linear double-stranded DNA.

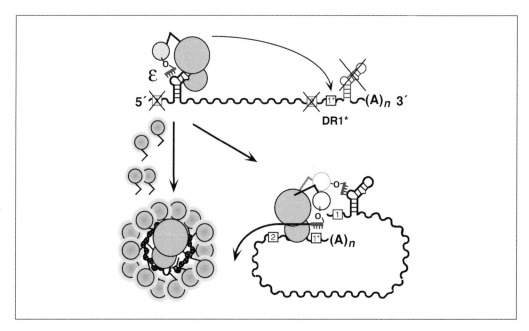

Fig. 6 The dual essential role of the ε/P protein interaction. P protein binding to 5' ε creates an as yet unidentified binding site for core protein dimers. Notably, this requires that the ε signal is also functional as (–)DNA primer template. Despite the presence of identical complements to the DNA primer sequence in all DR copies, translocation occurs specifically to DR1*. 3' ε cannot substitute for 5' ε, and its deletion neither affects RNA packaging nor (–)DNA synthesis. Though not experimentally proven, the pgRNA may be held in a circular conformation, e.g. by protein–protein interactions between polyA binding protein and cap-binding factors, that facilitates translocation of the P protein/primer complex from 5' ε to 3' DR1*.

levels of (–)DNA. The specific mutation was a C->G exchange at the 3' terminal position of the bulge (Fig. 5A). By primer extension, a complementary copy of the 5' terminal end of the (–)strand DNA was generated and enzymatically extended with a homopolymeric 3' tail to allow for PCR amplification. In this way, the 5' sequence of the (–)DNA could be determined although its 5' end is blocked by the linkage to P protein. This analysis revealed that the complement to the mutated bulge residue rather than to the wild-type nucleotide in the authentic 3' DR1* had been used as template. Hence 5'ε acts not only as encapsidation signal but also as origin of replication.

The transfection system also addressed the role of the second copy of ε at the 3' end of the terminally redundant pgRNA. Since mutations within or even complete deletion of 3' ε had no deleterious effects on (–)DNA synthesis this copy is not essential (64). Moreover, it cannot substitute for the 5' copy *in vivo*. This is in contrast to the DHBV *in vitro* system where either Dε copy can act as acceptor for P protein. The cause for the strict position-dependence of ε *in vivo* is not yet known.

The overall hepadnavirus strategy of P protein/ε-mediated pregenome encapsidation and initiation of reverse transcription as revealed by these genetic techniques outlined in Fig. 6. Evidently, however, it is extremely tedious if not impossible to derive further mechanistic details on the P protein/ε RNA interaction with this intact cell approach. Most of our current knowledge in this respect has been worked out using *in vitro* translated DHBV P protein—unfortunately, this approach is still not possible with the HBV enzyme.

3.3 Reconstituting hepadnavirus reverse transcription *in vitro*

Many laboratories have tried for years to heterologously express enzymatically active HBV P protein. There was no reason to believe that this reverse transcriptase should behave fundamentally different from the enzymes of retroviruses, many of which have been heterologously expressed and even are sold commercially to produce cDNA from mRNA. None the less, almost all attempts have failed with the hepadnavirus reverse transcriptases. The major reasons are two unusual properties of these proteins. First, they cannot, at least not under authentic conditions, initiate reverse transcription by extending the 3' end of a nucleic acid primer annealed to the RNA template as is the case for almost all other reverse transcriptases. Instead, initiation occurs via the side chain OH group of a Tyr residue in the protein (65, 66) and is restricted to the specific ε stem–loop structure as template. Only this self-synthesized primer is elongated after translocation of the covalent complex DR1* (Fig. 5B). Secondly, enzyme activity requires assistance from cellular chaperones, possibly in a similar fashion as steroid receptors are activated for hormone binding.

3.3.1 *In vitro translation of DHBV P protein*

A fortuitous observation marked the beginning of this new concept (67). DHBV P protein was *in vitro* translated in rabbit reticulocyte lysate programmed with an artificial mRNA containing just the P ORF and downstream sequences including the 3' copy of the Dε stem–loop (Fig. 7). Addition of dNTPs led to a limited synthesis of (–)DNA (but no (+)DNA) which, as in authentic replication, was covalently bound to P protein (this is conveniently followed by phenol extraction during which P protein bound DNA will partition into the organic phase). Primer extensions revealed one product that had the same apparent 5' end opposite DR1* as previously observed in transfection and infection experiments with DHBV. The initiation site of a second product, however, mapped in a region with Dε which we now know contains the characteristic bulge. This was the first evidence that reverse transcription was in fact not initiated *de novo* at DR1* but rather within the encapsidation signal. This mechanism was then confirmed by genetic experiments for both DHBV and HBV (Section 3.2.1). These results also indicated that, in contrast to *in vivo* replication, in the *in vitro* system the 3' copy of ε can serve as binding site for P protein. Fortunately, this relaxed position specificity *in vitro* has a significant experimental advantage. The ε stem–loop does not even need to be part of the P mRNA, it can be supplied as a separate moiety *in trans* (Fig. 7). Given suitable conditions, the protein will bind the ε RNA (61, 68) and use it as template. Short defined primers can be obtained by provision of appropriate mixtures of dNTPs, or inclusion of dideoxy derivatives. In addition, and similar to other polymerases, initiation can be greatly favoured over

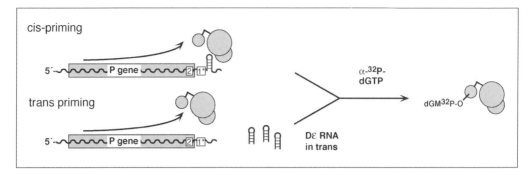

Fig. 7 Reconstitution of (–)DNA primer synthesis *in vitro*. P protein is *invitro* translated in rabbit reticulocyte lysate programmed with an artificial mRNA. If the RNA contains its own copy of ε at the 5′, or as shown at the 3′ end, P protein binds to it and, upon provision of radiolabelled dNTPs, initiates primer synthesis ('*cis*-priming'). This covalent transfer of the label to the protein provides a sensitive functional assay for a productive interaction between protein and RNA. If the P mRNA lacks a *cis* ε signal, it can also be supplied as a separate molecule *in trans*. The '*trans*-priming' assay also permits the use of synthetic ε analogues. Until now the assay works only with DHBV but not HBV P protein.

elongation by special buffer conditions (65, 69) that include Mn^{2+} instead of Mg^{2+}. If one of the dNTPs is labelled with ^{32}P at the α-position this will result in the covalent transfer of the label to the protein. This priming reaction (Fig. 7) provides a very sensitive functional assay that compensates for the minute amounts of protein that are produced in the *in vitro* system (in the range of a few ng per 25 μl translation mix). It has since been extensively used to investigate the roles of the Dε RNA, of P protein and cellular factors in a productive interaction.

As a basis for further detailed structure–function studies we first showed experi-mentally that wild-type Dε and its homologue from the heron HBV, Hε, have an over-all structure similar to that of HBV ε (70) with a lower and an upper stem, separated by a single-stranded bulge, and an apical loop (Fig. 3). The upper stem sequence of Hε is rather divergent from that of Dε, resulting in a much less base paired configuration. None the less, DHBV P protein accepts this natural variant of Dε rather efficiently as a priming template. A panel of artificial variants in which either the positions con-served or those variable between the two RNAs were mutated allowed to derive a set of rules for productive interaction with DHBV protein (71). One determinant is preservation of the overall secondary structure (which sometimes is dramatically influenced by even single mutations; Fig. 8B), another conservation of the base identity at a few strategic positions. While most of these data were compatible with a simple 'lock-and-key' model for the P protein–RNA interaction, there were a few notable exceptions. The first was a mutant that obviously violated the first criterion because it had a different structure from wild-type Dε yet still functioned as priming template (variant L5; Fig. 8B). The other was a mutant with a wild-type like structure that bound to P protein but could not serve as template. This suggested that the structure of the free RNA was not all that is required for a productive interaction but rather that this structure must be sufficiently flexible to adopt a new, priming-

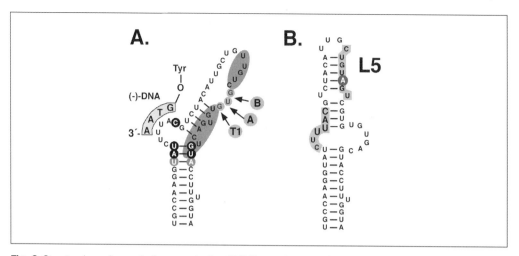

Fig. 8 Structural requirements for a productive RNA/P protein interaction derived from the *in vitro* translation system. (A) Structural changes induced in Dε upon complex formation with DHBV P protein. Arrows denote prominent cleavage sites in the protein-bound RNA that are completely inaccessible in free wild-type Dε. B, benzonase; A, RNase A; T1, RNase T. Grey ovals indicate regions protected in the complex from cleavage by Pb^{2+}, presumably caused by the bound protein. The encircled residues mark a principal 2' OH and base-dependent recognition element for productive interaction with P protein. All other residues, but not these, can be replaced by the corresponding deoxy derivatives. The base of the lower stem, except for the two to three base pairs at the tip acts as a scaffold to ensure formation of the bulge structure, but neither its sequence nor its ribose content is important. (B) A Dε mutant indicating the functional importance of a structural RNA rearrangement in the replication initiation complex. The mutant L5 carries a single C >A exchange (encircled A) that leads to a completely different structure in the apical region of free L5 RNA. The primer template and the loop region as present in Dε are highlighted by grey shading. None the less L5 is similarly active as template as wild-type Dε. Upon complex formation with P protein, L5 adopts a conformation that is virtually identical to that of protein-bound Dε. By contrast, some other point mutants with wild-type-like free state structures cannot adopt the priming-active new conformation and bind in an unproductive fashion to P protein.

competent conformation in the complex with the protein. In essence this indicated an 'induced-fit' mechanism which meanwhile has been demonstrated for a number of protein–RNA interactions. Proof of this model required secondary structure analyses of the P protein-bound RNA (72). Using a His-tagged protein version it was possible to completely separate the bound from the free RNA. Probing with nucleases and Pb^{2+} revealed a drastic conformational change in the upper stem region of Dε, during which previously base paired residues in the right half become highly accessible to single-strand specific reagents (Fig. 8A). On the other hand, several positions in the bulge and loop region accessible to Pb^{2+} mediated cleavage in the free RNA became protected, suggesting that here the RNA was shielded by protein contacts. The functional relevance of these structural changes was demonstrated using the exceptional variants mentioned above. The priming-active mutant L5 with a different free-state structure (Fig. 8B) ended up with essentially the same new structure as wt Dε in the P protein complex, while the one that was priming-inactive despite a wild-type like free-state structure did not change substantially upon binding to P protein. Together, these data strongly suggest a two-step mechanism for the P protein/ε recognition: the first discriminates binding-competent from other RNAs, the second involves a significant structural rearrangement during which the RT is properly positioned over the start site to initiate reverse transcription. Since replication initiation and encapsidation are mechanisticly linked via the ε signal, only RNAs which can be reverse transcribed will be packaged into capsids.

Another opportunity provided by the *in vitro* system is the use of synthetic nucleic acids, including variants with non-natural bases and/or backbone residues which cannot be produced inside a live cell. We have recently analysed a series of artificial oligonucleotides in which increasing parts of the ribonucleotide backbone have been replaced by the corresponding deoxy derivatives (S. G. Schaaf, J. Beck, and M. Nassal, submitted). While an all-DNA Dε analogue was completely priming-deficient, chimeras with as few as five ribose moieties were accepted as templates. These 2' OH dependent positions are the 3' terminal nucleotide of the bulge, i.e. the first template nucleotide, and the two base pairs at the tip of the lower stem directly underneath the bulge (Fig. 8A). At these positions also the base identities are important, since simple replacement of the Dε specific U–G pair by A–U, as present in HBV ε, renders such a variant completely inactive as template for DHBV P protein. These data demonstrate that the bulge region and its immediate vicinity contain a small, principal recognition element for P protein. The opening up of the double helix at the tip of the lower stem could widen the major groove which is narrow in A-form RNA and make it accessible for specific contacts with the protein (73). Alternatively, the essential 2' hydroxyls might be part of an 'indirect read-out' recognition mechanism (i.e. mediated via a specific backbone structure rather than the bases), or they could serve to ensure a specific three-dimensional structure of the bulge region, either by direct hydrogen bonding or by co-ordinating Me^{2+} ions. The major function of the lower stem, by contrast, is apparently that of a scaffold to ensure formation of the specific bulge structure. Therefore, neither 2' hydroxyls nor the nature of the bases involved are of importance. These data provide a basis for further biochemical studies with synthetic Dε analogues that should allow the contact regions between protein and RNA to be mapped. At some point, however, only direct biophysical studies will enable us to further elucidate the structural basis of the P protein/ε interaction. Probably, this will have to await reconstitution of the entire RNP complex from purified, heterologously expressed components to obtain the necessary quantities.

Not only the RNA template but also the P protein appears to undergo conformational changes upon complex formation (74, 75), as revealed by the distinct fragment patterns after partial proteolysis in the presence or absence of Dε. Since specific protein fragments correlate with the functional activity of Dε variants as primer templates, it is tempting to speculate that the structural alterations observed in the proteolysis assay are related to the above described rearrangement of the RNA into a replication-competent form. Accordingly, both the protein and the RNA adopt new structures that ensure initiation of DNA synthesis at the correct start site. This strategy is fundamentally different from nucleic acid primed DNA synthesis as employed in DNA replication and reverse transcription in retroviruses where the

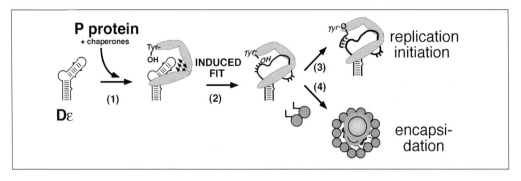

Fig. 9 Induced-fit activation of the ε RNA template for initiation of reverse transcription. P protein, assisted by chaperones, recognizes the overall structure of plus specific features within Dε RNA (1); the tip of the lower stem appears to be important for this initial event. Then RNA and protein undergo conformational changes during which the upper stem is molten and P protein is accurately positioned over the template region (2), allowing it to be copied into the short, protein-linked primer (3). The structural alteration is probably similarly important for the function of ε as encapsidation signal (4) since binding-competent but template-defective RNA sequences exhibit severe packaging defects in the context of complete DHBV genomes.

initiation site is simply defined by the 3' end of an appropriate complementary primer molecule. Recent evidence indicates (J. Beck and M. Nassal, unpublished) that also pgRNA encapsidation depends on this structural alteration (Fig. 9). Given that hepadnavirus (–)DNA synthesis from the ε template ceases after a few nucleotides it is likely that further structural rearrangements occur to prepare the complex for translocation to 3' DR1* and switching its activity from initiation to elongation mode.

In all, the above described studies have revealed that hepadnavirus replication initiation is a highly dynamic multi-step process. In this view it might be less surprising that this variety of ordered structural alterations is not realized in a simple two component ribonucleoprotein (RNP) complex but rather involves assistance by cellular factors, in particular chaperones which are the cell's professionals for modulating the structures of large biomolecules (Fig. 10).

The first evidence for the involvement of chaperones came from the observation that the priming activity of DHBV P protein translated in wheat germ extracts was very low but could be boosted by addition of rabbit reticulocyte lysate (76); the essential factor was probably the small co-chaperone p23 (see below). Similarly, immunodepletion of reticulocyte lysate with anti-Hsp90 or anti-p23 antibodies also rendered the P protein inactive, as did the treatment with the antibiotic geldana-mycin which prevents the interaction between Hsp90 and p23. This, and further evidence for the involvement of Hsp70, suggested that DHBV P protein may undergo a similar chaperone-mediated activation as steroid receptors (77–79). These receptors are unable to bind their cognate hormones unless they are present in a mature complex containing an Hsp90 dimer, p23, and one member of the prolyl-*cis-trans* isomerases of the immunophilin family. The mature complexes are generated in a dynamic cycle through sequential interactions with various chaperones that

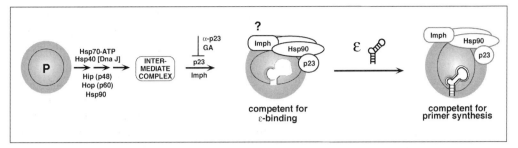

Fig. 10 Model for the chaperone-dependent activation of P protein. *In vitro* translated DHBV P protein can bind Dε RNA only when complexed with Hsp90 and the small co-chaperone p23, resembling the chaperone-mediated activation of steroid receptors. Based on this analogy, the complex should contain at least one member of the immunophilin family ('Imph'), and formation of the mature complex should occur via several intermediates involving a series of other chaperones. The functional relevance of complex formation is suggested by the sensitivity of the *in vitro* priming reaction to chaperone-specific antibodies, e.g. anti-p23, and small molecules like geldanamycin (GA) that specifically blocks Hsp90/p23 interaction. Conceivably, the chaperones are actively involved in the conformational alterations in P protein and Dε RNA that accompany initiation complex formation.

involve the Hsp70 system and accessory factors like Hip and Hop that mediate their subsequent association with Hsp90 (80). When hormone binds, the chaperones probably dissociate and the receptors enter the nucleus to activate transcription from the corresponding response elements. Otherwise, the mature complexes will fall apart and the receptor enters another activation circle. If the analogy holds true for P protein, the RNP complex should also contain an immunophilin, and the accessory factors should be important for generating a mature, replication-competent P protein complex. As yet, an involvement of these factors has not been experimentally demonstrated and, in contrast to the steroid receptors, the chaperones may not dissociate from P protein after binding of its ligand ε. In fact, based on the detectability of p23 virus particles it has been suggested that the entire RNP complex including the chaperones is encapsidated (81). However, the presence of Hsp90 signal in these preparations was not directly proven. While this may have technical reasons, e.g. the low affinity of the available anti-Hsp90 antibodies, the same study showed that p23, most of which is normally bound to Hsp90, can remain associated with P protein after release of Hsp90 from the complex. Hence only this fraction of p23 rather than the entire chaperone complex may be encapsidated. In any event these studies have shown a unique property of the hepadnavirus RT, and they explain the failures to detect enzymatic activity in preparations of isolated P protein. Despite the complexity of steroid receptor activation it has recently been possible to reconstitute hormone binding-competent receptors complexes using purified components (82). Though probably a very demanding goal it may therefore not be out of reach to accomplish a similar feat for DHBV P protein.

3.3.2 Why is HBV P protein translated *in vitro* inactive?

HBV P protein can be produced by *in vitro* translation in rabbit reticulocytes lysates in similar quantities as its DHBV counterpart but thus far all attempts to prove enzymatic activity have failed—the reason is still unclear. One speculation is that the avian virus P protein can productively interact with rabbit chaperones but that from the human virus cannot. Hence it might be possible to complement the rabbit cell extract with purified human chaperones. However, if the analogy to steroid receptor activation holds, a whole battery of chaperones and co-chaperones will sequentially be required to generate a functional P protein/ε complex—and each of them could be the limiting factor. As an alternative, we have recently prepared cytoplasmic extracts from HuH7 cells, the human hepatoma cell line that supports virus formation after transfection with cloned HBV DNA (J. Beck and M. Nassal, unpublished). These extracts are clearly competent for P protein translation; none the less, we could not observe any enzymatic activity of the translated protein. Surprisingly, even DHBV P protein was essentially inactive in these extracts, and addition of the hepatoma extract inhibited the protein's activity in reticulocyte lysate, indicating that some inhibitory factor or activity is present in the hepatoma cells. By fractionating the hepatoma extract and following the inhibitory effect of individual fractions it might be possible to trace that activity; whether this will hold the clue as to why the HBV enzyme is inactive in both extracts remains to be seen. An alternative concept would be to put the blame on the HBV ε RNA. As outlined above Dε undergoes a drastic conformational alteration upon productive binding to DHBV P protein that involves the opening of the upper stem (72). Since the stability of the upper stem region of HBV ε is much higher than that in Dε such a structural rearrangement may need assistance by some unwinding activity which is not present in sufficient amounts in the translation extracts. Given the increasingly acknowledged role of RNA binding proteins during the entire lifespan of an RNA molecule it might well be that the authentic pgRNA brings along such an unwinding activity from its site of synthesis in the nucleus which is lacking if the RNA is added from the outside to a cytoplasmic translation extract; a few proteins that bind to pgRNA have already been identified (20, 83, 84). This could be tested by adding nuclear extracts to the translation reaction. Alternatively, artificial ε-derivatives with lowered stability in the upper stem region might be found, by SELEX procedures (85), that can more easily undergo the conformational change. However, some caution is required since, as mentioned above, some Dε mutants have been found that bind to P but cannot act as templates for priming.

3.3.3 Baculovirus-mediated P protein expression in insect cells

Apart from *in vitro* translation, the only other system allowing at least for a restricted range of functional studies has been baculovirus-mediated expression of HBV protein in insect cells, which generates substantial amounts of protein. However, only a small fraction of it shows some enzymatic activity that, initially, appeared to resemble authentic priming and first strand DNA synthesis (86). Later, careful controls revealed that this activity was not dependent on ε (87), which is a hallmark of the authentic initiation process. It is still unclear whether some insect or baculovirus RNA can substitute for the authentic ε element or whether DNA synthesis results from a basal, poorly efficient ε-independent priming process which is only seen with the large quantities of P protein present in these experiments. None the less, several interesting data sets have been obtained with this system. The general optimal salt and pH conditions for the enzyme could be worked out (69), and by co-expression with the HBV core protein some fraction of P could apparently be incorporated into capsid-like particles (88). Surprisingly, this packaging worked poorly when P and core were produced from one mRNA, as in the authentic situation, but improved when the two proteins were generated from separate mRNAs. Again, the possible role of ε in this encapsidation reaction remains unclear as the presence of an ε element on the mRNAs seemed to stimulate packaging but was not essential. It is also unclear which of the RNAs was encapsidated and, if only one carried ε, whether this was preferentially packaged. Another interesting application of the system was to define intramolecular interactions within the P protein that gave some clues to its domain organization; by expressing the protein in two parts, with either one carrying a tag, it was shown by co-immunoprecipitation that they can form complexes that exhibit a similar activity as the full-length protein; association of the two parts was again stimulated by the presence of ε but did not strictly depend on it (89). Unfortunately, complex formation worked only upon co-expression of the two protein parts in one cell but not by mixing the extracts from separately infected cells, indicating that only during, or shortly after, translation is the protein in a conformation allowing it to enter an active state. Similarly, it has not been possible to reconstitute any activity by providing ε RNA to the purified protein or protein domains. This remains one of the biggest limitations as it precludes analyses involving the sequential addition of individual components, or the use of synthetic ε analogues, as they are possible with the *in vitro* translation system.

One possible reason for the poor specific activity of P protein in the insect cells is that the mammalian chaperones involved in its activation are only poorly substituted for by their insect counterparts. On the other hand, once all chaperones involved in P protein activation are identified it might be possible to utilize the insect cell system for the reconstitution of an authentic P protein/chaperone complex by co-expression of all components. Meanwhile baculovirus vectors carrying up to four independent expression cassettes are available, and co-infection with different baculoviruses is feasible; hence the chances for such an approach may not even be too slim.

4. Assembly of virus particles

4.1 Envelope containing particles

The three types of hepadnavirus particles are complete enveloped virions, or 'Dane particles', the nucleocapsids, or core particles, and empty envelopes called subviral, or S particles (Fig. 2). The S particles of the mammalian hepadnaviruses occur in two types, either spherical, or filamentous. The filaments, but not the spheres, contain substantial amounts of L protein which therefore may be responsible for the different morphologies. Since S particle formation as well as virion morphogenesis have been reviewed elsewhere in more detail (12) just a few key aspects are summarized here.

All hepadnaviruses produce such empty envelopes in large excess over virions. Like whole virions, they bud into the intracellular compartments of the secretory pathway (90, 91), hence hepadnavirus envelope proteins do not accumulate in the plasma membrane. S particles might enhance the infectivity of the virions (92), and/or they may act as decoys to trap antibodies directed against virus particles. The immunodominant epitope for neutralizing antibodies is located in an exterior lop of the S protein, and is accessible on both kinds of particles. This cross-reactivity is the reason that HBV S particles provide an effective vaccine against HBV.

Direct structural information about the envelope proteins remains scarce, except for the recent characterization of their N- and O-linked glycan moieties (93, 94). One major reason is that, as with every membrane protein, it is difficult to obtain crystals from the whole proteins that would be suitable for X-ray analysis. An alternative approach, that has been enormously successful with cellular membrane proteins as well as other viral env proteins (95, 96), e.g. the haemagglutinin of influenza virus and the env proteins of HIV, is to restrict analysis to the water soluble domains of the protein. The possibility to separately express the preS domains (31) may offer an opportunity for such an approach by X-ray or NMR analysis. If successful, it should provide important insights into the interactions with host cell factors and to the nucleocapsid.

An alternative approach that would also be applicable to whole virions (97, 98) is high resolution cryo-elecron microscopy combined with image reconstructions (Section 4.2.1). It does, however, require high relatively concentrations of purified, intact, and inherently symmetric particles. In addition, only few suitable microscopes are currently available and most of the facilities are not equipped to handle pathogenic organisms. This is particularly pitiful since large quantities of chronic carrier serum would be available as an abundant source for HBV while the supplies of animal sera containing high-titred mammalian hepadnaviruses are limited. Potential obstacles are a possibly non-symmetrical arrangement of the three different envelope proteins and the inward and outward directing preS domains. In addition, both virions and S particles contain substantial amounts of Hsc73, a constitutively expressed heat shock protein of the Hsp70 family (I. Swameye, M. Hild, U. Klingmüller, C. Kuhn, and H. Schaller, in preparation). Hsc73 binds to the preS domain of the L protein (99, 100), and also the M protein appears to interact with chaperones (101). These interactions may have an important influence on virion formation.

4.2 The nucleocapsid

The HBV capsid is currently by far the best characterized of the hepadnavirus particles, a result of the combined application of biochemical and biophysical techniques to recombinant capsids. Since the first observation that expression of the HBV core gene in *E. coli* yielded capsid-like particles (102), the inherent self-assembly ability of the core protein, which is independent of further viral gene products, has been exploited in a variety of heterologous expression systems. It should be noted, however, that it has as yet not been possible to assemble a complete nucleocapsid containing the RNA pregenome and the P protein *in vitro*. Many of the earlier studies have previously been reviewed (102, 103), hence only new structural aspects will be summarized here.

4.2.1 Structure of the HBV capsid

The actual assembly domain (Fig. 11A) resides in the N terminal 144 aa (104), while the Arg-rich C terminus is involved in nucleic acid binding (105) and is essential during reverse transcription of the RNA pregenome in authentic nucleocapsids (106). Replacement of the four authentic Cys residues by Ser or Ala defined a specific disulfide pattern (107, 108). The disulfide bridge that forms most easily is that between two Cys61 residues of neighbouring subunits. These disulfides stabilize the particles *in vitro*, but they are not required *per se* for particle formation. Even a mutant containing no Cys residue at all is able support virus formation (109, 110). An instrumental system for studying the assembly process was microinjection of an appropriate core protein mRNA into *Xenopus laevis* oocytes. Because of the relatively low amounts of protein produced the critical concentration for assembly could be estimated to be at around 1 μM (111, 112). Since no other intermediates were detected, dimers are probably the principal building blocks of the capsid (113).

A further advance was the development of an *in vitro* disassembly/reassembly system using purified particles consisting of C terminally truncated core protein variants. These can be dissociated into dimers under conditions of elevated pH, low salt, and modest concentrations of urea, and they reassemble spontaneously by restoring conditions of neutral pH, intermediate to high ionic strength, and removal of the urea (114). Unfortunately, this approach is not possible with the much more stable particles from full-length protein. Probably this extra stabilization is due to additional interactions with RNA inside the particles. In *E. coli*, this RNA is packaged by some non-specific process (104) that is different from the RNA packaging signal and P protein dependent encapsidation of the authentic genome in eukaryotic cells. Particle formation can be followed by size exclusion chromatography and, conveniently, by native agarose gel electrophoresis (104). As a consequence of their lower diffusion coefficients, particles migrate as much more sharp-edged bands than non-assembled proteins. The particles can be localized by subsequent Coomassie Blue staining, or more specifically by immunoblotting. Most easily, the gels are run in the presence of ethidium bromide which will penetrate the capsid shell and stain the packaged RNA, hence migration can be followed *in situ* (104). The same assay can also be applied to the much lower amounts of particles formed in transfected eukaryotic cells by visualizing the encapsidated nucleic acid by Southern/Northern hybridization.

Despite the availability of efficient expression systems, all attempts to crystallize the recombinant capsids were unsuccessful until very recently (115, 116). Structural analyses of the core protein and whole capsid were therefore confined to indirect biochemical approaches like antibody and protease accessibility studies (12). A major breakthrough came from recent progress in cryo-electron microscopy and advanced image reconstruction techniques. Detailed accounts of these techniques may be

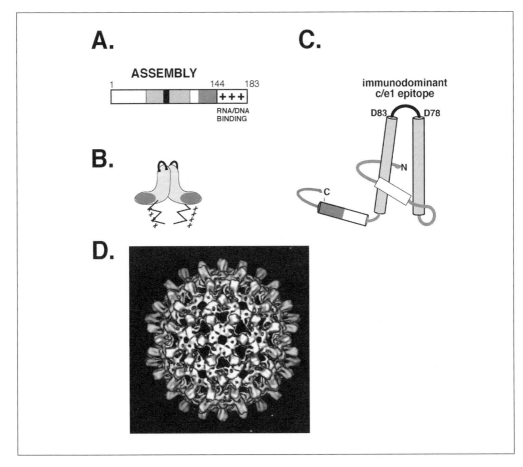

Fig. 11 Structural features of the HBV core protein and the capsid. (A) Primary sequence representation of the core protein. The domain structure and the homologous interaction regions were derived from biochemical analyses of heterologously expressed core protein. The sequence from the N terminus to about position 144 represents the assembly domain, the Arg-rich C terminus is required for nucleic acid binding. Light shading indicates regions involved in dimerization, darker shading regions important for multimerization of dimers. The black rectangle denotes the central immunodominant c/e1 epitope. (B) Low resolution structure of the core protein dimer. Earlier cryo-electron microscopic data indicated formation of a hammerhead-shaped dimer as the building block of the capsid, with the Arg-rich C terminus pointing towards the interior of the capsid shell. The shading of the interaction regions is as in (A). (C) Model for the fold of the core protein in a monomer. The model (120) is based on high-resolution image reconstructions (as in D) and biochemical evidence. It consists of four helices; the central ones are involved in a four-helix bundle that constitutes the dimer interface. Part of the C terminal helix and the following sequence mediates the contacts between the dimers in the complete capsid. The c/e1 epitope overlaps with the short connecting loop between the two central helices. (D) Image reconstruction of a complete, 240 subunit recombinant HBV capsid. The picture shows a surface representation of a T = 4 capsid at about 8 Å resolution (B. Böttcher, G. Beterams, and M. Nassal, unpublished data). The 120 spikes are generated by the four-helix bundles formed by each dimer. As in every icosahedrally symmetric capsid with a triangulation number > T = 1, the subunits are arranged in rings of five, and rings of six.

found elsewhere (117). Simply speaking, in the cryo-EM technique the specimen is embedded in non-crystalline ice without any heavy metal stains; hence the authentic structures are preserved much better than with conventional techniques. Secondly, electron densities are derived by scanning the micrographs. If the specimen consists of a number of identical inherently symmetric particles, and if the orientation of the individual particles can be determined, superimposing the electron densities from several particles will greatly improve the signal-to-noise ratio. Eventually, after several rounds of refinement, this results in a complete three-dimensional density map for the particle. In 1994 the technique revealed the principal architecture of recombinant HBV capsids (118). They occur in two different classes, characterized by the icosahedral triangulation numbers of T = 3, and T = 4. The triangulation number can be viewed as the multiple of 60 subunits required to build the corresponding icosahedral particle, i.e. 180 for T = 3, and 240 for T = 4. The surfaces of these particles showed 90, and 120, respectively, prominent protrusions, clearly indicating that the structures are based on the corresponding number of dimers; each dimer produces one of the visible spikes (Fig. 11B). This was in complete accord with the biochemical data described above. The same technique confirmed that, at least at this level of resolution (in the range of 25 Å), the recombinant capsid shells are not visibly distinct from authentic HBV capsids from human liver; the latter contain, however, a much higher inner electron density, plausibly owing to the packaged genome and P protein (119). Recently, the technique has been taken an important step further. Using highly concentrated samples of recombinant capsids several hundred or even thousands of particles could be included into one reconstruction, with a corresponding gain in resolution to below 10 Å (120, 121). At this level, secondary structure elements, in particular α-helices, can be directly visualized (Fig. 11D). It turned out that the prominent surface spikes are formed by a four-helix bundle in which each subunit of the dimer contributes two long helices; there is additional evidence for the existence of two further helices, one N- and one C-proximal. These data fit nicely to the large αhelix content of the protein determined by CD (114). Since most of the known capsids of other viruses adopt an eight-stranded antiparallel β-barrel conformation the structure of the HBV capsid represents a new class of capsid fold (122).

The next step is the assignment of individual residues of the primary sequence to these secondary structure elements. Based on the EM and previous biochemical data, Böttcher et al. (120) proposed a tentative model for the assembly domain (Fig. 11C). The validity of this model which has been further refined by additional EM data (123, 124) has recently been corroborated by independent approaches. The first study involved mapping of the homologous interaction regions by the two-hybrid system and the pepscan technique (125) and analysing the assembly phenotypes of mutants specifically designed according to these data (126). These experiments demonstrated the importance of central residues for dimerization, and of the C-proximal end of the assembly domain for multimerization. However, this clear-cut interpretation was only possible by the combination of the different approaches whilst the individual data sets derived from one technique alone could have led to misleading conclusions. For instance, mutation of several residues in the region that we now know for certain is part of the dimer interface did not detectably affect particle stability. Similarly, an internal deletion variant has been found which lacks almost half of the second central helix yet still forms particles when expressed in *E. coli* (P. Preikschat, H. Meisel, D. Krüger, and R. Ulrich, in preparation). Possibly, the unusually large contact area between the two helices from each monomer (122) maintains a sufficient number of favourable interactions even if local perturbations are introduced. The general lesson from these results is that as powerful as the interaction screening methods are the data produced with them, at least in structural terms, need to be interpreted very cautiously.

The second study used a large collection of partially random core protein variants carrying small deletions and sequence duplications throughout the primary sequence (127). A large fraction of the variants did not form particles in *E. coli*; in those that did mutations were tolerated in only few restricted regions. As these almost exclusively coincide with short loops connecting the proposed helices, these data strongly support the relevance of the EM-derived model.

In all, the heterologous expression systems combined a whole repertoire of genetic, biochemical, and biophysical techniques have thus given us a rather detailed picture of the architecture of the HBV capsid, and a very plausible model for the fold of its constituent core protein. Final confirmation of the structural model can be expected from X-ray analysis after the recent successful crystallization of recombinant core particles (115, 116) although the resolution appears to be limited to about 4 Å.

Despite these impressive results, several important aspects of HBV capsid assembly have not yet been adequately addressed in structural terms, and at present, preparing the starting material for such studies in sufficient quantities seems everything but trivial. Almost all data have been obtained with just recombinant core protein, or its assembly domain. The real replication-competent HBV nucleocapsid contains not only the specifically packaged RNA pregenome and the viral reverse transcriptase, it also provides the environment necessary for the complex process of reverse transcription. RNA and DNA containing particles must differ in some respects since only the mature DNA containing particles interact with envelope proteins to form a complete virion (128, 129). Furthermore, like any virus capsid, it has to be sufficiently stable to survive outside its host cell, yet must disintegrate at some step after entry into a new cell to allow uncoating of the packaged genome. Hence the nucleocapsid is not a static structure but rather a very dynamic assembly of changing composition. Probably the current data are just an invaluable reference to tackle in the future the much more demanding problem of analysing the core particle as a highly dynamic structure. Basically this will require preparation of complete core particles halted at different states of maturation.

4.2.2 Applications for recombinant HBV capsids

Because of the presence in its primary sequence of potent T helper epitopes and the repetitive display of regular structural features on its surface, the HBV capsid is an exceptionally efficient immunogen (130). It has therefore gained considerable interest as particulate carrier system for heterologous sequences that might induce potent B and T cell responses against the foreign epitopes (131, 132). Previous data had indicated that the strongest responses were obtained against foreign sequences inserted into the immunodominant c/e1 epitope of the core particle. Since this region is in the middle of the primary sequence however, it appeared that only rather short heterologous fragments could be inserted without interfering with formation of an assembly-competent structure. From the new structural data we now know that the c/e1 epitope overlaps with the small loop connecting the two central helices at the tips of the visible spikes. This suggested that the loop might be replaced by much larger foreign segments, provided these do not interfere with folding of the two separated core protein halves. Indeed we have recently been able to demonstrate that a complete foreign protein domain can be inserted in this site (133). As a model we used the green fluorescent protein (GFP); because GFP chromophore generation requires a native three-dimensional structure, production of fluorescent particles would directly indicate that both the core protein parts, and the inserted foreign domain are properly folded—and this was observed (Fig. 12). In addition, these particles elicited a strong humoral response against native GFP when injected into rabbits. This result has two major implications. First, it should be possible to insert, instead of GFP, protein domains from medically relevant organisms, and use the particles to efficiently induce immune responses against the displayed protein. Secondly, and given the foreign domains can be attached rigidly enough to the particle surface, they should be forced into the same highly symmetrical arrangement that allowed high-resolution EM techniques to be applied to the core protein itself. This would provide an alternative to study, by EM, the structure of small to medium-sized proteins that are refractory to crystallization or NMR.

Another application of recombinant HBV core particles with potential medical relevance was to use them as a probe to select capsid-binding peptide sequences from a phage display library (134). Indeed, a small peptide was found that binds tightly to the core protein (135). Moreover, its binding site at the tip of capsid spikes could recently be localized by cryo-EM techniques (136), suggesting that the bound peptide could interfere with envelopment. Hence it should be possible to develop recombinant HBV capsids into a useful tool for high-throughput screens aimed at finding small compound inhibitors of assembly and virion formation.

5. The enigma of HBx

Undoubtedly, new interaction screening methods represent a tremendous technical advance. However, the least understood of the mammalian hepadnavirus gene products, foresightedly called HBx, provides a telling example how the power of these screens can produce confusion rather than clarification. HBx is a regulatory protein that is conserved in all mammalian (but not the avian) hepadnaviruses. Though not essential in transfected cells (137), it is required to establish infection *in vivo*, at least in the woodchuck model (138, 139). This is one of few hard data on its physiological function. It is a modestly active, indirect transcription activator that acts on a whole variety of cellular and viral promoter/enhancer elements (140).

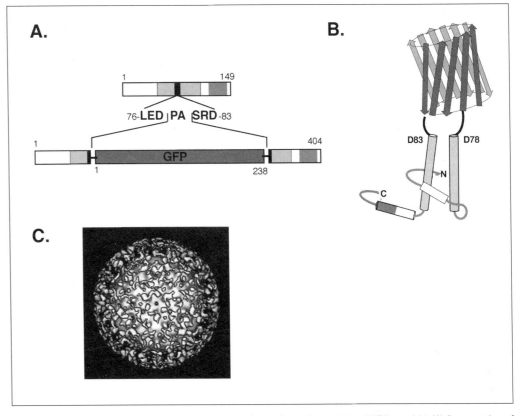

Fig. 12 Native display of foreign protein domains on the surface of recombinant HBV capsids. (A) Construction of a chimeric GFPcore gene. The complete gene for the 238 aa green fluorescent protein (GFP) was inserted into the central c/e1 epitope of the core protein's assembly domain. (B) schematic view of the structure of the chimeric GFPcore protein. GFP which forms an 11-stranded β-barrel replaces the short connecting loop between the central core protein helices. Formation of fluorescent particles upon expression of the chimeric protein in *E. coli* indicated that both the two halves of the core protein and the GFP had adopted native conformations. (C) Image reconstruction of recombinant GFPcore particles. The chimeric particles have a double-shelled appearance, with the inner shell closely resembling cores from wild-type core protein. The outer electron density originates from the GFP domains. The barrel structures are not visible because they are too flexibly attached to the particle surface. However, the presence of detectable density indicates that at least portions of the GFPs adopt positions with icosahedrally symmetric relation.

Much interest in HBx has also been spurred by its potential involvement in carcinogenesis (e.g. 141, 142, but also in this respect its role is not clear. Identical approaches, for instance cancer development in HBx transgenic mice, have led to contradictory results (143–145).

In specific *in vitro* cell culture settings, HBx may even act as an anti-oncogene (S. Schäfer and W. Gerlich, in preparation). Moreover, very plausible X-independent models for carcinogenesis can be imagined. First, although integration of HBV DNA into the host genome is not part of the normal replication cycle, liver tumours from chronic carriers frequently contain integrated HBV sequences that can act as insertional mutagens. Secondly, cancer may be the result of the permanent liver regeneration induced by decade-long attempts of the immune system to clear the organ from infected cells (146). The situation is even more complicated because HBx has been variously reported to act in the cytoplasm, or the nucleus, or both (e.g. 147), and by potential synergistic, or antagonistic, effects exerted by the apparent *trans*-activating potential of the preS-containing envelope proteins (148, 149), or truncated variants thereof (150, 151).

Viral regulatory proteins are notoriously multifunctional as illustrated, for instance, by SV40 T-antigen. Not too surprisingly, a variety of diverse cellular partners and pathways had earlier been suggested to be the target of HBx, for instance protein kinase C (152), an RNA polymerase II subunit (153), p53 (154, 155), or the protease tryptase TL2 (156). It was hoped that techniques like the two-hybrid system would soon disclose the true partners of HBx and hence elucidate its mode of action. However, it seems that in this case the new methods, rather than finding the needle, have increased the size of the haystack. There is an ever growing list of X associated or X interacting proteins (XAPs and XIPs) from diverse cellular playgrounds. Some of the more popular ones are the proteasome (157, 158), DNA repair proteins (159–162, PKC again (163), or p55sen, a protein associated with replicative senescence (164), as well as proteins of unknown function (165). Since most of the earlier proposed HBx interactions continue to be studied, e.g. with various transcription factors (166–168), RNA polymerase II (169, 170), or p53 (141), we are left with a tremendous amount of biochemically interesting data. However, for a skeptical viewer the elucidation of the true functions of HBx and its mechanism of action in the viral life cycle may appear farther away than before. This underscores dramatically the need for more feasible animal models of HBV infection to evaluate, *in vivo*, the physiological relevance of such *in vitro* derived data.

6. Conclusions and perspectives

The combination of a variety of experimental systems, from animal models and transfection of cloned virus genomes in cultured cells to *in vitro* translation and heterologous expression of individual gene products with diverse genetic, biochemical, and biophysical techniques has yielded a rather detailed picture of some of the molecular mechanisms underlying the hepadnavirus infectious cycle. However, several important aspects are not at all understood. For others we have a wealth of biochemical data without knowing their physiological relevance and, except for the capsid shell, very little is known about the structural basis of the interactions between viral and cellular components. A certainly incomplete list of important goals for the future would include the identification of the cellular receptor(s) for human HBV; the establishment of a feasible infection model for HBV or, at least, a virus that is more closely related to HBV than WHV or DHBV; an *in vitro* system for active HBV P protein; the reconstitution of an active replication initiation complex, or even of complete replication-competent nucleocapsids from purified components, at least for DHBV. With the combined efforts from various disciplines some of these goals may soon be within reach.

For lack of published data, one very powerful approach could not be adequately addressed in this review, i.e. combinatorial chemistry and high-throughput screening. With this methodology, which is mostly accessible to industry labs, inhibitors of specific interactions between viral, or viral and cellular factors, should be identified much faster than was previously possible. Because of low membrane permeability, toxicity, or other side-effects many of these compounds will probably not be suitable as therapeutic drugs; they could, however, be extremely useful tools for fundamental research on hepatitis B virus biology. It is therefore hoped that in the future the power of this technique can be exploited to its full extent by a more liberal exchange of materials and knowledge between industry and academia.

Acknowledgements

Work in the author's laboratory is supported by grants from the Deutsche Forschungsgemeinschaft (DFG), the Bundesministerium für Bildung und Forschung (BMBF), the Center for Clinical Research I of the University of Freiburg, the Land of Baden-Württemberg, and the Fonds der Chemischen Industrie. I thank the former and present members of my group for their commitment, and Dr Hubert E. Blum for providing a stimulating research environment at the University Hospital Freiburg.

References

1. Blumberg, B. S. (1997) Hepatitis B virus, the vaccine, and the control of primary cancer of the liver. *Proc. Natl. Acad. Sci. USA*, **94**, 7121.
2. Hoofnagle, J. H. (1998) Therapy of viral hepatitis. *Digestion*, **59**, 563.
3. Drees, B. L. (1999) Progress and variations in two hybrid and three hybrid technologies. *Curr. Opin. Chem. Biol.*, **3**, 64.
4. Vidal, M. and Legrain, P. (1999) Yeast forward and reverse 'n' hybrid systems. *Nucleic Acids Res.*, **27**, 919.
5. Nassal, M. (1999) Hepatitis B virus replication: Novel roles for virus-host interactions. *Intervirology*, **42**, 100.
6. Summers, J. and Mason, W. S. (1982) Replication of the genome of a hepatitis B like virus by reverse transcription of an RNA intermediate. *Cell*, **29**, 403.
7. Nassal, M. and Schaller, H. (1993) Hepatitis B virus replication. *Trends Microbiol.*, **1**, 2221.
8. Rothnie, H. M., Chapdelaine, Y., and Hohn, T. (1994) Pararetroviruses and retroviruses: a comparative review of viral structure and gene expression strategies. *Adv. Virus Res.*, **44**, 1.
9. Kiss-Laszlo, Z. and Hohn, T. (1996) Pararetro and retrovirus RNA: splicing and the control of nuclear export. *Trends Microbiol.*, **4**, 480.
10. Linial, M. L. (1999) Foamy viruses are unconventional retroviruses. *J. Virol.*, **73**, 1747.
11. Yu, S. F., Sullivan, M. D., and Linial, M. L. (1999) Evidence that the human foamy virus genome is DNA, *J. Virol.*, **73**, 1565.
12. Nassal, M. (1996) Hepatitis B virus morphogenesis. *Curr. Top. Microbiol. Immunol.*, **214**, 297.

13. Nassal, M. and Schaller, H. (1996) Hepatitis B virus replication an update. *J. Viral Hepat.*, **3**, 217.

14. Seeger, C. and Hu, J. (1997) Why are hepadnaviruses DNA and not RNA viruses? *Trends Microbiol.*, **5**, 447.

15. Kann, M., Sodeik, B., Vlachou, A., Gerlich, W. H., and Helenius, A. (1999) Phosphorylation dependent binding of hepatitis B virus core particles to the nuclear pore complex. *J. Cell Biol.*, **145**, 45.

16. Smith, G. J. R., Donello, J. E., Luck, R., Steger, G., and Hope, T. J. (1998) The hepatitis B virus post transcriptional regulatory element contains two conserved RNA stem loops which are required for function. *Nucleic Acids Res.*, **26**, 4818.

17. Donello, J. E., Loeb, J. E., and Hope, T. J. (1998). Woodchuck hepatitis virus contains a tripartite posttranscriptional regulatory element. *J. Virol.*, **72**, 5085.

18. Huang, J. and Liang, T. J. (1993) A novel hepatitis B virus (HBV) genetic element with Rev response element like properties that is essential for expression of HBV gene products. *Mol. Cell. Biol.*, **13**, 7476.

19. Huang, T., Wimler, K. M., and Carmichael, G. G. (1999). Intronless mRNA transport elements may affect multiple steps of pre mRNA processing. *EMBO J.*, **18**, 1642.

20. Heise, T., Guidotti, L. G., Cavanaugh, V. J., and Chisari, F. V. (1999). Hepatitis B virus RNA binding proteins associated with cytokine induced clearance of viral RNA from the liver of transgenic mice. *J. Virol.*, **73**, 474,

21. Lenhoff, R. J. and Summers, J. (1994). Coordinate regulation of replication and virus assembly by the large envelope protein of an avian hepadnavirus. *J. Virol.*, **68**, 4565.

22. Bruss, V. (1997) A short linear sequence in the pre S domain of the large hepatitis B virus envelope protein required for virion formation. *J. Virol.*, **71**, 9250.

23. Bruss, V., Gerhardt, E., Vieluf, K., and Wunderlich, G. (1996). Functions of the large hepatitis B virus surface protein in viral particle morphogenesis. *Intervirology*, **39**, 23.

24. Swameye, I. and Schaller, H. (1997). Dual topology of the large envelope protein of duck hepatitis B virus: determinants preventing pre S translocation and glycosylation. *J. Virol.*, **71**, 9434.

25. De Meyer, S., Gong, Z. J., Suwandhi, W., van Pelt, J., Soumillion, A., and Yap, S. H. (1997) Organ and species specificity of hepatitis b virus (HBV) infection: a review of literature with a special reference to preferential attachment of HBV to human hepatocytes. *J. Viral Hepat.*, **4**, 145.

26. Kuroki, K., Cheung, R., Marion, P. L., and Ganem, D. (1994) A cell surface protein that binds avian hepatitis B virus particles. *J. Virol.*, **68**, 2091.

27. Tong, S., Li, J., and Wands, J. R. (1995) Interaction between duck hepatitis B virus and a 170 kilodalton cellular protein is mediated through a neutralizing epitope of the pre S region and occurs during viral infection. *J. Virol.*, **69**, 7106.

28. Kuroki, K., Eng, F., Ishikawa, T., Turck, C., Harada, F., and Ganem, D. (1995) gp180, a host cell glycoprotein that binds duck hepatitis B virus particles, is encoded by a member of the carboxypeptidase gene family. *J. Biol. Chem.*, **270**, 15022.

29. Eng, F. J., Novikova, E. G., Kuroki, K., Ganem, D., and Fricker, L. D. (1998) gp180, a protein that binds duck hepatitis B virus particles, has metallocarboxypeptidase D like enzymatic activity. *J. Biol. Chem.*, **273**, 8382.

30. Breiner, K. M., Urban, S., and Schaller, H. (1998) Carboxypeptidase D (gp180), a Golgi resident protein, functions in the attachment and entry of avian hepatitis B viruses. *J. Virol.*, **72**, 8098.

31. Urban, S., Breiner, K. M., Fehler, F., Klingmüller, W., and Schaller, H. (1998) Avian

hepatitis B virus infection is initiated by the interaction of a distinct pre S subdomain with the cellular receptor gp180. *J. Virol.*, **72**, 8089.

32. Urban, S., Kruse, C., and Multhaup, G. (1999). A soluble form of the avian hepatitis B virus receptor. Biochemical characterization and functional analysis of the receptor ligand complex. *J. Biol. Chem.*, **274**, 5707.

33. de Roda Husman, A. M. and Schuitemaker, H. (1998) Chemokine receptors and the clinical course of HIV 1 infection. *Trends Microbiol.*, **6**, 244.

34. Marion, P. L. (1988) Use of animal models to study hepatitis B virus. *Prog. Med. Virol.*, **35**, 43.

35. Purcell, R. H. (1993) The discovery of the hepatitis viruses. *Gastroenterology*, **104**, 955.

36. Roggendorf, M. and Tolle, T.K. (1995) The woodchuck: an animal model for hepatitis B virus infection in man. *Intervirology*, **38**, 100.

37. Buendia, M. A. (1992) Mammalian hepatitis B viruses and primary liver cancer. *Semin. Cancer Biol.*, **3**, 309.

38. Mason, W. S., Cullen, J., Moraleda, G., *et al.* (1998) Lamivudine therapy of WHV infected woodchucks. *Virology*, **245**, 18.

39. Logrengel, B., Lu, M., and Roggendorf, M. (1998) Molecular cloning of the woodchuck cytokines: TNF alpha, IFN gamma, and IL 6. *Immunogenetics*, **47**, 332.

40. Obert, S., Zachman-Brand, B., Deindl, E., Tucker, W., Bartenschlager, R., and Schaller, H. (1996) A splice hepadnavirus RNA that is essential for virus replication. *EMBO J.*, **15**, 2565.

41. Schultz, U. and Chisari, F. V. (1999) Recombinant duck interferon gamma inhibits duck hepatitis B virus replication in primary hepatocytes. *J. Virol.*, **73**, 3162.

42. Lanford, R. E., Chavez, D., Brasky, K. M., Burns, R. B. R., and Rico-Hesse, R. (1998) Isolation of a hepadnavirus from the woolly monkey, a New World primate. *Proc. Natl. Acad. Sci. USA*, **95**, 5757.

43. Yan, R. Q., Su, J. J., Huang, D. R., Gan, Y. C., Yang, C., and Huang, G. H. (1996) Human hepatitis B virus and hepatocellular carcinoma. I. Experimental infection of tree shrews with hepatitis B virus. *J. Cancer Res. Clin. Oncol.*, **122**, 283.

44. Walter, E., Keist, R., Niederost, B., Pult, I., and Blum, H. E. (1996). Hepatitis B virus infection of tupaia hepatocytes *in vitro* and *in vivo*. *Hepatology*, **24**, 1.

45. Petersen, J., Dandri, M., Gupta, S., and Rogler, C. E. (1998) Liver repopulation with xeno-genic hepatocytes in B and T cell deficient mice leads to chronic hepadnavirus infection and clonal growth of hepatocellular carcinoma. *Proc. Natl. Acad. Sci. USA*, **95**, 310.

46. Ilan, E., Burakova, T., Dagan, S., *et al.* (1999) The hepatitis B virus trimera mouse: a model for human HBV infection and evaluation of anti HBV therapeutic agents. *Hepatology*, **29**, 553.

47. Akbar, S. K. and Onji, M. (1998) Hepatitis B virus (HBV) transgenic mice as an investi-gative tool to study immunopathology during HBV infection. *Int. J. Exp. Pathol.*, **79**, 279.

48. Chisari, F. V. (1996) Hepatitis B virus transgenic mice: models of viral immunobiology and pathogenesis. *Curr. Top. Microbiol. Immunol.*, **206**, 149.

49. Guidotti, L. G., Ishikawa, T., Hobbs, M. V., Matzke, B., Schreiber, R., and Chisari, F. V. (1996) Intracellular inactivation of the hepatitis B virus by cytotoxic T lymhocytes. *Immunity*, **4**, 25.

50. Cavanaugh, V. J., Guidotti, L. G., and Chisari, F. V. (1998) Inhibition of hepatitis B virus replication during adenovirus and cytomegalovirus infections in transgenic mice. *J. Virol.*, **72**, 2630.

51. Guidotti, L. G., Rochford, R., Chung, J., Shapiro, M., Purcell, R., and Chisari, F. V. (1999). Viral clearance without destruction of infected cells during acute HBV infection. *Science*, **284**, 825.

52. Takahashi, H., Fujimoto, J., Hanada, S., and Isselbacher, K. J. (1995) Acute hepatitis in rats expressing human hepatitis B virus transgenes. *Proc. Natl. Acad. Sci. USA*, **92**, 1470.

53. Delaney, W. E. T. and Isom, H. C. (1998) Hepatitis B virus replication in human HepG2 cells mediated by hepatitis B virus recombinant baculovirus. *Hepatology*, **28**, 1134.

54. Junker-Niepmann, M., Bartenschlager, R., and Schaller, H. (1990) A short cis acting sequence is required for hepatitis B virus pregenome encapsidation and sufficient for packaging of foreign RNA. *EMBO J.*, **9**, 3389.

55. Bartenschlager, R. and Schaller, H. (1992) Hepadnavirus assembly is initiated by polymerase binding to the encapsidation signal in the viral RNA genome. *EMBO J.*, **11**, 3413.

56. Berkowitz, R., Fisher, J., and Goff, S. P. (1996) RNA packaging. *Curr. Top. Microbiol. Immunol.*, **214**, 177.

57. Knaus, T. and Nassal, M. (1993). The encapsidation signal on the hepatitis B virus RNA pregenome forms a stem loop structure that is critical for its function. *Nucleic Acids Res.*, **21**, 3967.

58. Pollack, J. R. and Ganem, D. (1993) an RNA stem loop structure directs hepatitis B virus genomic RNA encapsidation. *J. Virol.*, **67**, 3254.

59. Calvert, J. and Summers, J. (1994). Two regions of an avian hepadnavirus RNA pregenome are required in cis for encapsidation. *J. Virol.*, **68**, 2084.

60. Rieger, A. and Nassal, M. (1995) Distinct requirements for primary sequence in the 5′ and 3′ part of a bulge in the hepatitis B virus RNA encapsidation signal revealed by a combined *in vivo* selection/*in vitro* amplification system. *Nucleic Acids Res.*, **23**, 3909.

61. Pollack, J. R. and Ganem, D. (1994) Site specific RNA binding by a hepatitis B virus reverse transcriptase initiates two distinct reactions: RNA packaging and DNA synthesis. *J. Virol.*, **68**, 5579.

62. Wang, G. H. and Seeger, C. (1993) Novel mechanism for reverse transcription in hepatitis B viruses. *J. Virol.*, **67**, 6507.

63. Nassal, M. and Rieger, A. (1996) A bulged region of the hepatitis B virus RNA encapsidation signal contains the replication origin for discontinuous first strand DNA synthesis. *J. Virol.*, **70**, 2764.

64. Rieger, A. and Nassal, M. (1996) Specific hepatitis B virus minus strand DNA synthesis requires only the 5′ encapsidation signal and the 3′ proximal direct repeat DR1. *J. Virol.*, **70**, 585.

65. Weber, M., Bronsema, V., Bartos, H., Bosserhoff, A., Bartenschlager, R., and Schaller, H. (1994) Hepadnavirus P protein utilizes a tyrosine residue in the TP domain to prime reverse transcription. *J. Virol.*, **68**, 2994.

66. Zoulim, F. and Seeger, C. (1994). Reverse transcription in hepatitis B viruses is primed by a tyrosine residue f the polymerase. *J. Virol.*, **68**, 6.

67. Wang, G. H. and Seeger, C. (1992) The reverse transcriptase of hepatitis B virus acts as a protein primer for viral DNA synthesis. *Cell*, **71**, 663.

68. Beck, J. and Nassal, M. (1996) A sensitive procedure for mapping the boundaries of RNA elements binding in vitro translated proteins defines a minimal hepatitis B virus encapsidation signal. *Nucleic Acids Res.*, **24**, 4364.

69. Urban, M., McMillan, D. J., Canning, G., Newell, A., Brown, E., Mills, J. S., *et al.* (1998) *In vitro* activity of hepatitis B virus polymerase: requirement for distinct metal ions and the viral epsilon stem loop. *J. Gen Virol.*, **79**, 1121.

70. Beck, J., Bartos., and Nassal, M. (1997) Experimental confirmation of a hepatitis B virus (HBV) epsilon like bulge and loop structure in avian HBV RNA encapsidation signals. *Virology*, **227**, 500.

71. Beck, J. and Nassal, M. (1997) Sequence and structure specific determinants in the interaction between the RNA encapsidation signal and reverse transcriptase of avian hepatitis B viruses. *J. Virol.*, **71**, 4971.

72. Beck, J. and Nassal, M. (1998) Formation of a functional hepatitis B virus replication initiation complex involves a major structural alteration in the RNA template. *Mol. Cell. Biol.*, **18**, 6265.

73. Draper, D. E. (1995) Protein RNA recognition. *Annu. Rev. Biochem.*, **64**, 593.

74. Tavis, J. E. and Ganem, D. (1996) Evidence for activation of the hepatitis B virus polymerase by binding of its RNA template. *J. Virol.*, **70**, 5741.

75. Tavis, J. E., Masse, B., and Gong, Y. (1998) The duck hepatitis B virus polymerase is activated by its RNA packaging signal, epsilon. *J. Virol.*, **72**, 5789.

76. Hu, J. and Seeger, C. (1996) Hsp90 is required for the activity of a hepatitis B virus reverse transcriptase. *Proc. Natl. Acad. Sci. USA*, **93**, 1060.

77. Smith, D. F. (1998) Sequence motifs shared between chaperone components participating in the assembly of progesterone receptor complexes. *Biol. Chem.*, **379**, 283.

78. Scheibel, T. and Buchner, J. (1998) The Hsp90 complex a super chaperone machine as a novel drug target. *Biochem. Pharmacol.*, **56**, 675.

79. Buchner, J. (1999) Hsp90 & Co. a holding for folding. *Trends Biochem. Sci.*, **24**, 136.

80. Frydman, D. and Hohfeld, J. (1997) Chaperones get in touch: the Hip Hop connection. *Trends Biochem. Sci.*, **22**, 87.

81. Hu, J., Toft, D. O., and Seeger, C. (1997) Hepadnavirus assembly and reverse transcription require a multi component chaperone complex which is incorporated into nucleocapsids. *EMBO J.*, **16**, 59.

82. Dittmar, K. D., Hutchison, K. A., Owens-Grillo, J. K., and Pratt, W. B. (1996) Reconstitution of the steroid receptor.hsp90 heterocomplex assembly system of rabbit reticulocyte lysate. *J. Biol. Chem.*, **271**, 12833.

83. Perri, S. and Ganem, D. (1997) Effects of mutations within and adjacent to the terminal repeats of hepatitis B virus pregenomic RNA on viral DNA synthesis. *J. Virol.*, **71**, 8448.

84. Perri, S. and Ganem, D. (1996) A host factor that binds near the termini of hepatitis B virus pregenomic RNA. *J. Virol.*, **70**, 6803.

85. Gold, L., Brown, D., He, Y., Shtatland, T., Singer, B. S., and Wu, Y. (1997) From oligonucleotide shapes to genomic SELEX: novel biological regulatory loops. *Proc. Natl. Acad. Sci. USA*, **94**, 59.

86. Lanford, R. E., Notvall, L., and Beames, B. (1995) Nucleotide priming and reverse transcriptase activity of hepatitis B virus polymerase expressed in insect cells. *J. Virol.*, **69**, 4431.

87. Lanford, R. E., Notvall, L., Lee, H., and Beames, B. (1997). Transcomplementation of nucleotide priming and reverse transcription between independently expressed TP and RT domains of the hepatitis B virus reverse transcriptase. *J. Virol.*, **71**, *2996.*

88. Seifer, M., Hamatake, R., Bifano, M., and Standring, D. N. (1998) Generation of replication competent hepatitis B virus nucleocapsids in insect cells. *J. Virol.*, **72**, 2765.

89. Lanford, R. E., Kim, Y. H., Lee, H., Notvall, L., and Beames, B. (1999) Mapping of the hepatitis B virus reverse transcriptase TP and RT domains by transcomplementation for nucleotide priming and by protein protein interaction. *J. Virol.*, **73**, 1885.

90. Roingeard, P and Sureau, C. (1998) Ultrastructural analysis of hepatitis B virus in HepG2 transfected cells with special emphasis on subviral filament morphogenesis. *Hepatology*, **28**, 1128.

91. Garoff, H., Hewson, R., and Opstelten, D. J. E. (1998) Virus maturation by budding. *Microbiol. Mol. Biol. Rev.*, **62**, 1171.

92. Bruns, M., Miska, S., Chassot, S., and Will, H. (1998) Enhancement of hepatitis B virus infection by noninfectious subviral particles. *J. Virol.*, **72**, 1462.

93. Tolle, T. K., Glebe, D., Linder, M., Schmitt, S., Geyer, R., *et al.* (1998) Structure and glycosylation patterns of surface proteins from woodchuck hepatitis virus. *J. Virol.*, **72**, 9978.

94. Schmitt, S., Glebe, D., Alving, K. *et al.* (1999) Analysis of the pre S2 N and O linked glycans of the M surface protein from human hepatitis B virus. *J. Biol. Chem.*, **274**, 11945.

95. Hughson, F. M. (1999) Structure snared at last. *Curr. Biol.*, **9**, R49.

96. Skehel, J. J. and Wiley, D. C. (1998) Coiled coils in both intracellular vesicle and viral membrane fusion. *Cell*, **95**, 871.

97. Cheng, R. H., Kuhn, R. J., Olson, N. H., Rossmann, M. G., Choi, H. K., Smith, T. J., *et al.* (1995) Nucleocapsid and glycoprotein organization in an enveloped virus. *Cell*, **80**, 621.

98. Fuller, S. D., Berriman, J. A., Butcher, S. J., and Gowen, B. E. (1995) Low pH induces swiveling of the glycoprotein heterodimers in the Semliki Forest virus spike complex. *Cell*, **81**, 715.

99. Prange, R., Werr, M., and Löffler-Mary, H. (1999) Chaperones involved in hepatitis B virus morphogenesis. *Biol. Chem.*, **380**, 305.

100. Löffler-Mary, H., Werr, M., and Prange, R. (1997) Sequence specific repression of cotranslational translocation of the hepatitis B virus envelope proteins coincides with binding of heat shock protein Hsc70. *Virology*, **235**, 144.

101. Werr, M. and Prange, R. (1998) Role for calnexin and N linked glycosylation in the assembly and secretion of hapatitis B virus middle envelope protein particles. *J. Virol.*, **72**, 778.

102. Cohen,. B. J. and Richmond, J. E. (1982) Electron microscopy of hepatitis B core antigen synthesized in *E. coli. Nature*, **296**, 677.

103. Ganem, D. (1991) Assembly of hepadnavirus virions and subviral particles. *Curr. Top Microbiol. Immunol.*, **168**, 61.

104. Birnbaum, F. and Nassal, M. (1990) Hepatitis B virus nucleocapsid assembly: primary structure requirements in the core protein. *J. Virol.*, **64**, 3319.

105. Hatton, T., Zhou, S., and Standring, D. N. (1992) RNA and DNA binding activities in hepatitis B virus capsid protein: a model for their roles in viral replication. *J. Virol.*, **66**, 5232.

106. Nassal, M. (1992) The arginine rich domain of the hepatitis B virus core protein is required for pregenome encapsidation and productive viralpositive strand DNA synthesis but not for virus assembly. *J. Virol.*, **66**, 4107.

107. Nassal, M., Rieger, A., and Steinau, O. (1992) Topological analysis of the hepatitis B virus core particle by cysteine cysteine cross linking. *J. Mol. Biol.*, **225**, 1013.

108. Zheng, J., Schodel, F., and Peterson, D. L. (1992) The structure of hepadnavirus core antigens. Identification of free thiols and determination of the disulfide bonding pattern. *J. Biol. Chem.*, **267**, 9422.

109. Zhou, S. and Standring, D. N. (1992) Cys residues of the hepatitis B virus capsid protein are not essential for the assembly of viral core particles but can influence their stability. *J. Virol.*, **66**, 5393.

110. Nassal, M. (1992) Conserved cysteines of the hepatitis B virus core protein are not required for assembly of replication competent core particles nor for their envelopment. *Virology*, **190**, 499.

111. Seifer, M., Zhou, S., and Standring, D. N. (1993) A micromolar pool of antigenically distinct precursors is required to initiate cooperative assembly of hepatitis B virus capsids in Xenopus oocytes. *J. Virol.*, **67**, 249.

112. Seifer, M. and Standring, D. N. (1995) Assembly and antigenicity of hepatitis B virus core particles. *Intervirology*, **38**, 47.

113. Zhou, S. and Standring, D. N. (1992) Hepatitis B virus capsid particles are assembled from core protein dimer precursors. *Proc. Natl. Acad. Sci. USA*, **89**, 10046.

114. Wingfield, P. T., Stahl, S. J., Williams, R. W., and Steven, A. C. (1995). Hepatitis core antigen produced in *Escherichia coli*: subunit composition, conformational analysis, and *in vitro* capsid assembly. *Bochemistry*, *34*, 4919.

115. Wynne, S. A., Leslie, A. G., Butler, P. J., and Crowther, R. A. (1999) Crystallization of hepatitis B virus core protein shells: determination of cryoprotectant conditions and preliminary X ray characterization. *Acta Crystallogr. D Biol. Crystallogr.*, **55**, 557.

116. Zlotnick, A., Palmer, I., Kaufman, J. D., Stahl, S. J., Steven, A. C., and Wingfield, P. T. (1999) Separation and crystallization of T = 3 and T = 4 icosahedral complexes of the hepatitis B virus core protein. *Acta Crystallogr. D Biol. Crystallogr.*, **55**, 717.

117. Mancini, E. J., de Haas, F., and Fuller, S. D. (1997) High resolution icosahedral reconstruction: fulfilling the promise of cryo electron microscopy. *Structure*, *5*, 741.

118. Crowther, R. A., Kiselev, N. A., Böttcher, B., Berriman, J. A., Borisova, G. P., Ose, V., *et al.* (1994) Three dimensional structure of hepatitis B virus core particles determined by electron cryomicroscopy. *Cell*, **77**, 943.

119. Kenney, J. M., von Bonsdorff, C. H., Nassal, M., and Fuller, S. D. (1995) Evolutionary conservation in the hepatitis B virus core structure: comparison of human and duck cores. *Structure*, **3**, 1009.

120. Böttcher, B., Wynne, S. A., and Crowther, R. A. (1997) Determination of the fold of the core protein of hepatitis B virus by electron cryomicroscopy. *Nature*, **386**, 88.

121. Conway, J. F., Cheng, N., Zlotnick, A., Wingfield, P. T., Stahl, S. J., and Steven, A. C. (1997) Visualization of a 4 helix bundle in the hepatitis B virus capsid by cryo electron microscopy. *Nature*, **386**, 91.

122. Zlotnick, A., Stahl, S. J., Wingfield, P. T., Conway, J. F., Cheng, N., and Steven, A. C. (1998) Shared motifs of the capsid proteins of hepadnaviruses and retroviruses suggest a common evolutionary origin. *FEBS Lett.*, **431**, 301.

123. Conway, J. F., Cheng, N., Zlotnick, A., Stahl, S. J., Wingfield, P. T., Belnap, D. M., *et al.* (1998) Hepatitis B virus capsid: localization of the putative immunodominant loop (residues 78 to 83) on the capsid surface, and implications for the distinction between c and e antigens. *J. Mol. Biol.*, **279**, 1111.

124. Conway, J. F., Cheng, N., Zlotnick, A., Stahl, S. J., Wingfield, P. T., and Steven, A. C. (1998) Localization of the N terminus of hepatitis B virus capsid protein by peptide based difference mapping from cryoelectron microscopy. *Proc. Natl. Acad. Sci. USA*, **95**, 14622.

125. Kramer, A. and Schneider-Mergener, J. (1998) Synthesis and screening of peptide libraries on continuous cellulose membrane supports. *Methods Mol. Biol.*, **87**, 25.

126. König, S., Beterams, G., and Nassal, M. (1998) Mapping of homologous interaction sites in the hepatitis B virus core protein. *J. Virol.*, **72**, 4997.

127. Köschel, M., Thomssen, R., and Bruss, V. (1999) Extensive mutagenesis of the hepatitis B virus core gene and mapping of mutations that allow capsid formation. *J. Virol.*, **73**, 2153.

128. Gerelsaikhan, T., Tavis, J. E., and Bruss, V. (1996) Hepatitis B virus nucleocapsid envelopment does not occur without genomic DNA synthesis. *J. Virol.*, **70**, 4269.

129. Wei, Y., Tavis, J. E., and Ganem, D. (1996) Relationship between viral DNA synthesis and virion envelopment in hepatitis B viruses. *J. Virol.*, **70**, 6455.

130. Milich, D. R., Chen, M., Schodel, F., Peterson, D. L., Jones, J. E., and Hughes, J. L. (1997) Role of B cells in antigen presentation of the hepatitis B core. *Proc. Natl. Acad. Sci. USA*, **94**, 14648.

131. Milich, D. R., Peterson, D. L., Zheng, J., Hughes, J. L., Wirtz, R., and Schodel, F. (1995) The hepatitis nucleocapsid as a vaccine carrier moiety. *Ann. NY Acad. Sci.*, **754**, 187.

132. Ulrich, R., Nassal, M., Meisel, H., and Kruger, D. H. (1998) Core particles of hepatitis B virus as carrier for foreign epitopes. *Adv. Virus Res.*, **50**, 141.

133. Kratz, P. A., Böttcher, B., and Nassal, M. (1999) Native display of complete foreign protein domains on the surface of hepatitis B virus capsids. *Proc. Natl. Acad. Sci. USA*, **96**, 1915.

134. Rodi, D. J. and Makowski, L. (1999) Phage display technology finding a needle in a vast molecular haystack. *Curr. Opin. Biotechnol.*, **10**, 87.

135. Dyson, M. R. and Murray, K. (1995) Selection of peptide inhibitors of interactions involved in complex protein assemblies: association of the core and surface antigens of hepatitis B virus. *Proc. Natl. Acad. Sci. USA*, **92**, 2194.

136. Böttcher, B., Tsuji, N., Takahashi, H., Dyson, M. R., Zhao, S., Crowther, R. A., *et al.* (1998) Peptides that block hepatitis B virus assembly: analysis by cryomicroscopy, mutagenesis and transfection. *EMBO J.*, **17**, 6839.

137. Blum, H. E., Zhang, Z. S., Galun, E., von Weizsäcker, F., Garner, B., Liang, T. J., *et al.* (1992) Hepatitis B virus X protein is not central to the viral life cycle *in vitro*. *J. Virol.*, **66**, 1223.

138. Chen, H. S., Kaneko, S., Girones, R., *et al.* (1993) The woodchuck hepatitis virus X gene is important for establishment of virus infection in woodchucks. *J. Virol.*, **67**, 1218.

139. Zoulim, F., Saputelli, J., and Seeger, C. (1994). Woodchuck hepatitis virus X protein is required for viral infection *in vivo*. *J. Virol.*, **68**, 2026.

140. Rossner, M. T. (1992) Review: hepatitis B virus X gene product: a promiscuous transcriptional activator. *J. Med. Virol.*, **36**, 101.

141. Feitelson, M. A. and Duan, L. X. (1997) Hepatitis B virus X antigen in the pathogenesis of chronic infections and the development of hepatocellular carcinoma. *Am. J. Pathol.*, **150**, 1141.

142. Cromlish, J. A. (1996) Hepatitis B virus induced hepatocellular carcinoma: possible roles for HBx. *Trends Microbiol.*, **4**, 270.

143. Reifenberg, K., Lohler, J., Pudollek, H. P., Schmitteckert, E., Spindler, G., Köck, J., *et al.* (1997) Long term expression of the hepatitis B virus core e and X proteins does not cause pathologic changes in transgenic mice. *J. Hepatol.*, **26**, 119.

144. Lee, T. H., Finegold, M. J., Shen, R. F., DeMayo, J. L., Woo, S. L., and Butel, J. S. (1990) Hepatitis B virus transactivator X protein is not tumorigenic in transgenic mice. *J. Virol.*, **64**, 5939.

145. Koike, K., Moriya, K., Iino, S., Yotsuyanagi, H., Endo, Y., Miyamura, T., *et al.* (1994) High level expression of hepatitis B virus HBx gene and hepatocarcinogenesis in transgenic mice. *Hepatology*, **19**, 810.

146. Cisari, F. V. and Ferrari, C. (1995). hepatitis B virus immunopathogenesis. *Annu. Rev. Immunol.*, **13**, 29.

147. Doria, M., Klein, N., Lucito, R., and Schneider, R. J. (1995). The hepatitis B virus HBx protein is a dual specificity cytoplasmic activator of Ras and nuclear activator of transcription factors. *EMBO J.*, **14**, 4747.

148. Hildt, E., Saher, G., Bruss, V., and Hofschneider, P. H. (1996). The hepatitis B virus large surface protein (LHBs) is a transcriptional activator. *Virology*, **225**, 235.

149. Rothmann, K., Schnolzer, M., Radziwill, G., Hildt, E., Moelling, K., and Schaller, H. (1998) Host cell virus cross talk: phosphorylation of a hepatitis B virus envelope protein mediates intracellular signaling. *J. Virol.*, **72**, 10138.

150. Hildt, E. and Hofschneider, P. H. (1998) The PreS2 activators of the hepatitis B virus: activators of tumour promoter pathways. *Recent Results Cancer Res.*, **154**, 315.

151. Henkler, F. F. and Koshy, R. (1996) Hepatitis B virus transcriptional activators: mechanisms and possible role in oncogenesis. *J. Viral Hepat.*, **3**, 109.

152. Kekule, A. S., Lauer, U., Weiss, L., Luber, B., and Hofschneider, P. H. (1993) Hepatitis B virus transactivator HBx uses a tumour promoter signalling pathway. *Nature*, **361**, 742.

153. Cheong, J. H., Yi, M., Lin, Y., and Murakami, S. (1995) Human RPB5, a subunit shared by eukaryotic nuclear RNA polymerases, binds human hepatitis B virus X protein and may play a role in X transactivation. *EMBO J.*, **14**, 143.

154. Feitelson, M. A., Zhu, M., Duan, L. X., and London, W. T. (1993) Hepatitis B x antigen and p53 are associated in vitro and in liver tissues from patients with primary hepatocellular carcinoma. *Oncogene*, **8**, 1109.

155. Wang, X. W., Forrester, K., Yeh, H., Feitelson, M. A., Gu, J. R., and Harris, C. C. (1994) Hepatitis B virus X protein inhibits p53 sequence specific DNA binding, transcriptional activity, and association with transcription factor ERCC3. *Proc. Natl. Acad. Sci. USA*, **91**, 2230.

156. Takada, S., Kido, H., Fukutomi, A., Mori, T., and Koike, K. (1994) Interaction of hepatitis B virus X protein with a serine protease, tryptase TL2 as an inhibitor. *Oncogene*, **9**, 341.

157. Huang, J., Kwong, J., Sun, E. C., and Liang, T. J. (1996) Proteasome complex as a potential cellular target of hepatitis B virus X protein. *J. Virol.*, **70**, 5582.

158. Seeger, C. (1997) The hepatitis B virus X protein: the quest for a role in viral replication and pathogenesis. *Hepatology*, **25**, 496.

159. Sitterlin, D., Lee, T. H., Prigent, S., Tiollais, P., Butel, J. S., and Transy, C. (1997) Interaction of the UV damaged DNA binding protein with hepatitis B virus X protein is conserved among mammalian hepadnaviruses and restricted to transactivation proficient X insertion mutants. *J. Virol.*, **71**, 6194.

160. Butel, J. S., Lee, T. H., and Slagle, B. L. (1996) Is the DNA repair system involved in hepatitis B virus mediated hepatocellular carcinogenesis? *Trends Microbiol.*, **4**, 119.

161. Lee, T. H., Elledge, S. J., and Butel, J. S. (1995) Hepatitis B virus X protein interacts with a probable cellular DNA repair protein. *J. Virol.*, **69**, 1107.

162. Jia, L., Wang, X. W., and Harris, C. C. (1999) Hepatitis B virus X protein inhibits nucleotide excision repair. *Int. J. Cancer.*, **80**, 875.

163. Cong, Y. S., Yao, Y. L., Yang, W. M., Kuzhandaivelu, N., and Seto, E. (1997) The hepatitis B virus X associated protein, XAP3, is a protein kinase C binding protein. *J. Biol. Chem.*, **272**, 16482.

164. Sun, B. S., Zhu, X., Clayton, M. M., Pan, J., and Feitelson, M. A. (1998) Identification of a protein isolated from senescent human cells that binds to hepatitis B virus X antigen. *Hepatology*, **27**, 228.

165. Melegari, M., Scaglioni, P. P., and Wands, J. R. (1998) Cloning and characterization of a novel hepatitis B virus x binding protein that inhibits viral replication. *J. Virol.*, **72**, 1737.

166. Choi, B. H., Park, G. T., and Rho, H. M. (1999) Interaction of hepatitis B viral X protein and CCAAT/enhancer binding protein alpha synergistically activates the hepatitis B viral enhancer II/pregenomic promoter. *J. Biol. Chem.*, **274**, 2858.

167. Haviv, I., Matza, Y., and Shaul, Y. (1998) pX, the HBV encoded coactivator, suppresses the phenotypes of TBP and TAFII250 mutants. *Genes Dev.*, **12**, 1217.

168. Lara-Pezzi, E., Armesilla, A. L., Majano, P. L., Redondo, J. M., and Lopez-Cabrera, M. (1998) The hepatitis B virus X protein activates nuclear factor of activated T cells (NF AT) by a cyclosporin A sensitive pathway. *EMBO J.*, **17**, 7066.

169. Lin, Y., Tang, H., Nomura, T., Dorjsuren, D., Hayashi, N., Wei, W., *et al.* (1998) The hepatitis B virus X protein is a co activator of activated transcription that modulates the transcription machinery and distal binding activators. *J. Biol. Chem.*, **273**, 27097.

170. Dorkjsuren, D., Lin, Y., Wei, W., Yamashita, T., Nomura, T., Hayashi, N., *et al.* (1998) RMP, a novel RNA polymerase II subunit 5 interacting protein, counteracts transactivation by hepatitis B virus X protein. *Mol. Cell. Biol.*, **18**, 7546.

2 | Protein–protein interactions in papillomavirus replication

DENNIS J. MCCANCE

1. Introduction

Certain members of the human papillomavirus genus are the etiological agents of lower genital tract cancers, in particular cervical cancer, and so an understanding of the protein–protein interactions involved in viral replication would be important if antiviral therapy were to become a reality. Viral DNA replication occurs in the upper half of stratified epithelium and this poses a problem for the virus. The viruses infect the basal epithelial cells and when these cells divide, one daughter cell moves up through the epithelium and undergoes terminal differentiation (Fig. 1A). This situation, if allowed to develop, will inhibit the ability of the virus to replicate, since differentiating cells will not contain the replicative machinery for DNA replication. Therefore papillomaviruses have acquired a strategy which involves stimulating cells into S-phase of the cell cycle where abundant replicative enzymes will be present. However, there is an important caveat to this scenario, that is, for viral DNA amplification to proceed, a certain level of keratinocyte differentiation is necessary for the switch from the early promoter to the late promoter and capsid protein production. Therefore the virus has the task of uncoupling S-phase progression from partial cell differentiation. In this chapter I will concentrate on what is known of the protein–protein interactions involved in the replication of HPV-16, but will refer to other HPV types in comparative terms or where little or nothing is known for this particular HPV type.

Human papillomaviruses replicate in either cutaneous surfaces or mucosal surfaces and it is at the latter sites that most of the cancers arise. The virus codes for three early proteins, E6, E7, and E5, which are responsible for altering the normal program of keratinocyte differentiation to facilitate replication of the viral DNA (Fig. 1B). There are two proteins, E1 and E2, which appear essential for the actual replication of the genome along with a number of cellular proteins, such as DNA polymerases. I will first discuss the roles of the early proteins in setting up the cellular environment, which allows the replication of HPV genomes.

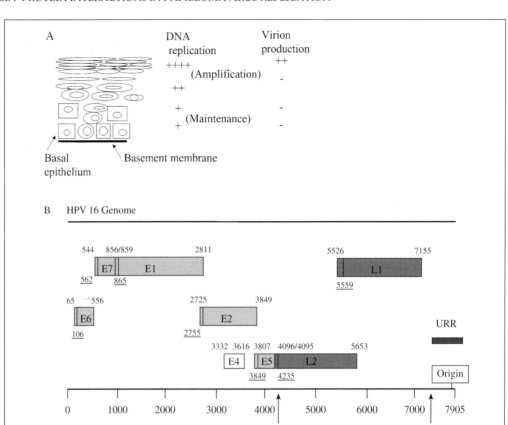

Fig. 1 (A) Schematic of stratifying epithelium showing the areas in the epithelium where the major viral DNA replication and virion formation occurs. (B) Diagram of the HPV-16 open reading frames (ORFs) indicating the base pairs of the ORDs and the starting ATG for each protein (underlined). The early (pA$_e$) and late (pA$_L$) polyadenylation signals are indicated. Also shown is the origin of replication in the upstream regulatory region (URR).

2. Early HPV proteins

The three early proteins all have properties, which are consistent with a role in altering the course of keratinocyte differentiation and stimulating cells into the S-phase of the cells cycle, as they exert their effects through the interaction with various cellular proteins, many of which are involved in cell differentiation and control of the cell cycle progression.

2.1 HPV-16 E6

E6 is 151 amino acids in length and contains two zinc fingers, which bind Zn^{2+} *in vitro* and, as will be seen later, are essential for most of the functions of the protein.

There is an upstream ATG, which if used would produce a protein of 158 amino acids, but it is not clear if both forms are produced *in vivo*. E6, is a multifunctional protein and can bind to at least four cellular proteins, E6 associated protein (E6AP) (51, 52), p 53, E6 binding protein (E6BP) (3), and Bak (4). It has been shown to activate transcription from some promoters and inhibit from others (5–7) and as we will see later this is partly through the ability of E6 to bind to E6AP and p53 in a trimer complex, resulting in the degradation of p53. E6 has been shown to immortalize human primary mammary gland cells (8), but not human primary keratinocytes, which are the natural host cell, although E6 does co-operate with E7 to immortalize keratinocytes (9). *In vivo* research on E6 has been hampered because in cells the protein is produced at low levels and detection is difficult, while purification for structural studies is hard because of the insolubility of the protein. Therefore apart from immortalization and transformation studies carried in epithelial cells and mouse fibroblasts, most of the research on E6 has been carried out *in vitro* using glutathione-*S*-transferase fusion proteins and *in vitro* transcription/translation-produced proteins. Consequently, it is difficult to determine how important these *in vitro* interactions are to the *in vivo* action of E6.

2.1.1   E6, p53, and E6AP interactions

The E6 from high-risk HPV types such as HPV-16 and -18 cause the degradation of p53 through the ubiquitin proteolytic pathway (10), which is a major mechanism for the breakdown of protein and amino acid recycling. While E6 can bind p53 directly, the binding is more efficient in the presence of the E6AP, which is an ubiquitin-protein ligase and an important component of the proteolytic cascade (2). The sequence of events is that E6 binds to E6AP, which then recruits p53, and since E6AP has ligase activity it mediates interaction of the trimeric complex with the E2 protein (part of the ubiquitin complex), which subsequently ubiquitinates lysines on p53 and leads to the degradation of the protein through the proteosome (11). The degradation of p53 is observed *in vivo* in cells expressing either E6 alone, or the full-length genome. While the low-risk types such as HPV-6 and -11 do not cause degradation of p53, there is some dispute as to whether E6 of these types bind to p53 (12, 13). One group observed that HPV-6 E6 bound to p53 at 40% the level of HPV-16 E6 (12), while another group failed to detect binding to HPV-6 E6 (13). HPV-11 was found to bind weakly to p53 (14). Although the differences have not been resolved, it is clear that the low-risk types do not cause degradation of p53.

The domains for HPV-16 E6 important for the interaction with p53 and E6AP have been mapped, although there are again inconsistencies in the data. Initial results suggested that the N terminal domain of HPV-16 E6 was responsible for the binding to p53 (12), but others have found that mutations throughout the length of the protein have profound effects on the ability of E6 to bind to E6AP (13). As Fig. 2 shows, two mutations F45Y-F47Y-D49H and C136G retained complete binding in one study, but the same mutations eliminate binding in another.

E6 has been shown to effect the transcriptional activity of genes controlled by p53. It was shown that promoters repressed by p53 could be derepressed and promoters

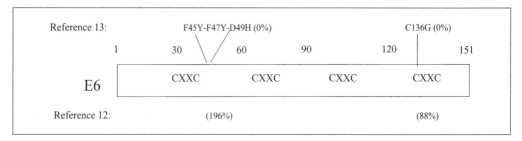

Fig. 2 Diagram of the E6 ORF showing the four CXXC motifs, which form the two zinc fingers. Also shown are examples of two mutations, which show different p53 binding characteristics in the two different studies indicated.

activated by p53 could be repressed by wild-type E6, but not by mutations which inhibited p53 degradation (15, 16). However, E6 has also been shown to transcriptionally activate genes, which are not known to be modulated by p53 indicating that there may be other cellular interactions of E6 that are yet to be uncovered.

2.1.2 Interaction of E6 and E6BP

Using the yeast two-hybrid system Chen *et al*. (3) showed that HPV-16 E6 bound to a cellular protein of 210 amino acids and molecular weigh of approximately 55 kDa. The protein was found to contain four potential calcium binding domains and an endoplasmic reticulum retention signal HDEL (one letter amino acid code). This protein was found to be homologous to ERC-55, which has recently been cloned and shown to be calcium binding protein with an EF-hand motif. The EF-hand motif was first described for paralbumin and consists of two perpendicularly placed α-helics (E and F) with an interhelical loop (for a review of EF-hand proteins see ref. 17). EF-hand containing proteins consist of single or multiple pairs of helix–loop–helix motifs, which change conformation on binding Ca^{2+}. Calcium regulates a wide variety of cellular processes including cell cycle control and differentiation. For instance, the differentiation of mouse and human keratinocytes can be induced by increasing levels of calcium. The binding site on E6BP for E6 is a helix–loop–helix domain, which is found in other E6 binding proteins such as E6AP (18). The fact that E6 binds to a cellular calcium binding protein and the fact that calcium can stimulate differentiation of keratinocytes suggests that E6 may alter differentiation by modulating the function of E6BP, although the mechanism is unclear.

2.1.3 E6 and Bak interaction

Apoptosis has been shown to be important during development and during events in the cell cycle which adversely effect DNA integrity and growth. It is therefore thought to be important in prevention of the dysregulation of the cell cycle and potential tumourigenesis. There are a number of proteins which are involved in stimulating or inhibiting apoptosis and tumour viral proteins have adapted to target or mimic the family of proteins involved in apoptosis. HPV-16 E6 has been shown *in vivo* in the lens of transgenic mice to inhibit apoptosis (19). Recently, evidence has been produced in tissue culture, which shows that E6 binds and induces the degradation of Bak, a pro-apoptotic protein (4). The E6 of HPV-18, a high-risk papillomavirus, can inhibit apoptosis induced by Bak in p53 null cell lines. The mechanism is thought to be through the binding of E6 to Bak and this complex then recruits E6AP resulting in the degradation of Bak and inhibition of apoptosis. Therefore E6 may target two proteins involved in the apoptotic pathways, p53 and Bak by a similar degradative mechanism. These functions may be important for survival of keratinocytes and the eventual replication of the viral genome.

2.1.4 Immortalization and transformation of cells by HPV-16 E6

Immortalization and transformation studies have been routinely carried out on viral proteins to indicate the ability of the proteins to disrupt differentiation and cell cycle control. HPV-16 E6 alone cannot immortalize primary human keratinocytes, but in combination with E7 immortalization is observed (20). However, E6 can immortalize human mammary gland cells (8), although the significance of this is unclear as there is no convincing evidence that papillomaviruses are involved in mammary cancer. It may be that mammary gland cells are very sensitive to the effects of E6 on p53 while keratinocytes need disruption of other pathways before cells are immortalized.

Therefore, in summary the fact that E6 binds and degrades p53, an important protein for control of the progression of the cell through G1 and a cellular protein involved in apoptosis, suggests that E6 stimulates S-phase progression by disrupting the normal function of p53 through binding to E6AP. When cells incur DNA damage or undergo the unwanted attentions of growth stimulators then they may apoptose. The fact that E6 can inhibit apoptosis induced by p53 and Bak may be important mechanisms to save the cells for viral DNA replication. The outcome of the interaction with E6BP is more problematic, but if it is involved in keratinocyte differentiation, then modulation of the activity may disrupt normal differentiation.

2.2 HPV-16 E7

The E7 protein plays a critical role in altering the cellular environment for the benefit of viral replication. It has a number of functions and interactions with cellular proteins that are consistent with it having a role in modulating differentiation and stimulating G1 progression. It appears to be the major transforming protein of HPV-16 in that it can transform rodent cell lines (21–24) and is essential for the immortalization of primary human keratinocytes (25). The protein has 98 amino acids and although it has a predicted molecular weight of 11 kDa, the HPV-16 E7 protein runs at 17 kDa on SDS–PAGE gel electrophoresis, due to charged residues in the N terminal ten amino acids (26, 27). Structural studies on E7 have been hampered by the relative insolubility of the protein. Initial NMR studies in solution have shown that E7 is very unstructured and in a 'floppy' state since no reference point could be determined during these experiments (Dr William Phelps, personal communication).

The protein can, for convenience, be broken down into three domains called CR1, 2, and 3 (Fig. 3). The N terminal domain, CR1 has important functions for tran-

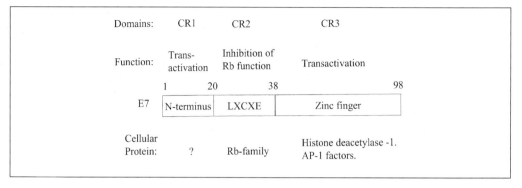

Fig. 3 Diagrammatic representation of the HPV-16 E6 protein, showing the functional domains CR1, CR2, and CR3 and the interacting cellular proteins are listed below.

scription and transformation, although at present no cellular protein has been shown to bind this region. The CR2 domain contains the LXCXE motif, which is important for the binding to the retinoblastoma protein (Rb) and has homology with the Rb binding domains of SV40 large T and adenovirus E1 a proteins (28). Finally, the CR3 domain contains the zinc finger, which has been shown to be important for the binding of AP-1 transcription factors (29), histone deacetylases (30; and Westbrook and McCance, unpublished data) and is necessary for the disruption of the Rb/E2F complexes (31). Both the N terminus and the zinc finger region have endogenous transcriptional activity when fused to the DNA binding domain of the GAL4 transcription factor. Dimerization of the protein is mediated through the zinc finger as disruption of either of the CXXC motifs forming the finger inhibits dimerization (32). These motifs and therefore the zinc finger are also important for transformation and immortalization, however, it is not clear if dimerization is important for these functions, since mutations in CXXC motif abrogate both functions and also dimerization. A more thorough mutational analysis of the CR3 domain is required to resolve these issues. The functions of E7 are mediated by interaction with cellular proteins, which are described next.

2.2.1 E7 and Rb and histone deacetylase interactions

The best documented E7 interaction is with Rb tumour suppressor protein. Rb regulates the progression of cells through G1 by binding and inhibiting a number of transcription factors, which are involved in the activation of S-phase genes. Rb inhibition is relieved as cells approach S-phase by the phosphorylation of Rb at a number of different sites in the molecule by cyclin E and D associated kinases (33). The E2F family of transcription factors is inhibited by Rb during G1, but when Rb is hyperphosphorylated, E2F is released and activates a number of genes important for DNA synthesis such as DNA polymerase-α, thymidine kinase, cyclin E, and dihydrofolate reductase. However, in the presence of E7 there is premature release of Rb inhibition of E2F and promotion of G1 progression and entry into S-phase.

The primary region for E7 binding to Rb is in the CR2 domain (34) and the motif

LXCXE in this region is essential for binding to the Rb B-pocket (between 649–772 amino acids). Certain mutations in the B-pocket severely reduce E7 binding (35–37). While this region is essential for binding there is a region in the CR3 domain that is required in addition to the CR2 domain for the release of E2F (31). This region in the CR3 domain also binds Rb, but further downstream from the binding site of CR2, and is in the carboxy terminus between amino acids 803 and 841 of Rb (31). The requirement for a region of E7 that interacts with the carboxyl terminal of Rb to displace E2F is consistent with the carboxy terminus of Rb playing a role in E2F binding. The low risk HPV types, such as HPV-6 and -11, infecting the genital tract do not bind Rb to the same level as the E7 from oncogenic viruses and the potential biological consequences are discussed below.

There are other proteins, which bind to the CR3 domain of E7 which may also be important for the repressive function of Rb. Recently, the histone deacetylase 1 protein (HDAC-1) was shown to bind to Rb in the B-pocket domain (30). Histone deacetylase proteins are important for repression of transcription by remodelling chromatin to make it difficult for transcription factors and the basic transcription machinery to access their DNA binding sites. Therefore the repressive activity of Rb on transcription may be due to the ability to bind HDAC. The activity of E7 in derepressing the inhibition of Rb on E2F may therefore need both the CR2 and CR3 domains to compete away Rb and HDAC-1 from transcription factors. Recent work has shown that E7 can bind a complex containing HDAC proteins in the absence of Rb, through binding to Mi2β (38), a protein found in the chromatin remodelling complex called NuRD (nucleosome remodelling histone deacetylase complex) (39, 40). While the full consequences of the interaction of E7 with Mi2β and HDAC-1 activity remain to be elucidated, it is possible that the targeting of acetylases and deacetylases may be a common feature of viral proteins that modulate transcription, especially those that do not bind directly to DNA.

The first interpretations of the results of the interaction between E7 and Rb were that the level of binding determined the potential of the virus strain to cause cancer. This potential could be measured by the ability of the viruses to immortalize primary human keratinocytes. Early on it was found that the HPV-16, but not the HPV-6 genome could immortalize keratinocytes (9, 41), however, the binding of E7 to Rb is not necessary for immortalization of keratinocytes, as mutations, which disrupt binding can immortalize cells in the context of the full-length genome (25). Nonetheless, E7 is necessary for immortalization, since mutations, which disrupt the zinc finger can not immortalize cells (25). Further evidence that Rb binding is not necessary for disease production comes from experiments with the cottontail rabbit papillomavirus (CRPV), which serves as a useful model of papillomavirus pathogenesis, since injection of viral DNA into the skin of the rabbit will produce warts and as the animals age about 75% will exhibit malignant conversion. Mutations in the E7 gene of the rabbit papillomavirus, which abrogate Rb binding were still able to produce warts, although at a reduced rate, indicating that Rb binding was not necessary for wart production (42). However, it is not clear if Rb binding was necessary for malignant conversion as the animals were sacrificed before carcinomas developed. Also

the E7 protein from HPV-1, a virus producing benign cutaneous warts binds Rb to 60–80% of HPV-16 E7, yet the genome from this virus cannot immortalize keratinocytes (43). On the other hand two virus types, HPV-8 and -47, which infect cutaneous surfaces and are associated with malignant disease in patients with a rare autosomal conditions called epidermodysplasia verruciformis have negligible transforming activity in tissue culture, and code for E7 proteins that bind Rb at low levels (43–45).

E7 of HPV-16 also binds to other Rb family members such as p107 and p130 (28). These members share considerable homology in the pocket region of the proteins, although they differ significantly in the N terminal region. The CR2 domain of E7 mediates binding to p107 and p130 (28, 46), although different amino acid mutations in the LXCXE motif have the capacity to bind to different Rb members. For instance mutations in the glutamic acid (E) abrogate binding to Rb but not to p107. The importance of this for the biology of HPV has still to be resolved.

2.2.2 E7 and AP-1 binding

The CR3 domain of E7 as well as binding HDAC-1 can also mediate binding to members of the AP-1 family of transcription factors (29). The members of this family are c-Jun, JunB, JunD, Fra1–6, c-Fos, and FosB1 and 2. The Jun members can homo- or heterodimerize to each other and to the Fos group. E7 appears to bind to all the members and activate transcription. AP-1 factors are involved in both proliferative responses and in differentiation of keratinocytes and myeloid cells. In addition, there are a number of genes whose products are involved in DNA synthesis or regulation of synthesis such as proliferating cell nuclear factor (PCNA), topoisomerase I, thymidylate transferase and thymidine kinase that are regulated by AP-1 factors. E7 enhances c-Jun induced transcription three- to fourfold over the levels of c-Jun alone, and mutations that abolish Rb binding have no effect on the *trans*-activation. Potentially important is the observation that the ability of the E7 mutations to enhance the *trans*-activation of c-Jun and to bind HDAC-1 paralleled their immortalization activity (25, 29, 30).

2.2.3 E7 and cyclin proteins

Cyclins and their dependent kinases are important at different stages of the cell cycle. Progression through G1 requires D and E type cyclins and their kinases cdk2, cdk4 and 6 and both are responsible for the hyperphopshorylation of Rb in mid to late G1 (33). In cells containing E7, cyclin D is reduced while cyclin E and cyclin A, a cyclin involved in G1/S-phase and passage through G2, are increased (47–51). The CR2 domain of E7 is required for the increase in cyclins A and E, though CR1 is also required for the up-regulation of cyclin A (51). In addition, the CR3 could conceivably also play a role in regulating cyclin A levels since the cdk inhibitor p27^{KIP1} has been shown to inhibit cyclin A gene expression (52) and E7 has been shown to relieve this inhibition by binding p27^{KIP1} through the CR3 region (53). Since cyclin E is involved in phosphorylating Rb into the inactive state, this may help E7 overcome Rb-mediated G1 arrest and stimulate S-phase entry. However, E7 also causes an increase in p21^{CIP}, which is a cyclin-dependent kinase inhibitor (54, 55), which on the surface would appear to be counterintuitive. Despite the high levels of the p21^{CIP}, the cyclin-dependent kinases are active, and although the mechanism underlying these observations is unclear, it appears that E7 may bind to p21^{CIP} and thus in some way inhibit it's activity (54, 55). All the above observations were carried out in the presence of E7 alone, even though in the life cycle of the virus there are other early viral proteins such as E6 and E5. When studies have been carried out in the presence of both E6 and E7 then the increased levels of p21^{CIP} are not observed, probably because E6 causes degradation of p53 a known regulator of the p21^{CIP} promoter (56; and Westbrook and McCance, unpublished data). In line with this is the observation that in the presence of E7 alone p53 and p21^{CIP} levels are high (Baglia and McCance, unpublished data).

The locus coding for the kinase inhibitor p16^{INK4a} also codes for a protein called p14^{ARF} through differential splicing (57). The two proteins have different exon 1 sequences but are then spliced to the same exon 2, but in different reading frames, producing two different proteins from the same DNA region. p14^{ARF} is not kinase inhibitor, but when overexpressed will cause cell cycle arrest (58). While the functions of p14^{ARF} are unclear it is able to bind to MDM2 and inhibit p53 degradation and so stabilizes p53 protein levels (57). In turn p53 has been shown to have an inhibitory effect on p14^{ARF} transcription although there are no p53 binding sites in the promoter region of p14^{ARF} (59). However, the promoter of p14^{ARF} contains three potential E2F binding sites and in fact can be activated by E2F demonstrating another way in which p53 and Rb pathways interact (60). Therefore it is anticipated that the presence of E7 will increase p14^{ARF} levels by increasing the levels of free E2F and that E6 would have an additive effect by degrading p53 and so removing the negative regulation of p14^{ARF} levels. However, since the function of p14^{ARF} is not known it is difficult at present to suggest a role in viral replication, although overexpression of p14^{ARF} will arrest cells and perhaps this is what happens in HPV infection, but the cells still move into S-phase. However, the cells do not divide and therefore act as a stable factory for viral particle production.

In summary, investigations point to the fact that E7 can act during the G1 phase of the cell cycle by up-regulating the levels of cyclins and inhibiting the activity of kinase inhibitors, with the result that substrates may be inappropriately phosphorylated late in G1 and cells stimulated to prematurely enter S-phase. These activities complement other E7 functions described above and result in the specific stimulation of the expression of genes whose products also encourage G1/S-phase transition. E7 of the high risk viruses also effects S-phase cells by up-regulating cyclin B and cdc2 through apparent translational and post-translational mechanisms, but does not disrupt cyclin B/cdc2 complexes. Since cyclin B is required for cells to exit mitosis, elevated levels could prolong mitosis and the period during which HPV DNA replication can precede. In addition, if the cells are prevented from dividing this would allow infectious virus to be produced without being diluted out by cell division. It should be remembered that the virus is trying to replicate in a very dynamic tissue with cells moving up the epithelium and being shed into the environment. The level of mature virus particles per cell is low and inhibiting division would concentrate the virus and since transmission of infected cells may be the main route of infection this would give a higher infectious dose per cell.

2.2.4 E7 and TBP and TAFs

While E7 can affect transcription with the release of E2F, it has also been shown that the protein can bind to the TATA-binding protein, TBP and to one of the associated proteins TAF-110 (61). Efficient interaction with TBP requires all three domains of E7 (62) and the interaction may not be stimulatory since it has been suggested that this interaction is responsible for the ability of E7 to repress p53 transcriptional activity. However, beyond this observation the consequences of the interaction are unknown.

2.2.5 E7 and pRB/c-Jun interactions

Rb has been implicated in the differentiation of certain cell types including muscle cells and adipocytes (63, 64) and the AP-1 factors have been shown to be up-regulated during keratinocyte and myeloid cell differentiation (65, 66). Since Rb can increase both DNA binding and transcription of c-Jun and E7 can bind to both proteins it was reasoned that E7 may effect the Rb/c-Jun interaction. The hypo-phosphorylated form Rb binds c-Jun through the B-pocket and the C terminus. Recall that E7 binds Rb at the B-pocket and the C terminus through the CR2 and CR3 domains. It was found that E7 could disrupt the Rb/c-Jun interaction probably through competition for the same binding regions. If Rb activation of c-Jun transcription is an important signal for keratinocyte differentiation, then disruption of this signal may inhibit differentiation and instigate a series of interactions which results in the cells moving into S-phase.

In summary, E7 binds to a number of cellular proteins, which are members of cellular pathways that interact during normal cell differentiation and growth. Most of the pathways act to control the progression of the cell through G1 into S-phase and since the virus is attempting to replicate in cells which are differentiating, these interactions are consistent with a role for E7 in preventing differentiation and stimulating S-phase entry.

2.3 HPV-16 E5

The E5 gene is located downstream of the E2 open reading frame (Fig. 1B), and is not well conserved at the DNA level between HPV types, although here is a conservation of the physico-chemical properties of the protein as they are all highly hydrophobic and membrane bound. The proteins from both human and animal viruses can trans-form rodent cells with varying degrees of efficiency (67–69). Most research has been carried out with the bovine papillomavirus type 1 (BPV-1) E5, mainly because this protein has a hydrophilic tail (70, 71), which gives the protein more solubility and is much easier to detect in transfected cells than the HPV E5 proteins, which do not contain any hydrophilic tail. The E5 protein of HPV-16 is composed of 84 amino acids, while that of BPV-1 is only 44 amino acids, making the latter one of the smallest oncoproteins known (72–76). HPV-16 E5 dimerizes, even under reducing conditions, although the amino acids mediating such an activity are unknown (77). The hydrophobic nature of the protein means that it is membrane bound and is found mostly in the endoplasmic reticulum and Golgi apparatus in transfected cells and transverses the membrane two or three times (78, 79). The hydrophobic nature also makes the protein very different to work with and production of good antibody reagents is difficult. However, for HPV-16 E5 there are antibodies directed against N and C terminal peptides (80).

2.3.1 E5 interactions with growth factor receptors and vacuolar ATPases

Both E6 and E7 interact with cellular proteins that are involved in growth suppression, such as p53 and Rb and abrogate the normal repressive function. E5 on the other hand, has activities mediated through the epidermal growth factor receptor (EGFR), which stimulates cell growth (68, 69, 81). E5 causes an increase in the signalling from the EGFR by inhibiting the degradation of the receptor after ligand stimulation and endocytosis of the receptor (81). In normal cells the receptor binds the epidermal growth factor (EGF), dimerizes, autophosphorylates itself, and is rapidly taken into cells in endosomes with 90% of the receptors internalized within ten minutes. The lumen of the endosomes are then acidified and they become late endosomes and lysosomes resulting in the degradation of the EGFR within an hour after stimulation (Fig. 4A). In the presence of E5 there is an inhibition, or delay in acidification of the endosomes and consequently a delay in the degradation of the receptor, resulting in the recycling of a substantial proportion of the receptors back to the surface of the cell, where they can bind more ligand causing another round of stimulation of the cell (82) (Fig. 4B). The receptors recycle back to the surface of the cell in the absence of the ligand. Inhibition of acidification is thought to be due to the binding of E5 to the 16 kDa pore subunit of the vacuolar ATPase complex and inhibition of the transfer of protons into the lumen of the endosome (78). The mechanism of the inhibition of proton transport is however, unknown. For example, it is not known if E5 simply blocks the pore preventing transport of proteins, or if E5 has an effect on the assembly of the complex or the enzymatic activity.

There is conflicting evidence on whether HPV-16 E5 binds to the EGFR itself (80, 83). In one study using anti-E5 antibodies, HPV-16 E5 was shown to bind the EGFR when both were expressed in COS-1 cells (54), while in another study no binding was observed (83). The results are difficult to reconcile except that in the latter study the E5 protein was epitope tagged to allow for the use of anti-tag antibodies in the immunoprecipitations, whereas in the other study anti-E5 antibodies were used. It is possible that the hydrophilic tag may be masked by the rest of the hydrophobic molecule, resulting in the protein not being recognized by the anti-tag antibodies.

Since EGF is a growth requirement for keratinocytes and is necessary for progression through the restriction point of G1, the activity of E5 indicates that it may act to stimulate cells through this phase of the cell cycle and into S-phase. This stimulation of cell growth is reflected in the fact that HPV-16 E5 can transform rodent fibroblasts and the rate of transformation is increased in the presence of EGF (68, 69, 81), but not platelet-derived growth factor (PDGF). The E5 protein alone cannot im-

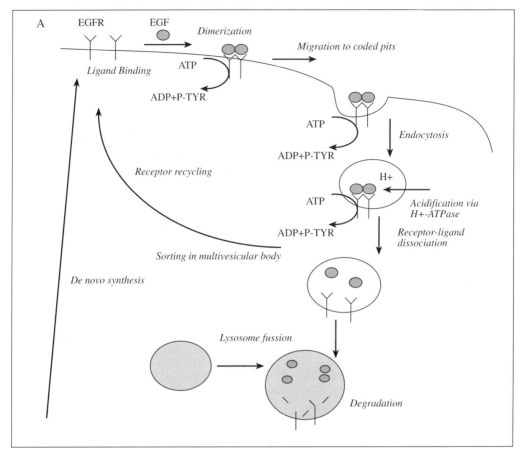

Fig. 4 (A) Schematic of the turnover of the epidermal growth factor receptor (EGFR) in the presence of epidermal growth factor (EGF). Normally in tissue culture cells after addition of EGF and secondary message stimulation 90% of the receptors are degraded through the lysosomal pathway. (B) In the presence of HPV-16 E5 turnover is normal but degradation is inhibited and up to 40% of receptors recycle to the surface. Therefore in the presence of E5 there is an increase in the number of receptors on the surface of cells (by permission of ASM Press).

mortalize primary human keratinocytes (69, 82), although a genome with a premature stop codon in E5 can only immortalize at 10% the efficiency of the wild-type genome, suggesting that E5 has an important co-operative role in immortalization (84).

2.4 Summary of the roles of E6, E7, and E5 in keratinocyte differentiation and cell cycle control

Since papillomaviruses contain few proteins dedicated to viral genome replication, infected keratinocytes must be stimulated into S-phase of the cell cycle instead of proceeding to terminal differentiation, to supply the necessary replicative machinery.

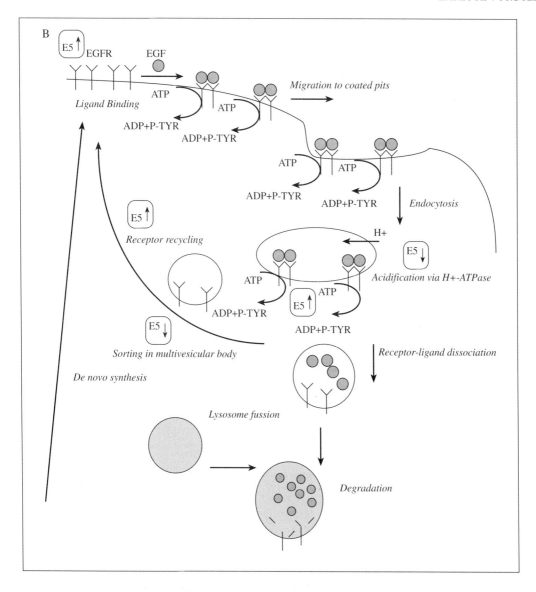

The shift into S-phase need not cause the cell to go through a complete cycle and in fact this may not be desirable as it would limit the replication time for viral DNA. Therefore the virus has to uncouple S-phase progression from differentiation and two of the viral proteins, E6 and E7 interact with cellular proteins that have important functions during G1 progression and S-phase entry. E6 binds and degrades p53 and also interacts with E6BP, which is a calcium binding protein and may have a role in cell differentiation pathways. p53 can cause cell cycle arrest in G1 or apoptosis if cells are exposed to a stress either externally or by infection, therefore reducing the effectiveness of this monitoring system the virus could more easily stimulate S-phase entry. In line with this activity, E6 also causes degradation of Bak, which has pro-apoptotoic activities. In co-operation with E6, E7 binds to Rb and abrogates the ability of this protein to cause G1 arrest. The third viral protein E5 up-regulates the activity of the EGFR by inhibiting degradation and initiation recycling of the receptor for further binding to the ligand. The result of this activity would be to stimulate G1-progression and complement the activities E6 and E7.

3. Viral proteins involved in viral DNA replication

HPV replication in epithelial cells can be broken down into three phases. First, the virus infects basal cells and the genome numbers increase to 50–100 copies very rapidly. The next phase is one of maintenance, where the 50–100 copies replicate in synchrony with chromosomal replication as cells divide in the basal layer. One of the daughter cells moves up through the epithelium and commences to differentiate at which stage amplification of the viral genomes within the cells is observed. It is now possible to follow these steps in primary keratinocytes and initial experiments using HPV-31 have been carried out (85, 86). Keratinocytes are transfected with circular HPV-31 DNA along with a drug resistance marker such as neomycin, the cells selected and cultured. The cultured cells retain a copy number of approximately 50 per cell. If the cell are then stimulated to differentiation by different methods, amplification of the genomes is observed in a population of the transfected cells (86, 87). These systems have not been used extensively to delineate the *cis*- and *trans*-elements important for replication, although a transient assay system, which was developed a number of years ago (88) and allows the replication of origin containing plasmids in the presence of *trans*-acting viral proteins, has elucidated elements involved in origin replication.

Using this assay two viral proteins, E1 and E2, were found to be involved in the replication of the viral genome in co-operation with the cellular machinery. Both the proteins bind to specific sequences in the origin region of the genome, which straddles the region around nucleotide number 1 (Fig. 5). There is a high level of conservation of the number and relative positions of E1 and E2 binding sites in the origin region between HPV types and the animal viruses such as BPV-1 (Fig. 5). E1 and E2 proteins are absolutely required *in trans* for origin replication for HPV-11 (89) and HPV-16 (90), but surprisingly E1 is only required for replication of the HPV-1 origin (91). The biological significance of this is unclear at present. These assays have also shown that E2 binding sites are required, but that the E1 site is dispensable (89, 92). Since E1 and E2 can bind together E2 may recruit E1 to the origin in the absence of an E1 binding site (89, 93). In fact recruitment of E1 to the origin may be an important function of E2, because E1 binds at a very low affinity to the E1 binding site in the absence of E2 (93–97). *In vitro* replication studies using BPV-1 and HPV-11 origins in nuclear extracts, reconstituted with E1 and E2 proteins has shown that for BPV-1 replication E1 only is required (97, 98), while both are required for HPV-11 replication (99). While these studies are carried out in the presence of

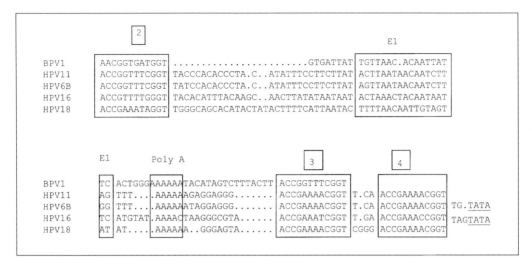

Fig. 5 The region of the origin of replication of four different human papillomavirus types and the bovine papillomavirus type 1. The origin of the human viruses has three E2 binding sites (E2BS) and one E1 binding site E1BS. The E2BS are numbered 2, 3, and 4 and surround the E1BS (E1). The polyA tract is just downstream of the E1BS. The fourth E2BS is right next to the TATA box (underlined for HPV-11 and HPV-16). Nucleotide 1 is in the middle of the E1BS.

excess proteins, they do indicate that E1 functions as the primary replicative protein and E2 as an auxiliary function.

There are three E2 binding sites (E2BS), which surround the single E1 binding site (E1BS) and the poly A-rich region (Fig. 5). The E2 binding sequence is a palindrome of $ACCN_6GGT$ and E2 has varying affinities depending on whether a G/C or A/T are at positions 4 and 9, with the former combination creating a high affinity site (100). The E1BS site of HPV-11 has a sequence of ACT*TAAT*AACAATCTTAG and mutation of the TAAT motif (*underlined*) to GCAG abrogated binding to the origin region (93). However, this mutated origin is replicated by E1 and E2, as mentioned above. The role of the polyA region is not clear since mutating the site to be G-rich did not eliminate replication, but the level of origin of replication was lower (89). The region may be important for initiation of replication since E1 has a helicase activity (see below) and this region may be targeted by such an enzymatic activity. While the transient assay system has given important insights into the replication of HPVs, it is a rather artificial assay, since the *trans*-acting proteins are overexpressed and there is no possibility of observing temporal effects of the proteins. This situation will be rectified now that HPV-31 DNA has been propagated and amplified in human keratinocytes.

3.1 E1 and E2 proteins

E1 binds to the origin at low affinity, but complex formation with E2 increases binding tenfold. E1/E2 complexes can form in solution and are DNA-independent.

E1, like SV40 large T (LT) has helicase and ATPase activity and has homology with the LT C-terminal region (101). While LT binds to Rb and p53, this function is found in E7 and E6 proteins respectively and not in E1. E1 appears to recruit cell replicative proteins to the origin such as DNA polymerase α (102) and E2 also binds other cellular proteins such as replication protein A (RPA) (103), indicating the importance of the complex for successful replication of the viral genome.

Mutational analysis has shown that the DNA binding region of HPV-11 E1 lies in a large area of the C-terminal portion of the protein between amino acids 346 and 649 (104). The domain for DNA binding overlaps this region and is between 186 and 649. Since the functional domains appear to overlap it has been difficult to determine if both DNA binding and E2 binding are essential for E1 to support replication. One piece of evidence that DNA binding by E1 *per se* is not essential, is the fact that the HPV-11 origin region with mutations in E1 binding site that abrogates binding is still replicated in the presence of both E1 and E2 (89). For BPV-1 E1 the DNA binding domain has been located between amino acids 180 and 270 (105), while the domain which binds E2 is more controversial. One group found binding to be between 458 and 605 amino acids at low temperatures and a more N-terminal domain at higher temperature (105), while another group observed a domain in the N-terminal region between amino acids 1 and 250 (106). A similar region was found for HPV-31 E1 (94). More refined mutational analysis needs to be carried out to determine the regions involved in DNA and E2 binding, but it could turn out that more than one domain is involved.

Mutational analysis of E2 has produced a clearer picture of the active domains of the protein. Sequences in the N-terminus of E2 are important for interaction with E1, and also for the trans-activation function, while C-terminal sequences are required for DNA binding (107–109). Within the N-terminal domain, mutations have shown that it is possible to separate the replicative from the *trans*-activation function of E2 and show that *trans*-activation is dispensable for replication (107, 109).

As well as a function in replication, E2 also has transcriptional control properties. As mentioned above it has a defined transcriptional activation domain in the N terminal region although it appears to have a repressive activity on the early promoter (110–112). This is in contrast to the activation properties of the BPV-1 E2 promoter on the bovine viral promoter. Repression may be due to the close proximity of one of the E2 sites (number 4, Fig. 5) to the TATA box resulting in steric hindrance of the binding of the TATA-binding protein (TBP) (112). A role in transcriptional activation for E2 has yet to be defined and the main function of E2, apart from a role in replication, may be to repress high expression of E6 and E7 to keep a balance between some level of differentiation and progression to S-phase. When HPV DNA integrates in malignant tissue, the E2 gene is often disrupted and the levels of E6 and E7 RNA increase. In addition to a transcriptional increase the stability of the RNA is increased, perhaps due to 3′ processing from a chromosomal site rather than the 3′ end of the early region (113). Since E1 and E2 bind together at the origin there may be a role for transcriptional activation of this complex, although this has not been confirmed.

3.2 Maintenance and amplification of HPV genomes

Most of the replicative research has been carried out using plasmids containing the origin region and E1 and E2 expressed *in trans* from exogenous promoters. While this will determine *cis-* and *trans*-elements that are important for replication of the origin, the method does not allow study of maintenance and amplification of the genome and control of replication from the natural promoter. Recent advances have made it possible to transfect cells with circular full-length HPV-31 genome and obtain stable episomal replication of 50–100 copies of the DNA in actively dividing keratinocytes (86). This replication might be considered the maintenance phase of replication, which takes place in the basal epithelial layer in the genital tract. After division of cells when segregation of the genomes takes place the viral DNA replicates to maintain a constant level of 50–100 copies per cell. Cell lines containing HPV-31 DNA in episomal form have been maintained in culture for over two years (L. A. Laimins, personal communication). There appears to be constitutive expression of the early open reading frames, which is only observed in keratinocytes and not in fibroblasts suggesting that there are cell-specific factors that may drive this transcription (111, 114, 115). Since there are a number of transcription factor sites in the URR, various factors have been implicated in the cellular specificity, but none have been well defined. For instance, YY1 and C/EBP complexes (116) have been suggested to determine specificity as has Sp1/Sp3 (117), although it has also been suggested that it is a combination of factors that determine specificity (118). At present it is unclear the nature of the specificity observed in the transcription of the early region genes. If the cells are stimulated to differentiation either by being placed in raft cultures or suspended in 1.6% methylcellulose then amplification of the genome occurs and late transcripts appear (86). Transcripts for the early proteins, E6, E7, E5, E1, and E2 are from the early promoters in the URR with initiation at 97 base pairs in HPV-16. The late promoter and initiation site for late message is in the middle of the E7 open reading frame at 742 base pairs. So there is a switch from the early to late promoter which is only observed in keratinocytes, indicating that there is cell-specific and differentiation-specific factors involved. However, searches for such factors have been unsuccessful.

4. Summary

Human papillomaviruses replicate in cells that are destined to differentiation and as such would not contain the replicative machinery necessary for viral DNA replication. Therefore the virus has to stimulate cells into synthesis phase to complete viral replication and there are three early proteins, E5, E6, and E7, which are involved in creating this environment. The viral proteins actually involved in the replication process are E1 and E2 and both bind at the origin of replication and can bind to each other even in the absence of DNA. The fact that HPV-31 can now be maintained episomally in keratinocytes and can be amplified when the cells are induced to differentiation, will allow the study of the elements important for maintenance and amplification of the genome and will facility studies of the temporal relationship between E1 and E2 during these replication stages.

Acknowledgements

I would like to thank members of my laboratory both past and present who contributed to work described here and to the National Institutes of Health, the American Cancer Society, the Medical Research Council, and the Cancer Research Campaign for funding over the years.

References

1. Huibregtse, J., Schneffer, M., and Howley, P. M. (1991) A cellular protein mediates association of p53 with E6 oncoprotein of human papillomavirus types 16 or 18. *EMBO Journal*, 10(13), 4129.
2. Huibregtse, J., Scheffner, M., and Howley, P. M. (1993) Localization of the E6-AP regions that direct human papillomavirus E6 binding, association with p53, and ubiquitination of associated proteins. *Molecular and Cellular Biology*, 13(8), 4918.
3. Chen, J. J., Reid, C. E., Band, V., and Androphy, E. J. (1995) Interaction of papillomavirus E6 oncoproteins with a putative calcium-binding protein. *Science*, 269(5223), 529.
4. Thomas, M. and Banks, L. (1998) Inhibition of Bak-induced apoptosis by HPV-18 E6. *Oncogene*, 17, 2943.
5. Etscheid, B. G., Foster, S. A., and Galloway, D. A. (1994) The E6 protein of human papillomavirus type 16 functions as a transcriptional repressor in a mechanism independent of the tumor suppressor protein, p53. *Virology*, 205(2), 583.
6. Lamberti, C., Morrissey, L. C., Grossman, S. R., and Androphy, E. J. (1990) Transcriptional activation by the papillomvirus E6 zinc finger oncoprotein. *EMBO Journal*, 9, 1907.
7. Sedman, S., Barbosa, M. S., Vass, W. C., Hubbert, N. I., Haas, J. A., Lowry, D. R., *et al* (1991) the full length E7 protein of human papillomavirus type 16 has transforming and transactivating activities and cooperates with E7 to immortalize keratinocytes in culture. *Journal of Virology*, 65, 4860.
8. Band, V., Zajchowki, D., Kulesa, V., and Sager, R. (1990) Human papillomvirus DNAs immortalize normal epithelial cells reduce their growth factor requirements. *Proceedings of the National Academy of Sciences, USA*, 87, 463.
9. Halbert, C., Demers, G. W., and Galloway, D. A. (1992) The E6 and E7 genes of human papillomavirus type 6 have weak immoralizing activity in human epithelial cells. *Journal of Virology*, 66, 2125.
10. Scheffner, M., Werness, B. A., Huibregtse, J. M., Levine, A. J., and Howley, P. M. (1990) The E6 oncoprotein encoded by human papillomavirus types 16 and 18 promotes the degradation of p53. *Cell*, 63(6), 1129.
11. Scheffner, M., Huibregtse, J. M., Vierstra, R. D., and Howley, P. M. (1993) The HPV-16 E6 and E6-AP complex functions as a ubiquitin-protein ligase in the ubiquitination of p53. *Cell*, 75, 495.
12. Crook, T., Tidy, J. A., and Vousden, K. H. (1991) Degradation of p53 can be targeted by HPV E6 sequences distinct from those required for p53 binding and trans-activation. *Cell*, 67(3), 547.

13. Foster, S. A., Demers, G. W., Etscheid, B. G., and Galloway, D. A. (1994) The ability of human papillomavirus E6 proteins to target p53 for degradation *in vivo* correlates with their ability to abrogate actinomycin D-induced growth arrest. *Journal of Virology*, 68(9), 5698.

14. Lechner, M. S. and Laimins, L. A. (1994) Inhibition of p53 DNA binding by human papillomavirus E6 proteins. *Journal of Virology*, 68(7), 4262.

15. Lechner, M. S., Mack, D. H., Finicle, A. B., Crook, T., Vousden, K. H., and Laimins, L. A. (1992) human papillomavirus E6 proteins bind p53 in viro and abrogate p53-mediated repression of transcription [published erratum appears in *EMBO J.* 1992 Nov; 11(11):4248]. *EMBO Journal*, 11(8), 3045.

16. Pim, D., Storey, A., Thomas, M., Massimi, P., and Banks, L. (1994) Mutational analysis of HPV-18 E6 identifies domains required for p53 degradation *in vitro*, abolition of p53 transactivation *in vivo* and immortalization of primary BMK cells. *Oncogene*, 9, 1869.

17. Ikura, M. (1996) Calcium binding and conformational response to EF-hand proteins. *Trends in Biochemical Science*, 21, 14.

18. Chen, J. J., Hong, Y., Rustamzedeh, E., Baleja, J. D., and Androphy, E. J. (1998) Identification of an alpha helical motif sufficient for association with papillomavirus E6. *Journal of Biological Chemistry*, 273, 13537.

19. Pan, H. and Griep, A. E. (1995) Temporally distinct patterns of p53-dependent and p53-independent apoptosis during mouse lens development. *Genes and Development*, 9(17), 2157.

20. Barbosa, M. S., Vass, W. C., Lowry, D. R., and Schiller, J. T. (1991) In vitro biological activities of the E6 and E7 genes vary among human papillomaviruses of different oncogenic potential. *Journal of Virology*, 65(1), 292.

21. Banks, L., Edmonds, C., and Vousden, K. H. (1990) Ability of the HPV16 E7 protein to bind RB and induce DNA synthesis is not sufficient for efficient transforming activity in NIH3T3 cells. *Oncogene*, 5(9), 1383.

22. Brokaw, J. L., Yee, C. L., and Munger, K. (1994) A mutational analysis of the amino terminal domain of the human papillomavirus type 16 E7 oncoprotein. *Virology*, 205(2), 603.

23. Chesters, P. M. and McCance, D. J. (1989) Human papillomavirus types 6 and 16 in cooperation with Ha-ras transform secondary rat embryo fibroblasts. *Journal of General Virology*, 70(Pt 2), 353.

24. Chesters, P. M., Vousden, K. H., Edmonds, C., and McCance, D. J. (1990) Analysis of human papillomavirus type 16 open reading frame E7 immortalizing function in rat embryo fibroblast cells. *Journal of General Virology*, 71(Pt 2), 449.

25. Jewers, R. J., Hildebrant, P., Ludlow, J. W., Kell, B., and McCance, D. J. (1992) Regions of human papillomavirus type 16 E7 oncoprotein required for immortalization of human keratinocytes. *Journal of Virology*, 66(3), 1329.

26. Armstrong, D. J. and Roman, A. (1993) The anomalous electrphoretic behavior of the human papillomavirus type 16 E7 protein is due to the high content of acidic amino acid residues. *Biochemical and Biophysical Research Communications*, 192(3), 1380.

27. Armstrong, D. J. and Roman, A. (1992) Mutagenesis of human papillomavirus types 6 and 16 E7 open reading frames alters the electrophoretic mobility of the expressed proteins. *Journal of General Virology* 73(Pt 5), 1275.

28. Dyson, N., Guida, P., Munger, K., and Harlow, E. (1992) Homologous sequences in adenovirus E1A and human papillomavirus E7 proteins mediate interaction with the same set of cellular proteins. *Journal of Virology*, 66(12), 6893.

29. Antinore, M. J., Birrer, M. J., Patel, D., Nader, L., and McCance, D. J. (1996) The human papillomavirus type 16 E7 gene product interacts with and trans-activates the AP1 family of transcription factors. *EMBO Journal*, 15(8), 1950.

30. Brehm, A., Miskka, E. A., McCance, D. J., Reid, J. L., Bannister, A. J., and Kouzarides, T. (1998) Retinoblastome protein recruits histone deacetylase to repress transcription. *Nature*, 391, 597.

31. Patrick, D. R., Oliff, A., and Heimbrook, D. C. (1994) Identification of a novel retinoblastoma gene product binding site on human papillomavirus type 16 E7 protein. *Journal of Biological Chemistry*, 269(9), 6842.

32. McIntyre, M C., Frattini, M. G., Grossman, S. R., and Laimins, L. A. (1993) Human papillomavirus type 18 E7 protein requires intact Cys-X-X-Cys motifs for zinc binding, dimerization, and transformation but not for Rb binding. *Journal of Virology*, 67(6), 3142.

33. Lundberg, A. S. and Weinberg, R. A. (1998) Functional inactivation of the retinoblastome protein requires sequential modification by at least two distinct cyclin-cdk complexes. *Molecular Cell Biology*, 18, 753.

34. Phelps, W. C., Munger, K., Yee, C. L., Barnes, J. A., and Howley, P. M. (1992) Structure-function analysis of the human papillomavirus type 16 E7 oncoprotein. *Journal of Virology*, 66(4), 2418.

35. Hu, Q. J., Dyson, N., and Harlow, E. (1990) The regions of the retinoblastoma protein needed for binding to adenovirus E1A or SV40 large T antigen are common sites for mutations. *EMBO Journal*, 9(4), 1147.

36. Huang, S., Wang, N. P., Tseng, B. Y., Lee, W. H., and Lee, E. H. (1990) Two distinct and frequently mutated regions of retinoblastoma protein are required for binding to SV40 T antigen. *EMBO Journal*, ((6), 1815.

37. Kaelin, W. G., Jr., Ewen, M. E., and Livingston, D. M. (1990) Definition of the minimal simian virus 40 large T antigen- and adenovirus E1A-binding domain in the retinoblastoma gene product. *Molccular and Cellular Biology*, 10(7), 3761.

38. Brehm, A., Nielsen, S. J., Miska, E. A., McCance, D. J., Reid, J. L., Bannister, A. J., *et al.* (1999) The E7 protein associates with Mi2 and histonre deacetylase activity to promote cell growth. *EMBO Journal*, 18, 2449.

39. Tong, J. K., Hassig, C. A., Schnitzler, G. R., Kingston, R. E., and Schreiber, S. L. (1998) Chromatin deacetylase by an ATP-dependent nucleosome remodelling complex. *Nature*, 395, 217.

40. Zhang, Y., LeRoy, G., Seelig, H.-P., Lane, W. S., and Reinberg, D., (1998) The dermatomyositis-specific autoantigen Mi2 is a component of a complex containing histone deacetylase and nucleosome remodelling activities. *Cell*, 95, 279.

41. McCance, D. J., Kopan, R., Fuchs, E., and Laimins, L. A. (1988) Human papillomavirus type 16 alters human epithelial cell differentiation *in vitro*. *Proceedings of the National Academy of Sciences, USA*, 85(19), 7169.

42. Defeo-Jones, D., Vuocolo, G. A., Haskell, K. M., Hanobik, M. G., Kiefer, D. M., McAvoy, E. M., *et al.* (1993) Papillomavirus E7 protein binding to the retinoblastoma protein is not required for viral induction of warts. *Journal of Virology*, 67(2), 716.

43. Schmit, A., Harry, J. B., Rapp, B., Wettstein, F. O., and Iftner, T. (1994) Comparison of the properties of the E6 and E7 genes of low- and high-risk cutaneous papillomaviruses reveals strongly transforming and high Rb-binding activity for the E7 protein of the low-risk human papillomavirus type 1. *Journal of Virology*, 68(11), 7051.

44. Hiraiwa, A., Kiyono, T., Segawa, K., Utsumi, K. R., Ohashi, M., and Ishibashi, M. (1993)

Comparative study on E6 and E7 genes of some cutaneous and genital papillomaviruses of human origin for their ability to transform 3Y1 cells. *Virology*, 192(1), 102.

45. Iftner, T., Bierfelder, S., Csapo, Z., and Pfister, H. (1988) Involvement of human papillomavirus type 8 genes E6 and E7 in transformation and replication. *Journal of Virology* 62(10), 3655.

46. Davies, R., Hicks, R., Crook, T., Morris, J., and Vousden, K. (1993) Human papillomavirus type 16 E7 associates with a histone H1 kinase and with p107 through sequences necessary for transformation. *Journal of Virology*, 67(5), 2521.

47. Foster, S. A. Galloway, D. A. (1996) Human papillomavirus type 16 E7 alleviates a proliferation block in early passage human mammary epithelial cells. *Oncogene*, 12(8), 1773.

48. Lukas, J., Muller, H., Bartkova, J., Spitkovsky, D., Kjerulff, A., Jansen-Durr, P., *et al.* (1994) DNA tumor virus oncoproteins and retinoblastoma gene mutations share the ability to relieve the cell's requirement for cyclin D1 function in G1. *Journal of Cell Biology, 125*(3), 625.

49. Pei, X. F. (1996) The human papillomavirus E6/E7 genes induce discordant changes in the expression of cell growth regulatory proteins. *Carcinogenesis*, 17(7), 1395.

50. Pusch, O., Soucek, T., Wawra, E., Hengstschlager-Ottnad, E., Bernaschek, G., and Hengstschlager, M. (1996) Specific transformation abolishes cyclin D1 fluctuation throughout the cell cycle. *FEBS Letters*, 385(3), 143.

51. Zerfass, K., Schulze, A., Spitkovsky, D., Friedman, V., Hengelin, B., and Jansen-Durr, P. (1995) Sequential activation of cyclin E and cyclin A gene expression by human papillomavirus type 16 E7 through sequences necessary for transformation. *Journal of Virology*, 69(10), 6389.

52. Schulze, A., Zerfass-Thome, K., Berges, J., Middendorp, S., Jansen-Durr, P., and Henglein, B. (1996) Anchorage-dependent tanscription of the cyclin A gene. *Molecular and Cellular Biology*, 16(9), 4632.

53. Zerfass-Thome, K., Schulze, A., Zwerschke, W., Vogt, B., Helin, K., Bartek, J., *et al* (1997) p27kip1 blocks cyclin E-dependent transactivation of cyclin A gene expression. *Molecular and Cellular Biology*, 17(1), 407.

54. Funk, J. O., Waga, S., Harry, J. B., Espling, E., Stillman, B., and Galloway, D. A. (1997) Inhibition of CDK activity and PCNA-dependent DNA replication by p21 is blocked by interaction with the HPV-16 E7 oncoprotein. *Genes and Development*, 11(16), 2090.

55. Jones, D. L., Alani, R. M., and Munger, K. (1997) The human papillomavirus E7 oncoprotein can uncouple cellular differentiation and proliferation in human keratinocytes by abrogating p21 Cip10mediated inhibition of cdk2. *Genes and Development*, 11(16), 2101.

56. Ruesch, M. and Laimins, L. A. (1998) Human papillomavirus oncoproteins alter differentiation-dependent cell cycle exit on suspension in semisolid medium. *Virlogy* 250, 19.

57. Stott, F. J., Bates, S., James, M. C., McConnell, B. B., Starborg, M., Brookes, S. *et al.* (1998) The alternative product from human CDKN2A locus, p14ARF, participates in a regulatory feedback loop with p53 and MDM2. *EMBO Journal*, 17, 5001.

58. Kamijo, T., Zindy, F., Roussel, M. F., Quelle, D. E., Downing, J. R., Ashmun, R. A., *et al.* (1997) Tumor suppression at the mouse INK4a locus mediated by alternative reading frame product p19ARF. *Cell*, 91, 649.

59. Robertson, K. D. and Jones, P. A. (1998) The human ARF cell cycle regulatory gene promoter is a CpG island which can be silenced by DNA methylation and down-regulated by wild-type p53. *Molecular Cell Biology*, 18, 6457.

60. Bates, S., Philips, A. C., Clark, P. A., Stott, F., Peters, G., Ludwig, R. L. *et al.* (1998) p14ARF links the tumor suppressors Rb and p53. *Nature*, 395, 124.

61. Mazzarelli, J. M., Atkins, G. B., Geisberg, J. V., and Ricciardi, R. P. (1995) The viral oncoproteins Ad5 E1A, HPV16 E7 and SV40 TAg bind a common region of the TBP-associated factor-110. *Oncogene*, 11(9), 1859.

62. Phillips, A. C. and Vousden, K. H. (1997) Analysis of the interaction between human papillomavirus type 16 E7 and the TATA-binding protein, TBP. *Journal of General Virology*, 78(Pt 4), 905.

63. Chen, P. L., Riley, D. J., Chen, Y., and Lee, W. H. (1996) Retinoblastoma protein positively regulates terminal adipocyte differentiation through direct interaction with C/EBPs. *Genes and Development*, 10(21), 2794.

64. Gu, W., Schneider, J. W. Condorelli, G., Kaushal, S., Mahdavi, V., and Nadal-Ginard, B. (1993) Interaction of myogenic factors and the retinoblastoma protein mediates muscle cell commitment and differentiation. *Cell*, 72(3), 309.

65. Angel, P. and Karin, M. (1991) The role of Jun, Fos and the AP-1 complex in cell-proliferation and transformation. *Biochimica et Biophysica Acta*, 1072(2–3), 129.

66. Nead, N., Antinore, M. J., Baglia, L., Ludlow, J. W., and McCance, D. J. (1998) Rb binds c-Jun and activates transcription. *EMBO Journal*, 17, 2342.

67. Bouvard, V., Matlasheqski, G., Gu, Z. M., Storey, A., and Banks, L. (1994) The human papillomavirus type 16 E5 gene cooperates with the E7 gene to stimulate proliferation of primary cells and increases viral gene expression. *Virology*, 203(1), 73.

68. Leechanachi, P., Banks, L., Moreau, F., and Matlashewski, G. (1992) The E5 gene from human papillomavirus type 16 is on oncogene which enhances growth factor-mediated signal transduction to the nucleus. *Oncogene*, 7, 19.

69. Straight, S. W., Hinkle, P. M., Jewers, R. J., and McCance, D. J. (1993) The E5 oncoprotein of human papillomavirus type 16 transforms fibroblasts and effects the down regulation of the epidermal growth factor receptor in keratinocytes. *Journal of Virology*, 67(8), 4521.

70. Bubb, V., McCance, D. J., and Schlegal, R. (1998) DNA sequence of the HPV-16 E5 ORF and the structural conservation of its encoded protein. *Virology* 163(1), 243.

71. Burkhardt, A., Willingham, M., Gay, C., Jeang, K. T., and Schlegal, R. (1989) The E5 oncoprotein of bovine papillomavirus is oriented asymmetrically in Golgi and plasma membranes. *Virology*, 170(1), 334.

72. Burkhardt, A., DiMaio, D., and Schlegal, R. (1987) Genetic and biochemical definition of the bovine papillomavirus E5 transforming protein. *EMBO Journal*, 6(8), 2381.

73. DiMaio, D., Guralski, D., and Schiller, J. T. (1986) Translation of open reading frame E5 of bovine papillomavirus is required for its transforming activity. *Proceedings of the National Academy of Sciences, USA*, 83(6), 1797.

74. Rabson, M. S., Yee, C., Yang, Y. C., and Howley, P. M. (1986) Bovine papillomavirus type 13' early region transformation and plasmid maintenance functions. *Journal of Virology*, 60, 626.

75. Schiller, J. T., Vass, W. C., Vousden, K. H., and Lowy, D. R. (1986) E5 open reading frame of bovine papillomavirus type 1 encodes a transforming gene. *Journal of Virology*, 57(1), 1.

76. Schlegel, R., Wade-Glass, M., Rabson, M. S., and Yang, Y. C. (1986) The E5 transforming gene of bovine papillomavirus encodes a small, hydrophobic polypeptide. *Science*, 233(4762), 464.

77. Kell, B., Jewers, J. R., Cason, J., Pkarian, F., Kaye, J. N., and Best, J. M. (1994) Detection of E5 oncoprotein in human appillomavirus type 16-positive cervical scrapes using antibodies raised to synthetic peptides. *Journal of General Virology*, 75, 2451.

78. Conrad, M., Bubb, V. J., Schlegel, R. (1993) The human papillomavirus type 6 and 16 E5 proteins are membrane-associated proteins which associate with the 16-kilodalton pore-forming protein. *Journal of Virology*, 67(10), 6170.

79. Sparkowski, J., Anders, J., and Schlegel, R. (1995) E5 oncoprotein retained in the endo-plasmic reticulum/cis Golgi still induces PDGF receptor autophosphorylation but does not transform cells. *EMBO Journal*, 14(13), 3055.

80. Hwang, E. S., Nottoli, T., and Dimaio, D. (1995) The HPV16 E5 protein: expression, detection, and stable complex formation with transmembrane proteins in COS cells. *Virology*, 211(1), 227.

81. Pim, D., Collins, M., and Banks, L. (1992) Human papillomavirus type 16 E5 gene stimulates the transforming activity of the epidermal growth factor receptor. *Oncogene*, 7, 27.

82. Straight, S. W., Herman, B., and McCance, D. J. (1995) The E5 oncoprotein of human papillomavirus type 16 inhibits the acidification of endosomes in human keratinocytes. *Journal of Virology*, 69(5), 3185.

83. Conrad, M., Goldstein, D., Andresson, T., and Schlegel, R. (1994) The E5 protein of HPV-6, but not HPV-16, associates efficiently with cellular growth factor receptors. *Virology*, 200(2), 796.

84. Stoppler, M. C., Straight, S. W., Tsao, G., Schlegel, R., and McCance, D. J. (1996) The E5 gene of HPV-16 enhances keratinocyte immortalization by full-length DNA. *Virology*, 223(1), 251.

85. Frattini, M., Lim, H., and Laimins, L. A. (1996) *In vitro* synthesis of oncogenic human papillomaviruses requires episomal templates for differentiation-dependent late expression. *Proceedings of the National Academy Sciences, USA*, 91, 3062.

86. Ruesch, M. N., Studenrauch, F., and Laimins, L. A. (1998) Activation of papllomavirus late gene transcription and genome amplification upon differentiation in semisolid medium is coincident with expression of involcrin and transglutaminase but not keratin-10. *Journal of Virology*, 72, 5016.

87. Meyers, C., Frattini, M. G., Hudson, J. B., and Laimins, L. A. (1992) Biosynthesis of human papillomavirus from a continuous cell line upon epithelial differentiation. *Science*, 257, 971.

88. Ustav, M. and Stenlund, A. (1991) Transient replication of BPV-1 requires two poly-peptides encoded by E1 and E2 open reading frames. *EMBO Journal*, 10m 449.

89. Lu, J. Z.-J., Sun, Y.-N., Rose, R. C., Bonnex, W., and McCance. D. J. (1993) Two E2 binding sites (E2BS) alone or one E2BS plus an A/T-rich region are minimal requirements for the replication of the human papillomavirus type 11 origin. *Journal of Virology*, 67, 7131.

90. Ciang, C.-M., Ustav, M., Stenlund, A., Ho, T. F., Broker, T. R., and Chow, L. T. (1992) Viral E1 and E2 proteins support replication of homologous and heterologous papillomaviral origins. *Proceedings of the National Academy of Sciences, USA*, 89, 5799.

91. Gopalakrishan, V. and Kahn, S. A. (1994) E1 protein of human papillomavirus type 1a is sufficient for viral DNA replication. *Proceedings of the National Academy of Sciences, USA*, 91, 9597.

92. Sverdrup, F. and Kahn, S. A. (1995) Two E2 binding sites alone are sufficient to function as the minimal origin of replication of human papillomavirus type 18 DNA. *Journal of Virology*, 69, 1319.

93. Sun, Y.-N., Lu, J. Z.-J., and McCance, D. J. (1996) Mapping of HPV-11 E1 binding site and determination of other important cis-elements for replication of the origin. *Virology*, 216, 219.

94. Frattini, M. G. and Laimins, L. A. (1994) The role of the E1 and E2 proteins in the replication of human papillomavirus type 31b. *Virology*, 204, 799.

95. Lusky, M. and Fontaine, E. (1991) Formation of the complex of BPV-1 E1 and E2 proteins is modulated by E2 phosphorylation and depends upon sequences. *Proceedings of the National Acadey Sciences, USA*, 88, 6363.

96. Mohr, I. J., Clark, R., Sun, S., Androphy, E., MacPherson, P., and Botchan, M. R. (1990) Targeting the E1 replication factor to the papillomavirus origin of replication by complex formation with the E2 transactivator. *Science*, 250, 1694.

97. Seo, T.-S., Muller, F., Lusky, M., Gibbs, E., Kim, H.-Y., Phillips, B., *et al.* (1993) BPV-1 encoded E2 proteins enhance binding of E1 to the BPV origin. *Proceedings of the National Academy of Sciences, USA*, 90, 2865.

98. Tang, L., Mohr, I. J., Li, R., Nottoli, S., Sun, S., and Botchan, M. R. (1993) The E1 protein of BPV-1 is an ATP-dependent DNA helicase. *Proceedings of the National Academy of Sciences, USA*, 90, 5086.

99. Kuo, S.-R., Liu, J.-S., Broker, T. R., and Chow, L. T. (1994) Cell-free replication of human papillomavirus DNA with homologous viral E1 and E2 proteins and human cell extracts. *Journal of Biological Chemistry*, 269, 24058.

100. Androphy, E., Lowry, D. R., and Schiller, J. T. (1987) Bovine papillomavirus E2 trans-activating gene binds to specific sites in papillomavirus DNA product. *Nature,* 325, 70.

101. Clertant, P. and Seif, I. (1984) A common function for polyoma virus large T-antigen and papillomavirus E1 proteins? *Nature*, 311, 276.

102. Park, P., Copeland, L., Yang, L., Wang, T., Botchan, M. R., and Mohr, I. J. (1994) The cellular DNA polymerase alpha-primase is required for papillomavirus replication and association with the viral E1 helicase. *Proceedings of the national Academy of Sciences, USA*, 91, 8700.

103. Li, R. and Botchan, M. R. (1993) The acidic transcriptional activation domains of VP16 and p53 bind the replication protein A and stimulate *in vitro* BPV0-1 replication. *Cell*, 73, 1207.

104. Sun, T.-N., Han, H., and McCance, D. J. (1998) Active domains of human papillomavirus type 11 E1 protein for origin replication. *Journal of General Virology*, 79, 1651,

105. Thorner, L. K., Lim, D. A., and Botchan, M. R. (1993) DNA-binding domain of bovine paplomavirus type 1 E1 helicase: structural and functional aspects. *Journal of Virology*, 67, 6000.

106. Bensen, J. D. and Howley, P. M. (1995) Amino-terminal domains of the bovine papillomavirus type 1 E1 and E2 protein participate in complex formation. *Journal of Virology*, 69, 4364.

107. Brokaw, J. L., Blanco, M., and McBride, A. A. (1996) Amino acids critical for the functions of the bovine papillomavirus type 1 E2 transactivator. *Journal of Virology*, 70, 23.

108. Grossel, M., Baraoum, J., Prakash, S., and Endrophy, E. A. (1996) The BPV-1 E2 DNA-contact helix is required for transcriptional activation but not replication in mammalian cells. *Virology*, 217, 301.

109. Sakai, H., Yasugi, T., Benson, J. D., Dowhanick, J. J., and Howley, P. M. (1996) Targeted mutagenesis of human papillomavirus type 16 E2 transactivation domain reveals separable transcriptional activation and DNA replication functions. *Journal of Virology*, 70, 1602.

110. Benard, B. A., Bailly, C., Lenoir, M.-C., Darmon, M., Thierry, F., and Yaniv, M. (1989). The human papillomavirus type 18)HPV18) E2 gene product is a repressor of the HPV18 regulatory region in human keratinocytes. *Journal of Virology*, 63, 4317.

111. Cripe, T. P., Haugen, T. H., Turk, J. P., Tatabai, F., Schmid, P. G., Durst, M., *et al.* (1987) Transcriptional regulation of the HPV 16 E6/E7 promoter by a keratinocyte-dependent enhancer and by E2 transactivator and repressor gene products: implications for cervical cancer. *EMBO Journal*, 6, 3745.

112. Tan, S.-H., Leong, E. C., Walker, P. A., and Bernard, H.-U. (1994) The human papillomavirus type 16 E2 transcription factor binds with low cooperativity to two flanking sites and represses the E6 promoter through displacement of Sp1 and TFIID. *Journal of Virology*, 68, 6411.

113. Jeon, S. and Lambert, P. F. (1995) Integration of human papillomavirus type 16 DNA into the human genome leads to increased stability of E6 and E7 mRNAs: implications for cervical carcinogenesis. *Proceedings of the National Academy of Sciences, USA*, 92(5), 1654.

114. Gloss, B., Bernard, H. U., Seedork, K., and Klock, G. (1987) The upstream regulatory region of the human papillomavirus-16 contains an E2 protein-independent enhancer which is specific for cervical carcinoma cells and is regulated by glucocorticoid hormones. *EMBO Journal*, 6, 3735.

115. Steinberg, B. M., Auborn, K. J., Brandsma, J. L., and Taichman, L. B. (1989) Tissue site-specific enhancer function of the upstream regulatory region of human papillomavirus type 11 in cultured keratinocytes. *Journal of Virology*, 63, 957.

116. Bauknecht, T., See, R. H., and Shi, Y. (1996) A novel C/EBP beta-YY1 complex controls the cell-type specific activity of the human papillomavirus type 18 upstream regulatory region. *Journal of Virology*, 70, 7695.

117. Apt, D., Watts, R., Suske, G., and Bernard, H.-U. (1996) High Sp1/Sp3 levels in epithelial cells during differentation and cellular transformation correlate with activation of the HPV16 promoter. *Virology*, 224, 281.

118. Chang, T., Chan, W., and Bernard, H.-U. (1989) Transcriptional activation of HPV16 by nuclear factor 1, AP-1, steroid receptors and possibly a novel transcription factor. *Nucleic Acids research*, 18, 465.

3 | Molecular interactions in herpes simplex virus DNA replication

NIGEL D. STOW

1. Introduction

The family *Herpesviridae* comprises a diverse and interesting group of viruses characterized by their distinctive virion morphology and the possession of large linear double-stranded DNA genomes which range in size from approximately 120–250 kb (for review see refs 1, 2). Many vertebrate species are natural hosts to herpesviruses including fish, amphibians, reptiles, birds, marsupials, and mammals, and eight viruses that infect man have been identified. The family has historically been subdivided into the *Alpha-*, *Beta-*, and *Gammaherpesvirinae* subfamilies on the basis of biological properties; a classification largely (but not completely) supported by our rapidly advancing understanding of the genetic relationships amongst these viruses based upon DNA sequencing studies. The prototype and most extensively studied herpesvirus is herpes simplex virus type 1 (HSV-1), a human pathogen belonging to the *Alphaherpesvirinae* which is most frequently responsible for cold-sore lesions around the lips and mouth. In common with most other herpesviruses, HSV-1 is able to enter a latent state in its natural host. The latent viral genomes reside in a quiescent form within neuronal cells of the sensory ganglia but a variety of stimuli can trigger sporadic reactivation of viral replication giving rise to the recurrent symptoms of disease.

Early investigations of herpesvirus DNA replication were stimulated by the expectation that these large DNA viruses would encode the majority of the enzymes required for their own DNA synthesis and hence provide useful general models for eukaryotic DNA replication. Although subsequent work has confirmed that the herpesvirus replicative machinery is largely virus-coded, it is the smaller DNA viruses such as SV40 which have probably contributed most greatly to our present knowledge of the mechanisms of eukaryotic DNA replication. Nevertheless, numerous interesting features of herpesvirus DNA synthesis have emerged and much present interest centres around the possibility that a detailed understanding of the process will facilitate the development of novel and effective antiviral treatments.

The essential challenge that faces investigators of HSV-1 DNA replication is to determine the route by which a single genome is reproduced to yield multiple identical copies. In order to achieve this aim it is necessary to define both the pathway of replicative intermediates and the functions and mechanisms of action of the components which participate in DNA synthesis. During the 1980s major advances were made in identifying the *cis*-acting signals required for the synthesis of HSV-1 DNA and the encapsidation of the viral genome into virus particles. In addition the set of virus-coded proteins that perform direct and essential roles in HSV-1 DNA synthesis was defined and their major enzymatic activities elucidated. A major focus of more recent studies has been understanding how these proteins act together to bring about genomic replication.

Two excellent reviews of HSV-1 DNA replication have recently been published which describe in detail the advances leading to our present understanding of the viral origins of DNA replication and the functions and biochemical properties of the virus encoded proteins involved in DNA synthesis (3, 4). The present article will summarize this knowledge within the context of more detailed descriptions of the molecular interactions which occur between the replication components and the possible mechanisms of HSV-1 DNA synthesis. Space limitations prevent the presentation of a comprehensive bibliography, and I have therefore tended to cite mainly review articles and more recent papers which will hopefully provide interested readers with further information and refer them back to the relevant earlier work. I apologize to those who feel that this has resulted in their own contributions to the field being unfairly neglected.

2. Identification of the *cis*-acting signals and *trans*-acting proteins involved in HSV-1 DNA synthesis

Like other herpesvirus genomes, that of HSV-1 is characterized by the presence of reiterated regions. The 152 kb virion DNA consists of two covalently linked components, L and S, each comprising a unique region, U_L and U_S respectively, flanked by a set of inverted repeats (Fig. 1). The terminal and inverted repeats flanking U_L are referred to as TR_L and IR_L and the corresponding repeats flanking U_S as TR_S and IR_S. In addition the genome contains a true terminal redundancy of approximately 400 bp known as the *a* sequence that is also present in inverted orientation at the junction between the L and S segments. Additional copies of the *a* sequence may occur at the junction between the L and S segments and also at the terminus of the L, but not S component. The presence of the inverted repeats allows the two segments to invert relative to one another, so that virion DNA comprises an equimolar mixture of four genomic isomers distinguishable by the relative orientation of their L and S segments. Although physically distinct, the four isomers appear to be biologically equivalent (5, 6).

Compelling evidence regarding the presence and approximate location of specific *cis*-acting signals involved in the initiation of DNA synthesis was initially provided

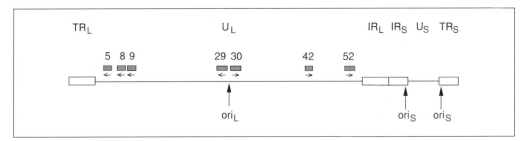

Fig. 1 Structure of the HSV-1 genome showing the inverted repeats as open boxes flanking the unique regions (thin lines). The locations of the replication origins, ori_S and ori_L, are indicated together with the positions and orientations of the seven ORFs encoding the proteins essential for viral DNA replication (hatched boxes).

by the characterization of defective HSV-1 genomes (7, 8). These studies indicated that separate signals that appear to function as origins of DNA replication reside within the S component and near the middle of the U_L region and these are now referred to as ori_S and ori_L, respectively. The development of transient replication assays in the early 1980s allowed the precise definition of the sequences involved in origin activity and facilitated subsequent investigations of the proteins that participate in viral DNA synthesis (9–11).

The transient assays for HSV-1 origin-dependent DNA synthesis are based upon observations that plasmids containing cloned fragments of HSV-1 DNA replicate autonomously in transfected tissue culture cells if a functional HSV-1 origin of DNA synthesis is present within the insert and the appropriate *trans*-acting viral proteins are supplied. In order to identify the *cis*-acting signals required for the initiation of viral DNA synthesis, test plasmids were transfected into permissive cells and the necessary helper functions were provided by superinfection with wild-type virus. Total cellular DNA was prepared and amplification of the transfected plasmid was assessed by Southern blot hybridization using a probe comprising the plasmid vector sequences (9). The sensitivity of the assay was increased by using restriction enzymes that can distinguish between unreplicated input plasmid DNA molecules and DNA that has been replicated following transfection (12, 13). For example, plasmids that have been propagated in *dam*+ strains of *E. coli* contain GATC motifs in which the A residues are methylated, but this modification is not present following replication of these sequences in a eukaryotic host. As a consequence replicated plasmid DNA is resistant, and non-replicated DNA sensitive, to cleavage by the enzyme *DpnI*. Although never directly proven, it is presumed that the *cis*-acting sequences which confer autonomous replication correspond to the sites at which synthesis is initiated, i.e. these signals correspond to origins of DNA replication. They will therefore be referred to as origins throughout the remainder of this chapter.

The transient replication assay allowed confirmation that the S component contains a functional replication origin, and fragments mapping entirely within the flanking IR_S and TR_S repeats were shown to contain all the *cis*-acting signals necessary for activity. The HSV-1 genome therefore contains two identical copies of ori_S which are located within the IR_S and TR_S regions (Fig. 1). The sequences required for ori_S function were subsequently shown to reside in the intergenic region between the 5' ends of divergently transcribed immediate early (IE) genes and a minimal functional region of approximately 90 bp (the core origin) was identified (14).

Although the presence of ori_S enabled plasmid amplification, the replication products were not packaged into virus particles and hence could not be serially passaged in the presence of the helper virus. To allow this a separate *cis*-acting packaging signal was required within the origin containing plasmid, and the necessary sequences were shown to reside entirely within the HSV-1 *a* sequence (15). Plasmids containing both a functional viral origin and packaging signal can be serially propagated as defective genomes in the presence of an appropriate helper virus and are generally referred to as amplicons (8). They have attracted much interest, particularly within the gene therapy field, as vectors for the delivery and expression of foreign genes (16).

Characterization of ori_L was hampered by the fact that the region of U_L found within CLASS II defective genomes is prone to deletion when cloned in *E. coli*. Significantly, when an amplicon containing a cloned class II repeat unit was serially propagated in the presence of wild-type helper virus, the deleted sequences were invariably repaired by recombination with the helper suggesting that the deletion prone sequences were important for ori_L function (8). The DNA sequence of the deletion prone region was obtained directly from virion DNA and revealed the presence of a long palindromic sequence that was probably responsible for cloning instability. Direct demonstration of ori_L function in a transient assay has now been achieved by cloning fragments from this region of the HSV-1 genome in an undeleted form in yeast vectors or recombination deficient *E. coli* strains (10, 17). The sequences which specify ori_L function lie within the intergenic region between the divergently transcribed early genes encoding the viral DNA polymerase and major single-stranded DNA binding protein (SSB) very close to the centre of U_L (Fig. 1). HSV-1 mutants lacking a functional copy of ori_L or in which both copies of ori_S have been inactivated have been described and show no obvious defects in tissue culture (18, 19). This indicates that at least *in vitro* there is a degree of redundancy between ori_S or ori_L and that either can suffice for genome replication.

The *trans*-acting viral proteins required for HSV-1 DNA synthesis were identified by a combination of studies using temperature sensitive (*ts*) mutants and transient DNA replication assays (for review see refs 3, 20). Systematic replacement of wt helper virus in the transient replication assay with series of plasmids encoding various HSV-1 functions allowed identification of a set of seven viral proteins that are both necessary and sufficient for origin-dependent DNA synthesis (11, 21). These seven proteins are encoded by genes which all lie within the U_L region of the HSV-1 genome; namely UL5, UL8, UL9, UL29, UL30, UL42, and UL52 (Fig. 1). (Throughout this article genes and their protein products are referred to interchangeably using the ULx nomenclature) (22). Both ori_S and ori_L containing plasmids are dependent upon the same set of replication proteins for their amplification in transient assays. Fine-mapping of the lesions within DNA negative *ts* mutants and the construction of specific null mutants demonstrated that all seven gene products are also required for the replication of genomic DNA (20). The seven DNA replication genes all belong to the early gene class and the expression of viral IE proteins is normally required to activate their transcription (11). However, when placed under the control of constitutive promoters the seven replication genes alone are sufficient to activate a viral replication origin (23).

In 1988, when the set of replication proteins was first identified, their functions were relatively poorly understood. UL30 was known to encode a DNA polymerase and the UL29 and UL42 genes to specify DNA binding proteins with preference for single- and double-stranded DNA respectively. Intensive interest subsequently focused on the biochemical activities of these proteins which have now all been purified from HSV-1 infected cells and overexpressed using a variety of vectors, of which the baculovirus *Autographa californica* nuclear polyhedrosis virus (AcMNPV) has probably been the most useful. The seven replication gene products are now known to comprise four distinct proteins or protein complexes with well characterized biochemical activities (Table 1). UL29 encodes a protein (ICP8) that binds co-operatively but without sequence specificity to single-stranded DNA. The UL9 protein exists as a homodimer in solution and functions both as a sequence-specific origin-binding protein (OBP) and a DNA helicase. A heterodimeric DNA polymerase with associated 3'–5' exonuclease and RNase H activities consists of a catalytic subunit encoded by UL30 and an accessory subunit encoded by UL42. The UL5, UL8, and UL52 proteins constitute a heterotrimeric complex that exhibits both DNA helicase and DNA primase activities (3, 4).

It should be noted that although the transient assays described above have permitted a full identification of the viral proteins which participate in HSV-1 origin-dependent DNA synthesis there remains the possibility that host proteins may also play essential roles in the process. Indeed, consideration of the biochemical activities of the HSV-1 DNA replication proteins reveals that at least two activities that would be expected to be required for bidirectional replication of double-stranded DNA are not represented, namely a topoisomerase and a DNA ligase, both of which are

Table 1. Viral proteins and their complexes involved in HSV-1 DNA replication.

Complex	Protein	Predicted size (Da)	Functions
Single-stranded DNA binding protein (monomer)	UL29 (ICP8)	128342	Binds co-operatively to single-stranded DNA, interacts with and stimulates activites of other complexes
DNA polymerase (heterodimer)	UL30 (Pol)	136413	DNA polymerase, 3'-5' exonuclease, RNase H
	UL42	51156	Processivity factor
DNA helicase-primase (heterotrimer)	UL5	98710	Contains helicase active site
	UL8	79921	Modulation of helicase and primase, interactions with UL9 and Pol
	UL52	114416	Contains primase active site
Origin-binding protein (homodimer)	UL9	94246	Sequence-specific origin-binding, DNA helicase

assumed to be provided by the host cell. The ability of the seven DNA replication proteins to support HSV-1 origin-dependent DNA synthesis in *Spodoptera frugiperda* (*Sf*) insect cells when expressed from recombinant baculoviruses nevertheless suggests that any host activities necessary for HSV-1 DNA synthesis must also be operative in infected *Sf* cells although they may be specified by either the cellular or AcMNPV genome (24).

Two cellular proteins have recently been implicated as having indirect roles in HSV-1 DNA synthesis. An inhibitor of host topoisomerase II inhibited viral DNA synthesis at early but not late times (25) possibly by blocking decatenation of daughter molecules rather than through an effect at the replication fork. Studies on a *ts* cell line conditionally able to support viral growth suggested that the host RCC1 (regulator of chromosome condensation) protein might function during circular-ization of the linear genome prior to its expression and replication (26). Although similar approaches will probably identify other important host proteins, a complete description of the factors that contribute to origin-dependent HSV-1 DNA synthesis is ultimately likely to require the development of a faithful cell-free system utilizing purified components.

The functions required for HSV-1 genome replication *in vivo* may also differ from those necessary in tissue culture cells. Although the major components of the syn-thetic machinery have undoubtedly been identified, it is noteworthy that HSV-1 encodes at least two enzymes involved in nucleotide metabolism, namely a thymidine kinase and a ribonucleotide reductase, that are essential for DNA synthesis in non-dividing cells *in vivo*. Additionally, virus-specified uracil N-glycosylase and deoxyuridine triphosphatase activities appear to play important roles during viral replication in the intact animal (4).

3. Models for HSV-1 DNA replication

Although many aspects of the pathway by which the linear HSV-1 genome is repro-duced remain unclear, overwhelming evidence indicates that it exists in an 'endless' form during most of the replicative cycle, thereby circumventing the problems that are associated with faithfully copying the ends of linear DNA molecules. Following infection input HSV-1 genomes are rapidly circularized, presumably by host proteins and/or components of the virus particle (27, 28). As noted above the cellular protein RCC1 appears to play a role in the process (26). Circularization could occur either by direct ligation of the termini (which have complementary single over-hanging 3′ nucleotides) or by homologous recombination between terminal *a* sequences (29). The latter process is incompatible with the known sequences of many herpesvirus genomes which have no terminal redundancy, and additionally would result in the loss of a copy of the *a* sequence.

Double-stranded circular DNA templates are predominantly replicated by either a theta-form or rolling-circle mechanism. A frequently cited model for HSV-1 DNA replication (reviewed in ref. 4) bears many similarities to the strategy adopted by bacteriophage λ (reviewed in ref. 30). An initial early phase is proposed in which the circular templates are amplified by a theta-form mechanism, followed by a later phase of rolling-circle DNA synthesis which produces long concatemers consisting of tandem head-to-tail duplications of the viral genome. Finally the concatemers are cleaved at specific sites to generate the unit length genomes packaged into virus particles. Analysis of the replicative intermediates produced during HSV-1 infection has, however, been hampered by their large size and fragility and the fact that recombination events may also be occurring. Convincing data in support of several aspects of the above model is consequently lacking and recent observations suggest that the overall process may be somewhat more complicated.

The occurrence of an early phase of theta-form replication is consistent with several observations. First, viral DNA accumulates rapidly and with non-linear kinetics following the initiation of HSV-1 DNA synthesis (31). Secondly, the observed inhibition of DNA synthesis at early but not late times by ICRF-193, an inhibitor of the cellular topoisomerase II, suggests a possible early requirement for decatenation of circular progeny molecules (25). Finally, the association of a DNA helicase but not a strand nicking activity with the viral OBP (UL9), and the overall symmetrical structure of the replication origins (Section 5) are features common to bidirectional theta-form synthesis. The rolling-circle mode, in contrast, requires a DNA nick for initiation, proceeds in one direction and predicts linear kinetics of DNA accumulation (30). Although circularized input HSV-1 genomes can be detected by pulsed-field gel electrophoresis (28), the accumulation of circular monomers predicted by the theta-form model has not been observed during the early stages of viral DNA synthesis. This may be because recombination between circular molecules generates structures that fail to enter the gel, or could indicate that a different mechanism is operational during these early stages.

At later times after infection replicating HSV-1 DNA sediments much more rapidly than unit length DNA and restriction enzyme analysis is consistent with the generation of concatemeric structures consisting of tandem repeats of the genome (32–34). The structure of these concatemers has recently been analysed by pulsed-field gel electrophoresis (31, 35–37). The data indicate that within concatemers, adjacent genomes can carry their L segments in opposite orientations, the four possible relative orientations being present in approximately equimolar amounts. These inversion events are probably a result of homologous intra- or intermolecular recombination between the inverted repeat regions. Also of considerable significance is the observation that even when replicative intermediates are digested with a restriction enzyme that cleaves each genome once, a significant proportion of the DNA fails to enter the gel. The most likely explanation is that the DNA which fails to enter the gel is highly branched due to the occurrence of intermolecular recombination. Although high frequency recombination between infecting HSV-1 genomes has also been demonstrated by conventional genetic analysis (38), it remains unclear whether recombination events represent an obligatory feature of the replication mechanism. In this context it may be relevant that homologous recombination events within concatemers could give rise to new circular template molecules enabling viral

DNA to accumulate with non-linear kinetics without the necessity of invoking an initial phase of theta-form replication.

The formation of concatemers is entirely consistent with a rolling-circle mechanism, but other possibilities cannot be excluded. Such structures might, for example, be generated by homologous recombination involving monomeric replication intermediates. Indeed, in bacteriophage T4 genetic recombination between molecules has been proposed as an important mechanism for establishing new replication forks (30). As indicated above, the establishment of a rolling-circle requires a DNA nick. Experiments with cell-free systems have shown that the HSV-1 replicative machinery can utilize circular plasmid molecules, or circular molecules containing a preformed replication fork, as templates for rolling-circle synthesis independent of the presence of a viral replication origin (39–41). The factors which convert a duplex circular molecule into a form suitable for rolling-circle replication have not been identified, and it remains unclear whether this mechanism operates *in vivo*.

Of the other mechanisms for HSV-1 DNA synthesis that have been suggested, one based on the yeast *Saccharomyces cerevisiae* 2μ circle mode of replication is of particular interest since it also involves a circular template, DNA segment inversion, concatemer production, and high frequency recombination. The 6.3 kb plasmid comprises two unique regions separated by inverted copies of a repeated 599 bp sequence, and can occur in two forms dependent on the relative orientation of the unique regions. It has been proposed that replication proceeds bidirectionally from a specific origin and that intramolecular recombination between one replicated and one unreplicated copy of the repeat region can give rise to the situation in which the two replication forks, rather than converging as in a theta-form model, move in the same direction around the circular template. This 'double rolling-circle' generates concatemers as the template is copied. Further recombination events may be involved in the reorientation of the replication forks to terminate replication and the resolution of the concatemer into monomeric circles (30). Such a model appears consistent with most of the features of HSV-1 replicative intermediates and the properties of the viral origins and OBP (31). However, viable viruses with non-inverting genomes have been described, and the symmetrical positioning of the viral origins (i.e. within the TR_S/IR_S regions and close to the centre of U_L) may preclude the initial type of recombination event necessary for establishing the 'double rolling-circle'.

Whatever the mechanism of replication, at least two processes are involved in the maturation of the replication products into linear double-stranded genomes. An 'alkaline exonuclease', encoded by gene UL12, has been implicated in removing branched structures prior to encapsidation, and null mutants appear to be impaired in the 'quality' of the DNA they package (42). Several other viral proteins are required for the cleavage of concatemers into monomeric genomes, a process tightly coupled to encapsidation of the DNA (43, 44). If initial circularization of the HSV-1 genome occurs by direct ligation of the ends, the simplest mechanism for generating unit length genomes from concatemers would be for sequence-specific cleavage events to occur at the novel junctions formed upon joining of the terminal *a* sequences (45). The ends of the genome would thus be faithfully copied as part of the overall replication process without having to invoke any specific mechanism for their duplication. More complicated mechanisms involving repair synthesis have been suggested which would enable cleavage between two genomes separated by a single *a* sequence to generate two termini each containing an *a* sequence. Such a process would appear to be an obligatory event if *a* sequences are lost during circularization by homologous recombination unless wastage of DNA occurs (46, 47).

4. Molecular interactions involved in HSV-1 DNA synthesis

During HSV-1 DNA replication, interactions occur between single- or double-stranded DNA and proteins, between proteins and nucleotides, and between various combinations of protein molecules. The macromolecular interactions, upon which the remainder of this chapter focuses, are likely to play key roles in the assembly of initiation complexes at the viral origins of replication and in regulating the activities of the various enzymes involved in copying the parental DNA chains. Those that have been described to date are listed in Table 2.

Evidence for protein–DNA and protein–protein interactions has been provided by a variety of approaches. In many instances the physical interaction has been directly demonstrated, e.g. by purification and sizing of protein complexes, affinity chromatography, co-immunoprecipitation, or electrophoretic mobility shift assays for protein–DNA complexes. In other cases the evidence is less direct, e.g. one protein influencing the intracellular localization or biochemical activity of another. A major experimental objective is not only to describe the interactions that occur amongst the components of the replicative machinery but also to analyse their overall contribution to DNA synthesis. To assess the significance of a particular interaction, mutated proteins specifically impaired in association should ideally be examined in functional assays and tested for their ability to support HSV-1 DNA synthesis. As will become apparent in subsequent sections, although numerous interactions have already been identified, our understanding of their roles remains far from complete.

5. Interactions involved in unwinding the replication origins

Initiation of HSV-1 DNA synthesis is assumed to involve the initial sequence-specific recognition of the origins, localized unwinding of the duplex DNA, and the recruitment of the proteins required for synthetic activity. The unwinding of the origin region appears to result from a series of molecular interactions involving the DNA sequences that comprise the origin, the sequence-specific OBP (UL9), and viral SSB (ICP8). Purified UL9 protein shares several properties with other DNA replication initiator proteins such as T-antigen of simian virus 40 and the E1 proteins of bovine

Table 2. Macromolecular interactions involved in HSV-1 DNA replication (see text for details).

INTERACTING COMPONENTS	POSSIBLE ROLE
Interactions within major complexes	
UL9-UL9	Formation of UL9 dimer
Pol-UL42	Formation of DNA polymerase holoenzyme
UL5-UL8, UL5-UL52 and UL8-UL52	Formation of helicase-primase complex
Protein-nucleic acid interactions	
UL9 with binding sites in replication origins	Origin recognition
ICP8 with single-stranded DNA	Coating of single-stranded DNA
UL9 with single-stranded DNA	Origin unwinding
UL5-UL8-UL52 complex with single-stranded DNA	Unwinding at replication fork
Pol-UL42 with primer-template junction	Chain elongation
Co-operative interactions	
Between UL9 dimers	Bending, looping distortion of origins
Between ICP8 monomers	Coating of single-stranded DNA
Other protein-protein interactions	
UL9-ICP8	Unwinding of origin
UL9-UL8	Recruitment of helicase-primase complex to origins
UL9-UL42	Recruitment of DNA polymerase to origins
UL9-DNA polymerase α	Possible role in initiation of DNA synthesis
UL8-Pol	Recruitment of DNA polymerase to origins, co-ordination of helicase-primase and DNA polymerase activities at replication fork
UL8-ICP8	Regulation of helicase-primase function
ICP8-UL42, ICP8-Pol	Regulation of DNA polymerase activity

and human papillomaviruses. These include sequence-specific binding to the origins, the ability to function as an energy-dependent DNA helicase, and to induce energy-independent structural alterations within the origin region (3, 4, 30).

5.1 ICP8

ICP8, encoded by gene UL29, is an abundant, zinc-containing protein of 1196 residues that functions in both viral DNA synthesis and gene regulation. The protein is monomeric in solution and binds single-stranded DNA with a fivefold higher affinity than duplex DNA (48). Binding to single-stranded DNA is rapid, highly co-operative, and independent of sequence with an estimated binding site of 12–22 bases per molecule (49). Like many other SSBs, ICP8 exhibits helix-destabilizing properties. Short duplex DNA fragments can be melted in the absence of ATP and the renaturation of complementary single strands promoted (50, 51). ICP8 also facilitates strand transfer, a property very probably associated with homologous viral recombination (52). Studies on ICP8 mutants have demonstrated that the protein can exert both positive and negative effects on transcription of viral genes during the course of a lytic infection (53, 54).

The helix-destabilizing properties of ICP8 probably play essential roles in facilitating the initial opening up of the duplex DNA prior to initiation of synthesis and in maintaining single-stranded DNA at the replication forks in a configuration suitable for use by the viral DNA polymerase. In addition to interacting with both itself (55) and the DNA template there is also abundant evidence for physical and functional interactions between ICP8 and other components of the replicative machinery. These will be considered in subsequent sections.

5.2   HSV-1 DNA replication origins

As outlined in Section 2, the minimal sequences required for ori_S function map to an approximately 90 bp region whilst a 425 bp fragment specifying ori_L activity has been identified (10, 14, 17, 56). Because of the tendency of ori_L sequences to delete when propagated in *E. coli*, more detailed analysis has been performed on ori_S. Nevertheless, sequence comparisons reveal that the two origins are closely related and are likely to function similarly.

The major features of the origins are illustrated in Fig. 2. Both origins contain long palindromic DNA sequences, an imperfect 45 bp palindrome in ori_S and a perfect one of 144 bp in ori_L. The minimal core ori_S sequence comprises the palindrome flanked by approximately 30 bp on one side and shows over 80% sequence identity to the corresponding region of the ori_L palindrome. At the centre of the ori_S and ori_L palindromes are 18 bp and 20 bp, respectively, containing exclusively A and T residues (AT regions). The sequence-specific OBP, UL9, binds efficiently to the sequence YGYTCGCACT (where Y is a pyrimidine) and strong binding sites occur in opposite orientation on the arms of the palindromes separated by approximately 30 bp. In the case of ori_S the two binding sites differ approximately fivefold in their affinity for UL9 (57–60). Site I, the higher affinity site, has the sequence CGTTCGCACT, and site II, TGCTCGCACT. In contrast, both UL9 binding sites within ori_L have the site I sequence. A third closely related element (site III) contains the sequence CGTTCTCACT and occurs once within ori_S and twice within ori_L. When tested in isolation the affinity of this sequence for UL9 is approximately 1000-fold lower than that of site I (59, 60). Nevertheless, in the context of a complete ori_S, co-operative interactions between UL9 molecules facilitate binding to site III (61). Analysis of ori_S mutants has shown that both sites I and II are essential for efficient origin function whilst removal of site III reduces replicative ability by approximately fivefold (12, 13, 62).

A binding site for an as yet unidentified cellular protein, OF-1, partially overlaps ori_S site I and mutagenesis experiments suggest that binding of this factor is necessary for full origin function (63). A weaker interaction of OF-1 was also detected with probes including the site II and III sequences. Analysis of the sequence requirements for binding suggested that OF-1 may be related to the cellular CCAAT-binding protein CP2. OF-1 interacts similarly with sequences overlapping binding sites I and III of ori_L (17).

Ori_S and ori_L appear to be functionally equivalent in tissue culture cells and in

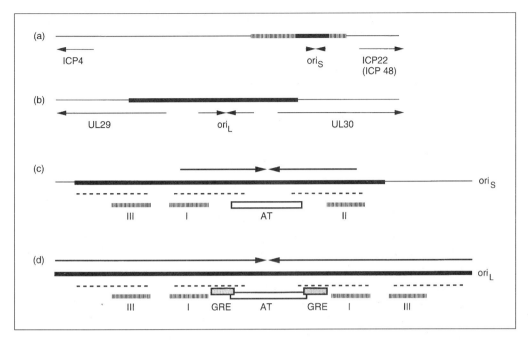

Fig. 2 Position and structure of the HSV-1 replication origins. (a) and (b) represent 900 bp from the regions containing ori_S and ori_L, respectively. Fragments that exhibit origin activity are shown as thickened lines. In the case of ori_S the core origin is represented by the solid line and flanking sequences which influence origin activity (ref. 65) are shaded. The ori_S and ori_L palindromes are indicated by inverted arrows and the positions of the 5' ends of the flanking transcripts and their protein products are shown. (c) and (d) show expanded versions of 110 bp from ori_S and ori_L, respectively. The core ori_S region and the central part of the ori_L palindrome are shown as solid thickened lines with the palindromic regions indicated by arrows. I, II, and III indicate UL9 binding sites, and the positions of the central AT regions and the two halves of the putative glucocorticoid receptor element (GRE) are indicated. The dotted lines show probable sites of interaction for the cellular factor OF-1 which overlap binding sites I, II, and III (Section 5.2).

transient assays exhibit similar kinetics of replication (17). It was noted, however, that one of the consequences of the minor sequence differences was the occurrence of a putative glucocorticoid-response element (GRE) unique to ori_L. Although the sequence arrangement and spacing of the two half sites of this motif differed significantly from the published consensus it was nevertheless able to interact with the glucocorticoid receptor DNA binding domain. The putative GRE was necessary for a specific enhancement of ori_L activity by the synthetic glucocorticoid, dexamethasone, in a cell line of neural origin (PC12) that had been terminally differentiated with nerve growth factor (64). Both nerve growth factor, which plays a role in the survival and function of the neuronal cells in which HSV-1 establishes latency, and glucocorticoids are induced in response to stress. It is therefore tempting to speculate that a component of the viral reactivation process may involve the specific stimulation of ori_L function. However, in a murine model for HSV-1 latency the presence of ori_L was not required for reactivation from explanted ganglia (18).

Further information relating to the identification of OF-1 and the mechanisms by which it and the glucocorticoid hormone–receptor complex influence origin activity are eagerly awaited.

For maximum activity both ori_L and ori_S require the presence of additional sequences which flank the minimal or core regions (17, 56, 65). Those surrounding the core ori_S have been investigated in greatest detail. Ori_S lies within an intergenic region of 697 bp between the 5' ends of the IE mRNAs encoding ICP4 in TR_S and IR_S and either ICP22 or ICP48 on opposite sides of U_S. In transient assays inclusion of additional sequences on either side of the core region resulted in an enhancement of replicative activity. For reasons that remain unclear the magnitude of this effect varied from about fourfold (14) to 50-fold (56, 65). The auxiliary flanking sequences contain numerous binding sites for transcription factors involved in the expression and regulation of the two genes, but neither IE gene transcription *per se* nor the presence of the IE gene-specific TAATGARAT promoter elements appeared to play a major role (14, 65). A significant part of the stimulatory activity could be explained by short elements immediately adjacent to either side of the core region which contain binding sites for cellular proteins. Proteins binding to the element on the ICP4 side included transcription factors Sp1 and Sp3 whilst those binding to the other element remain unidentified (65). The mechanism by which the cellular OBPs enhance replication remains unclear; one attractive possibility is that their presence facilitates assembly of the viral replicative machinery at the origin. Intriguingly, although the origin regions of the closely related herpes simplex virus type 2 (HSV-2) are very similar to those of HSV-1, no significant stimulation of core origin activity was observed in the presence of flanking sequences (66). A transcript running across the HSV-1 ori_S region has been described, but since the sequences specifying the two ends are dispensable for origin function in transient replication assays it is unclear whether it plays any role in viral DNA synthesis (67).

5.3 Interaction of UL9 with the viral origins

The viral OBP, UL9, is an 851 amino acid polypeptide that exists as a dimer in solution (68, 69) and comprises two essential functional domains (70, 71). Residues 1–534 specify a DNA helicase and associated DNA-dependent ATPase activity (72) while amino acids 535–851 are sufficient for sequence-specific binding to the replication origins (73) and contain a region important for interaction with ICP8 (74). The origin-binding domain (amino acids 535–851) is monomeric in solution (61), suggesting that residues required for dimerization are located in the N terminal portion of the protein. A leucine zipper motif within the N terminal region plays an essential role in co-operative binding to sites I and II, presumably by facilitating protein–protein interactions between UL9 dimers (75). Genetic evidence indicates that for productive UL9–UL9 interactions to occur bound molecules must be in the appropriate relative orientation on the same side of the helix. When increasing numbers of AT dinucleotides were introduced into the AT region of ori_S an oscillating effect on origin function was observed with maxima corresponding to the insertion of an integral number of full helical turns (66).

A key question regarding the stoichiometry of binding of UL9 to its recognition sequences remains unresolved, namely whether a single binding site interacts with one or two UL9 molecules. It has been pointed out that the UL9 recognition site exhibits a degree of symmetrical character with the presence of overlapping penta-nucleotide motifs, GT(T or G)CG, on the two strands suggesting the possibility of recognition by a dimeric protein (58). Evidence consistent with this has been presented by Stabell and Olivo (76) and by Fierer and Challberg (77). Binding of the monomeric C terminal recognition domain was also shown to be highly co-operative such that complexes formed with a single DNA recognition sequence invariably contained two protein monomers (77). This co-operativity suggests the existence of as yet uncharacterized protein–protein interactions in the C terminal region of the protein. In contrast, Martin *et al.* (78) and Gustafsson *et al.* (79) provided similarly compelling data indicating that only one UL9 molecule can bind per recognition site, allowing one UL9 dimer to potentially interact with two recognition sites. Resolution of this conflict, which may ultimately require crystallographic analysis of complexes between UL9 and its recognition sequence, is clearly crucial to an understanding of the initiation mechanism.

Only limited information is available concerning the molecular basis for sequence-specific recognition by UL9. The major contacts with DNA appear to align on one face of the DNA and form a recognition surface in the major grove (58, 60, 80). Site-directed mutagenesis experiments and the characterization of a monoclonal anti-body against UL9 that blocks sequence-specific binding suggest that amino acids directly involved in recognition lie within the region comprising amino acids 712–818 of UL9 (81). Secondary structure predictions suggest that three segments within this region have a high probability of forming α-helices, but it remains to be determined whether such structures participate in sequence recognition.

The co-operative interaction of full-length dimeric OBP with ori_S results in structural changes to the AT region between binding sites I and II. Distortion of the DNA helix in this region was revealed by specific hypersensitivity to DNase I and altered reactivity with the modifying agent potassium permanganate (59, 69, 82). The structural changes were independent of ATP but altered reactivity with potassium permanganate depended upon the free energy of supercoiled DNA. These data have led to the suggestion that interactions between UL9 dimers bound to sites I and II result in the intervening DNA being held in a loop (82).

Further evidence for the assembly of a complex nucleoprotein structure at ori_S has come from electron microscopic examination of DNA–protein complexes (83, 84). In the absence of ATP, interactions between dimers bound to sites I and II induced an approximately 86° degree bend in the DNA presumably accounting for the observed distortion of the central AT region (84). When ATP was present the UL9 helicase was able to unwind up to 1 kb of DNA whilst maintaining the interactions between the bound dimers resulting in the extrusion of unwound DNA from the base of a UL9 complex. The extruded single-stranded loops were visualized as a four-stranded condensed rod-like structure (84). As discussed below, during the initiation of DNA synthesis this origin-specific unwinding activity of UL9 is likely to be exploited through interactions involving other DNA replication proteins.

5.4   DNA helicase activity of UL9 and interaction with ICP8

A detailed analysis of the UL9 amino acid sequence revealed that the protein contains a set of six motifs shared by a superfamily of proteins (SF2) that includes a number of known or putative DNA and RNA helicases (85). Subsequently both DNA helicase and DNA-dependent ATPase activities were found to be associated with the purified protein (68). The DNA helicase activity is demonstrable on partially single-stranded molecules and a variety of model substrates have been used to show a 3′–5′ directionality and the requirement for a single-stranded DNA loading site. No sequence specificity was apparent in the unwinding reaction (69, 86). The helicase activity is coupled to the hydrolysis of ATP although other ribonucleotide and deoxyribonucleotide triphosphates are also capable of acting as energy sources *in vitro*. The introduction of specific mutations within the helicase motifs resulted in the generation of UL9 proteins that were unable to support viral DNA synthesis suggesting that the helicase function is essential for DNA synthesis (70). Helicase activity is independent of the presence of the sequence-specific DNA-binding domain, and a single-stranded DNA-binding activity has been shown to reside with the N terminal 535 amino acids of UL9 (72).

The sequence-independent helicase activity is stimulated by ICP8 which increases both the rate and extent of the DNA unwinding reaction (69, 86). Recent analysis has shown that ICP8 functions to increase the processivity of the enzyme and that maximum stimulation requires equimolar amounts of ICP8 rather than quantities sufficient to coat the single-stranded DNA template. Stimulation by ICP8 appears specific since human and *E. coli* SSBs did not enhance the UL9 helicase activity (87). These observations are consistent with the two proteins forming a complex that translocates along the DNA as it unwinds the two strands. Electron microscopic examination suggests that the binding of free ICP8 additionally facilitates the melting of the DNA duplex (88).

The physical interaction between UL9 and ICP8 has also been demonstrated directly. UL9 protein was specifically retained on an ICP8–agarose column and could be eluted with increasing salt concentrations and vice versa (49). The C terminal DNA-binding domain of UL9 is sufficient for the interaction which exhibits a 1:1 stoichiometry (49, 79). It was further shown that deletion of the C terminal 27 amino acids of UL9 destroyed its ability to interact with ICP8 but not its origin-binding or DNA helicase activities (74). The truncated protein was no longer stimulated by ICP8 but rather surprisingly exhibited basal levels of DNA-dependent ATPase and DNA helicase approximately eightfold greater than those of full-length UL9. The deleted protein was also greatly reduced in its ability to support viral origin-dependent DNA synthesis suggesting that formation of a complex between UL9 and ICP8 plays a significant role during DNA replication (74).

Although unwinding of a duplex origin region by UL9 has been observed in electron microscopy studies (84) there has been no direct biochemical demonstration to date. Sequence specificity in DNA unwinding was, however, noted with model substrates containing the box I sequence and was dependent on the presence of ICP8 (89). The model substrates contained a duplex box I sequence and required a single-stranded DNA tail on the side corresponding to the AT region of an intact origin. These results suggest that unwinding of the origin region may depend upon a combination of the sequence-specific DNA binding activity of UL9 and the single-stranded DNA binding activity of ICP8 provided by a protein complex.

5.5 Unwinding of the replication origins

The above features of UL9 and ICP8 have been incorporated into a tentative model describing the mechanism by which opening of the origin region may occur prior to the initiation of DNA synthesis (4) (Fig. 3). The initial stages involve the co-operative binding of a pair of UL9 dimers to sites I and II which results in the bending, looping, and distortion of the intervening AT region. The recruitment of ICP8 to the origin

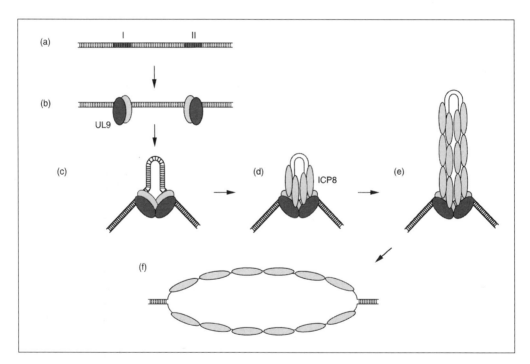

Fig. 3 Model for the unwinding of the HSV-1 replication origins. (a) Schematic diagram showing binding sites I and II in HSV-1 ori$_S$. (b and c) Pairs of UL9 dimers bind to sites I and II and through co-operative interactions involving the N termini loop and distort the intervening AT region. (d) ICP8 binds to UL9 and the distorted DNA. (e) Energy-dependent unwinding of the DNA, whilst maintaining interactions through the UL9 N termini, results in the extrusion of single-stranded regions that become coated with ICP8. The structure shown in (e) can potentially be converted to a bubble (f) through displacement of UL9, or disruption of the N terminal interactions.

probably occurs by virtue of its association with UL9 although it is not clear whether the UL9–ICP8 complex forms prior or subsequent to the sequence-specific binding by UL9. The observations of Lee and Lehman (89) suggest that in order for UL9 to function as a DNA helicase the AT region must display sufficient single-stranded characteristics to allow binding of ICP8. Following these initial interactions, the two UL9–ICP8 complexes are proposed to unwind the DNA in an energy-dependent reaction during which they remain associated through interactions involving the N termini of the UL9 proteins. Single-stranded DNA is consequently extruded and the unwound region stabilized by ICP8. It has been noted that the UL9–ICP8 interaction is disrupted in the presence of single- but not double-stranded DNA and that this may facilitate the ordered coating of single-stranded DNA by ICP8 as unwinding proceeds (79). Recruitment of DNA primase and DNA polymerase activities is proposed to occur on this structure allowing DNA synthesis to be initiated (Section 6). An associated displacement of UL9, or disruption of its co-operative interactions, would then allow the region to adopt a replication-bubble configuration and the establishment of bidirectional replication forks.

It should be stressed that many aspects of this model remain to be tested and that the biological relevance of some of the observations on which it is based is uncertain. Further significant advances in our understanding of the roles of UL9 and ICP8 in initiation probably await the development of a cell-free system for HSV-1 origin-dependent DNA synthesis.

5.6 Regulation of origin activity

Several lines of evidence suggest that the interaction of UL9 with the HSV-1 origins of replication may represent a site at which viral DNA synthesis can be regulated. Overexpression of wild-type UL9 protein in permissive cells resulted in a severe inhibition of HSV-1 growth (71, 90) and a similar effect has been observed on ori_S function in a transient assay in insect cells (74; N. Stow, unpublished results). The mechanism of inhibition is unknown but could involve a direct effect of UL9 on origin function. Alternatively, excess UL9 protein might sequester other viral or host components of the replicative machinery making them unavailable to participate in DNA synthesis. This finding not only suggests that regulation of UL9 levels may be important during viral infection but also indicates that caution should be exercised in interpreting experimental effects that require high UL9 concentrations.

Expression of the C terminal DNA binding domain of UL9 resulted in a potent dominant inhibition of viral DNA replication, probably due to its ability to bind to the origins and block their subsequent participation in initiation events (71, 91). A related polypeptide, OBPC, that consists of the C terminal 486 residues of UL9 and is encoded by an mRNA that initiates within the UL9 ORF, has been described in HSV-1 infected cells (92, 93). OBPC was detected in nuclear extracts with kinetics lagging slightly behind those of UL9 and was able to inhibit viral growth and DNA synthesis suggesting that it might function as a regulator of viral DNA replication. Inactivation of origin function by OBPC may play a role in establishing the rolling-circle mode of replication or prior to packaging of DNA (93).

6. Interactions involved in the initiation of DNA synthesis

Following interaction of UL9 and ICP8 with the HSV-1 origins of replication, initiation of DNA synthesis requires the recruitment of DNA primase and DNA polymerase activities. These functions are performed by two complexes which together comprise the remaining five HSV-1 proteins required for viral DNA synthesis.

6.1 The helicase–primase complex

Both a DNA helicase and a DNA primase activity were found to be associated with a trimeric complex purified from HSV-1 infected cells comprising subunits encoded by genes UL5, UL8, and UL52 (94). The helicase has associated DNA-dependent ATPase and GTPase activities, and experiments using model substrates indicate that it translocates in a 5′–3′ direction and requires a 3′ tail on the strand to be displaced (95). The primase function is responsible for the synthesis of short (6–13 bases) RNA primers on single-stranded DNA templates (96, 97). The enzyme exhibits a strong preference for certain template sequences, for example primer synthesis is initiated at the underlined C residue in the sequence 3′-AG<u>C</u>CCTCCCA, the predominant template site on φX174 virion DNA (98). A trimeric complex essentially indistinguishable from that isolated from HSV-1 infected cells can be generated in insect cells infected with recombinant baculoviruses (99).

Although the UL5 subunit contains a number of conserved ATP-binding and DNA helicase motifs (85) and a short motif in the UL52 protein exhibits homology to known DNA primases (100, 101) none of the subunits, expressed in isolation, exhibits detectable enzymatic activity. Site-directed mutagenesis experiments have confirmed that the active sites for the helicase and primase functions contain residues from the UL5 and UL52 subunits, respectively, and that both activities are essential for viral DNA synthesis (100–103). Nevertheless certain mutations within UL5 and UL52 have been shown to affect both the helicase and primase activities demonstrating that the two functions are not completely partitioned between these subunits (104–107).

The functional interdependence of the UL5 and UL52 subunits is also illustrated by the formation of a stable dimeric subassembly that retains both DNA helicase and DNA primase activities when the two proteins are co-expressed by recombinant baculoviruses (108, 109). This finding initially led to some puzzlement regarding the role of UL8 in the complex, particularly since it also exhibited no nucleic acid binding activity (110). However, subsequent biochemical comparisons of the trimeric helicase–primase and the UL5–UL52 subassembly indicate a role for UL8 in both enzyme activities. When homopolymeric templates were replaced by natural single-stranded DNAs, UL8 was shown to stimulate synthesis of RNA primers approximately tenfold (111). An effect on the helicase is suggested by the observed requirement for UL8 for efficient leading strand synthesis in a strand-displacement replication assay utilizing a preformed replication fork (J. Gottlieb and M. Challberg, unpublished results cited in ref. 3). Consistent with these biochemical studies, immunoprecipitation experiments performed using extracts of cells infected with recombinant baculoviruses showed that UL8 could interact separately with both the UL5 and UL52 subunits (112). The UL5–UL8 and UL52–UL8 complexes have not, however, been further analysed.

The UL8 subunit also appears to exhibit several other properties that may be related to its essential role in viral DNA synthesis. Immunofluorescence experiments showed that UL8, although not translocated into the nucleus when expressed alone, was required for efficient nuclear uptake of the UL5 and UL52 subunits (113, 114). In addition, direct physical interactions of UL8 with both UL9 and UL30 have been demonstrated, and a functional interaction with ICP8 is strongly suggested by biochemical data. These interactions will be described in detail in subsequent sections.

6.2   Effect of ICP8 on the helicase–primase complex

Several recent studies have examined the effects of ICP8 on the helicase–primase complex. Comparisons of the UL5–UL52 subassembly with the trimeric holoenzyme demonstrated that the presence of UL8 was required for efficient helicase and primase activities on templates in which single-stranded regions were coated with ICP8 (115, 116). In addition, the primase, DNA-dependent ATPase and helicase functions of the holoenzyme were stimulated by ICP8, but not other SSBs, when present in amounts much lower than required to saturate the substrate DNA (116, 117). Surface plasmon resonance measurements showed that the trimeric helicase primase complex, but neither the UL5–UL52 subassembly nor UL8 protein alone, could bind to immobilized ICP8 (115). Taken together, these results strongly suggest that ICP8 can establish a specific physical interaction with the trimeric helicase–primase complex that is mediated, at least in part, through the UL8 subunit. Moreover, stimulation of the enzymatic activities appears to be due to this physical interaction rather than to non-specific effects of ICP8 on the DNA template. During viral DNA synthesis the interaction possibly serves to promote the binding of the helicase–primase complex to regions of single-stranded DNA that are coated with ICP8, though other mechanisms may additionally contribute to enhanced helicase and primase activity. The rate of DNA unwinding *in vitro* by the HSV-1 helicase–primase in the presence of ICP8 is 60–65 base pairs/second (118), very similar to the value of 50 base pairs/second estimated for replication fork movement *in vivo* for another alphaherpesvirus, pseudorabies virus (119). Were bidirectional replication forks to be established at all three HSV-1 origins, the viral genome could be completely replicated in less than ten minutes.

6.3   The HSV-1 DNA polymerase holoenzyme

The interaction between the UL30 and UL42 components of the heterodimeric HSV-1 DNA polymerase (120–122) represents the best studied example of a protein–protein interaction amongst the viral replicative machinery. The effect on biochemical activity has been elucidated and genetic approaches have demonstrated a key role in viral DNA synthesis and growth. In addition, the interacting region of one of the subunits has been localized and its secondary structure subjected to biophysical examination. The biochemistry, genetics, and role of the HSV-1 DNA polymerase as a target for antiviral drugs have recently been reviewed (123).

The larger subunit, encoded by the 1235 codon UL30 ORF, functions as a DNA polymerase when expressed alone and is the site of numerous mutations that result in a thermolabile activity or resistance to inhibitory drugs (124–128). The catalytic UL30 subunit is hence commonly referred to as Pol. The Pol subunit also specifies intrinsic 3'–5' exonuclease and RNase H activities that are presumed to function in proof-reading and the removal of RNA primers, respectively (121, 128, 129). The UL30 polypeptide contains a number of conserved regions which are shared with other prokaryotic and eukaryotic DNA polymerases belonging to the cellular DNA polymerase α family, and three motifs characteristic of the 3'–5' exonuclease domain of other polymerases.

The 488 amino acid UL42 subunit exhibits high affinity but sequence-independent binding to double-stranded DNA (120) and can form functional complexes with free Pol. UL42 is present in approximately 20-fold greater amount than Pol in HSV-1 infected cells (122) but it is unclear whether the uncomplexed form has a specific replicative function since a transformed cell line expressing only 5% wt levels of UL42 can support efficient replication of UL42 null mutants (130).

The key role of UL42 appears to be related to the great increase in processivity on singly-primed single-stranded DNA templates which it confers upon Pol (122, 131). Under certain assay conditions (e.g. at high salt and limiting concentration of primer ends) this can be reflected as general stimulation of DNA polymerase activity (132). DNase protection experiments utilizing a model primer–template junction containing a 5' single-stranded tail showed that Pol alone protected 14 bp of the duplex region and 18 bases of single-stranded DNA. In the presence of UL42 the footprint was found to extend a further 5–14 bp into the duplex region suggesting that the double-stranded DNA-binding activity of UL42 might act to anchor the polymerase to newly synthesized DNA at the primer–template junction (133). The dimeric complex has a 10–15-fold higher affinity than Pol alone for the primer–template and dissociates more slowly without causing any reduction in the rate of strand elongation (133, 134). This mechanism of action contrasts markedly with that of other well characterized DNA polymerase processivity factors, including the eukaryotic replication protein PCNA, the β subunit of *E. coli* DNA polymerase III, and bacteriophage T4 gene 45 product, which lack double-stranded DNA binding activity but function as multimeric complexes which encircle the DNA duplex (for review see ref. 135).

Mutational analysis has shown that approximately 338 amino acids from the N terminus of UL42 are sufficient for viral DNA synthesis and growth, and that this region of the protein retains the ability to bind to and stimulate Pol (136–139). Although no smaller subdomain capable of interaction has been identified, insertion mutants which specifically disrupt the binding of UL42 to either Pol or DNA have been generated (138, 140). Characterization of these mutants indicated that both activities were necessary for processivity and provided strong evidence that the interaction between Pol and UL42 was essential for viral DNA synthesis and growth.

The Pol sequences important for interaction with UL42 have been located at the C terminus and appear to constitute a discrete functional domain unique to herpesvirus DNA polymerases. A fragment comprising the C terminal 228 amino acids of the protein was sufficient for interaction with UL42 (141), whilst at least 59 amino acids could be removed without affecting the catalytic activity of Pol (142). Further analysis indicated that deletion mutants lacking approximately 19–27 amino acids from the C terminus of Pol were unable to bind to or be stimulated by UL42, support viral DNA synthesis in a transient assay or allow viral growth (143, 144). These observations again provide strong evidence for the Pol–UL42 interaction being essential *in vivo* and have stimulated the search for drugs that might act by disrupting the complex or preventing its formation.

The importance of the C terminus of Pol in interacting with UL42 has been confirmed by several independent approaches. In a competition ELISA assay the binding of UL42 to immobilized Pol was inhibited approximately fourfold less efficiently by a deleted form of Pol lacking the C terminal 27 amino acids than by the full-length molecule suggesting that these amino acids contribute at least 75% of the binding energy (145). In addition anti-idiotype antibodies that recognized the UL42 subunit were detected in antisera raised against peptides representing the C terminal 12 or 15 amino acids of Pol. The generation of such antibodies, which arise as a consequence of reactivity against the combining region of antibodies to the initial peptide immunogen, provides direct evidence for structural complementarity between the C terminus of Pol and a region of UL42 (145). Finally, peptides representing the C terminal 18, 27 and 36 amino acids of Pol blocked the physical interaction with UL42 and inhibited its effects on DNA polymerase activity (145, 146). Circular dichroism studies on C terminal peptides indicated that two separate regions within the C terminal 36 amino acids exhibited α-helical character in solution (146). Peptides representing these two regions each self-associated in solution compatible with the their predicted amphipathic natures (143). In contrast, a peptide corresponding to the C terminal 36 amino acids also exhibited α-helical characteristics but remained monomeric with increasing concentration. These observations led to the proposal that in the 36 residue peptide the two α-helical regions might interact to form a hairpin structure (146). Since mutational analysis of UL42 failed to identify a discrete region capable of binding Pol it is possible that the interacting regions of UL42 are non-contiguous. This conclusion is supported by the failure to identify peptides from UL42 which specifically inhibit Pol binding or to identify single UL42 peptides that react with the anti-Pol anti-idiotype antibodies described above (145, 147). Dimerization of the DNA polymerase holoenzyme may thus involve the interaction of a discrete terminal protrusion of one subunit (Pol) with a possibly depressed region on the surface of the other (UL42). Interestingly, the C terminal regions of UL9 and UL8 that participate in interactions with ICP8 and Pol, respectively (Sections 5.4 and 6.4), are also strongly predicted to contain α-helices.

6.3.1 Effect of ICP8 on HSV-1 DNA polymerase

HSV-1 DNA polymerase activity is stimulated in the presence of ICP8 (48, 131). However, in contrast to its effects on UL9 and the helicase–primase complex, the effect on the polymerase appears largely non-specific since ICP8 can be substituted by other SSBs (148), suggesting that stimulation is at least partly due to the elimination of secondary structure from single-stranded template DNA. It is also possible that a direct physical interaction between the proteins may be important. Although such an interaction has not been characterized in detail, evidence from affinity chromatography suggests that ICP8 can interact directly with the UL42 subunit (149). Moreover, *ts* mutants with lesions in ICP8 have been described that at the permissive temperature show altered sensitivity to drugs that act on the Pol subunit (150).

6.4 Interaction between UL8 and UL9

A specific interaction between UL9 and the UL8 subunit of the helicase–primase complex was detected in immunoprecipitation experiments using extracts from insect cells mixedly infected with recombinant baculoviruses (112). The N terminal domain of UL9 (amino acids 1–535) was initially shown to be sufficient for the interaction (112), and subsequent analysis of UL9 mutants suggests that sequences essential for association are present between residues 131 and 211 (G. McLean, unpublished observations). It seems probable that the UL8–UL9 interaction plays an important role in the recruitment of the helicase–primase to the origin (Section 7).

6.5 Interactions of the HSV-1 DNA polymerase with UL8 and UL9

Two physical interactions have recently been described that may be relevant to the recruitment of the viral DNA polymerase to the replication origins: the catalytic UL30 (Pol) subunit is able to bind the UL8 component of the helicase–primase complex (151), and the accessory UL42 subunit can associate with the OBP, UL9 (152).

The UL8–Pol interaction was first identified in immunoprecipitation experiments using a monoclonal antibody against the UL8 protein. The antibody specifically co-precipitated Pol from extracts of insect cells mixedly infected with baculoviruses expressing the UL8 and Pol proteins. An ELISA assay was subsequently developed which allowed the interaction to be characterized in greater detail. Two lines of evidence implicate the C terminus of UL8 in the interaction. First, when a panel of anti-UL8 monoclonal antibodies was tested for their ability to block the interaction, three of the five inhibitory antibodies were found to recognize an epitope within the C terminal 29 amino acids of UL8. Secondly, a peptide representing the region spanning 13–32 amino acids from the UL8 C terminus inhibited the UL8–Pol interaction with an IC50 of approximately 5 μM (151). A UL8 mutant lacking 33 amino acids from its C terminus was unable to support ori_S-dependent DNA synthesis in a transient replication assay (114). However, it was not possible to conclude that this defect was a direct consequence of its inability to interact with Pol since nuclear uptake of a helicase–primase complex containing the truncated protein was also impaired (114).

A specific interaction between the UL9 and the UL42 proteins has also recently been reported (152). The two proteins were efficiently co-precipitated from extracts of insect cells in which they were co-expressed by recombinant baculoviruses. The presence of ethidium bromide or prior incubation of extracts with DNase I or micrococcal nuclease had no effect, excluding the possibility that the observed co-precipitation was a result of mutual association with DNA. *In vitro* transcribed and translated UL9, and a fragment corresponding to the N terminal 533 amino acids, additionally bound to a UL42–glutathione-*S*-transferase affinity column. It seems probable that the UL9–UL42 interaction, like that of UL8 and UL30, plays a role in recruitment of DNA polymerase to the origins. However, the association of the two proteins may be only transient since it is thought that only the DNA polymerase functions at the replication fork.

6.6 Interaction of host DNA polymerase α with UL9

An interaction between purified UL9 protein and the purified 180 kDa catalytic subunit of human DNA polymerase (pol α) was observed in immunoprecipitation and ELISA assays. The 180 kDa pol subunit also affected the mobility of UL9 complexes with an ori_S-containing fragment in gel shift assays, although its presence in the complexes was not proven. In addition, purified UL9 could stimulate the catalytic activity of the 180 kDa pol α subunit (153).

These observations suggest the interesting possibility that the DNA polymerase α–primase complex might be involved in the initiation of HSV-1 DNA synthesis and further characterization of its interaction with UL9 is awaited with interest. Genetic evidence (reviewed in ref. 4), however, suggests that the interaction is unlikely to play an obligatory role during viral DNA synthesis. Specifically, HSV-1 mutants have been isolated encoding altered UL30 (Pol) proteins that permit virus replication in the presence of drugs such as aphidicolin or the nucleoside analogue, araA, that are able to inhibit the intracellular activities of the DNA polymerase α–primase complex.

7. A model for the establishment of the HSV-1 replication fork

The above interactions involving the viral helicase–primase complex and DNA polymerase are postulated to facilitate recruitment of these enzymes to an unwound origin structure (Section 5.5) thereby allowing initiation of DNA synthesis (Fig. 4). Since both the helicase–primase complex and DNA polymerase are able to interact with UL9, recruitment of these enzymes may also serve to disrupt co-operative

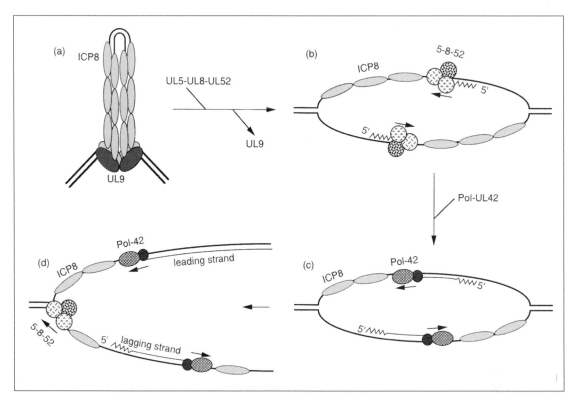

Fig. 4 Model for the establishment of HSV-1 replication forks. (a) Recruitment of the helicase–primase complex to the unwound origin region results in dissociation of UL9, (b) the synthesis of RNA primers (jagged lines), and the formation of a bubble structure. (c) The RNA primers are extended by the Pol–UL42 complex to establish leading strand synthesis. (d) Further unwinding at the replication fork by the helicase–primase allows the synthesis of RNA primers and their extension by Pol–UL42 on the lagging strand. Template and newly synthesized DNA strands are shown by thick and thin continuous lines, respectively. The direction of movement of protein complexes is indicated by arrows.

interactions between UL9 dimers and facilitate the conversion of the initial unwound structure into a replication bubble. It is therefore of particular interest to note that the region of UL9 required for the interaction with UL8 (between amino acids 131 and 211) includes the leucine zipper implicated in co-operative UL9 binding (75).

The sites of the initiating primers have not been mapped and it is not known whether recruitment of DNA polymerase occurs before or after their synthesis. Oligonucleotides representing the entire core $oris$ sequence did not serve as templates for the viral primase activity (98), but sites which meet the criteria for template recognition identified by Challberg (3) do lie within 150 bp on either side of this region. It seems quite probable that unwinding of the origin in the presence of UL9 and ICP8 could expose these sites and allow synthesis of the initiating primers. The RNA primers would then be extended by DNA polymerase to establish bidirectional leading strand DNA synthesis. Single-stranded regions of the opposite strand generated by this process would become coated with ICP8 prior to serving as templates for primer synthesis and lagging strand DNA replication. As a possible alternative pathway, it has been proposed that host DNA polymerase α–primase complex might be recruited to the origin by UL9 and perform the initial synthesis prior to displacement by the viral DNA polymerase and helicase–primase complex and the establishment of replication forks(4).

Irrespective of the mechanism of initiation, it seems unlikely that UL9 function is required at the replication forks and the protein is probably displaced from the replicative machinery, although data addressing this have not been presented. If the interactions of UL8 with UL9 and Pol are mutually exclusive, establishment of the UL8–Pol interaction may play a role in this process.

The remaining six HSV-1 DNA replication proteins (UL5–UL8–UL52, Pol–UL42, and ICP8) are presumed to function together as a multiprotein complex, or replisome, at the replication fork to allow co-ordinated synthesis on the two strands, and the various interactions amongst them are likely to be involved in forming and regulating this assembly. Consistent with the formation of such a complex, a high molecular weight assembly ($M_r > 10^6$) comprising the six fork proteins has been isolated from insect cells infected with recombinant baculoviruses and shown to be able to perform rolling-circle type DNA synthesis on circular plasmid templates (40).

Current models for eukaryotic replication forks (154) propose that the lagging strand is looped in such a way that the leading and lagging strand DNA polymerases can both move along the DNA in the same direction as the replication fork. In the case of HSV-1 such an arrangement might allow a single UL5–UL8–UL52 trimer to both unwind the duplex DNA and prime lagging strand DNA synthesis (Fig. 5). The

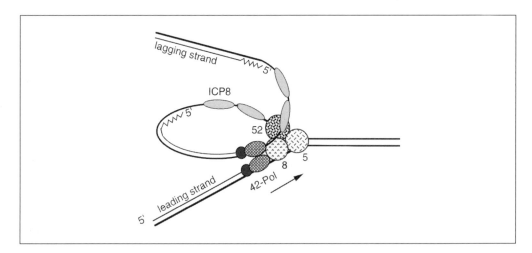

Fig. 5 Model for an HSV-1 replication fork. The diagram shows how looping of the lagging strand template might allow DNA polymerase complexes on both strands to move in the direction of the replication fork (arrow) and a single helicase primase complex to be positioned to unwind DNA at the fork, prime lagging strand DNA synthesis, and establish interactions with the leading and lagging strand DNA polymerases. RNA primers are shown as jagged lines, and the template and newly synthesized DNA strands as thick and thin continuous lines, respectively.

UL8–Pol interaction is envisaged to play important roles in co-ordinating leading strand DNA synthesis with template unwinding, and lagging strand DNA synthesis with primer generation. Since highly processive DNA synthesis occurs on the leading but not the lagging strand, a key question remains as to whether the Pol–UL42 complex is responsible for the replication of both strands. The possibility that the less processive free Pol performs lagging strand synthesis seems unlikely since the UL42 subunit is present in substantial molar excess and virtually all the viral DNA polymerase extracted from HSV-1 infected cells is purified in the form of a Pol–UL42 complex (122). Processivity of a Pol–UL42 complex on the lagging strand might be modulated through its interactions with other proteins of the replication complex (e.g. UL8 and/or ICP8) or possibly by altered degrees of phosphorylation of the UL42 subunit. Of the host-specified activities that are presumed to be required for viral DNA synthesis, it seems probable that a DNA ligase might also be associated with the replication fork proteins whilst a topoisomerase would function ahead of the fork.

The six HSV-1 replication fork proteins are members of a set of approximately 40 core genes that are conserved amongst members of the alpha-, beta-, and gamma-herpesvirus subfamilies suggesting that the mechanisms of DNA strand elongation are likely to be similar throughout the virus family and that similar sets of protein–protein interactions contribute to the process. In marked contrast, the structure of the replication origins and the presence of UL9 homologues are features that are largely specific to the alphaherpesvirus subfamily. Two notable exceptions are human herpesviruses 6 and 7 which, although they are betaherpesviruses, also contain UL9 homologues and replication origins with similar structure to HSV-1 (155, 156). Thus, whilst initiation of DNA synthesis probably occurs by similar mechanisms throughout the alphaherpesviruses, the process may differ significantly in the beta- and gammaherpesviruses (for reviews see refs 3, 157). Both human cytomegalovirus (HCMV, a betaherpesvirus) and Epstein–Barr virus (EBV, a gammaherpesvirus) encode viral proteins unique to their respective subfamilies which appear to perform essential origin-specific roles in viral DNA synthesis. In the case of EBV this protein is the transcriptional activator Zta which, like UL9, is a sequence-specific DNA binding protein capable of interacting with specific sites within the origin region (158). The analogous HCMV protein is encoded by gene UL84 and is possibly involved in a functional interaction with an origin-specific transcript (159, 160). Zta and HCMV UL84 possibly contribute in a similar fashion to UL9 to recruiting the other components of the replicative machinery prior to the initiation of DNA synthesis, though possibly involving different interactions and additional host or viral proteins. This hypothesis is supported by the recent demonstration that Zta interacts with at least two subunits of the EBV helicase-primase homologous to HSV-1 UL5 and either UL8 or UL52 (161).

8. Other interactions involving HSV-1 ICP8

The viral SSB, ICP8, has been reported to interact with at least two viral proteins that do not perform direct roles in HSV-1 DNA synthesis. The first of these is the viral alkaline exonuclease, a deoxyribonuclease encoded by gene UL12 that appears to be involved in removing branched structures from DNA replication intermediates prior to packaging of the viral genome (42). The interaction between the two proteins was detected by affinity chromatography and immunoprecipitation with monoclonal antibodies (162). Although the two proteins appear to co-localize in replication compartments (Section 9), UL12 exhibits a more diffuse intranuclear distribution (162, 163). The interaction between the two proteins may therefore be only transient, and it remains unclear what role it plays during the viral life cycle.

The second interaction of ICP8 is with the UL37 protein, a component of the tegument layer between the virion capsid and envelope (164). This interaction, which accounted for the previously reported ability of UL37 to bind to single-stranded DNA (165), has not been examined in further detail and its function is unknown.

9. Intranuclear localization of HSV-1 DNA synthesis

Although it has been known for many years that the DNA replication proteins and replicating genomes are present within discrete regions of the nuclei of HSV-1 infected cells, the basis for this distribution is only now becoming apparent. The recent advances in our understanding have been associated with an increasing use of confocal immunofluroescence microscopy to locate the viral DNA replication proteins in infected and transfected cells, and the identification of a cellular site to which parental viral genomes localize.

The large globular structures that represent the sites of viral DNA synthesis and contain all seven DNA replication proteins are referred to as replication compartments (166–169). Interestingly, virtually identical patterns of replication compartments were seen in the nuclei of HSV-1 infected binucleate or daughter cells (170, 171). This provides compelling evidence that rather than being established at random sites, the location of the replication compartments is determined through interactions between viral components and a pre-existing feature of the nuclear architecture. Confocal microscopy has demonstrated that replication compartments have a complex organization with late viral transcription, DNA synthesis, and capsid assembly occurring within partially overlapping but distinct domains (172).

When HSV-1 DNA synthesis is blocked by phosphonoacetic acid, which inhibits the viral DNA polymerase, the replication proteins accumulate in smaller punctate structures which have been referred to as pre-replicative sites (168, 169). Several host proteins involved in DNA replication (DNA polymerase α, DNA ligase-1, PCNA, and RF-A) as well as anti-oncogene proteins Rb and p53 co-localized with ICP8 in replicative compartments and pre-replicative sites (173). Although none of these proteins has a proven role in viral DNA synthesis it is interesting to note that DNA polymerase α has been reported to interact with UL9 (Section 6.5) whilst a host DNA ligase activity is predicted to be necessary. Similarly, in support of an essential role in HSV-1 DNA synthesis (25), the p170 form of host topoisomerase II redistributes to replication compartments where it interacts with progeny viral DNA (174). The assembly of pre-replicative site structures has also been described in cells infected with HSV-1 mutants in the Pol and UL42 genes, and following transfection with genes encoding the three components of the helicase–primase complex and ICP8 (168). Based on these observations Liptak *et al.* (168) proposed a model in which UL9, ICP8, and the helicase–primase complex assembly at pre-replicative sites followed by the recruitment of the Pol–UL42 heterodimer. This model is consistent with the known physical interactions amongst these proteins and the mechanism for the initiation of viral DNA synthesis proposed above.

Transfection experiments indicated that replication compartments could form in cells transfected with an ori_S containing plasmid and the seven DNA replication genes in the absence of any other viral gene product (175, 176). Furthermore, neither the presence of UL9 or ori_S was absolutely necessary for their formation (176, 177). The DNA synthesis occurring in the latter case is probably largely non-physiological and represents amplification of either cellular or transfected plasmid DNA in the presence of high concentrations of the viral DNA replication proteins. It is of interest to note that in the absence of UL9, the remaining six DNA replication proteins are also able to induce SV40 DNA replication provided a functional SV40 origin and T-antigen are present (23). T-antigen is able to unwind the SV40 origin region and it appears that the HSV-1 replication fork proteins can utilize such a structure for the initiation of DNA synthesis. Similar structures in the DNA of S-phase cells may also have the potential to serve as sites for DNA synthesis mediated by the HSV-1 replicative machinery.

Much confusion has until recently existed regarding the nature of pre-replicative sites, probably because this name was used to refer to structures with a similar punctate appearance formed under different experimental conditions. A clearer picture is now beginning to emerge in which it appears that two distinct classes of pre-replicative site can form. One class accounts for numerous (> 20 per nucleus) punctate regions seen in S-phase cells. These correspond to regions of cellular DNA synthesis and/or single-stranded DNA to which components of the viral replicative machinery can be recruited (177, 178) and the term 'pre-replicative' site therefore seems inappropriate.

The second class appears to represent the true pre-replicative sites of HSV-1 genomes and the precursors of replication compartments. They are fewer in number, their establishment is independent of the state of host cell DNA synthesis, and their locations may correspond to cellular ND10 domains (177, 178). The ND10 sites are ultrastructurally defined bodies with diameters of 0.3–0.5 μm that are associated with the nuclear matrix and contain a number of important cellular regulatory proteins. Although their precise role in HSV-1 replication remains unclear it is interesting to note that like human adenovirus 5 and SV40 (179), input HSV-1 genomes are preferentially localized to the periphery of ND10 domains (171). HSV-1 transcription is initiated at these sites and one of the viral IE proteins, ICP0, rapidly causes the dispersal of ND10-associated proteins (171, 180). At least one of the dispersed ND10 proteins (PML) then relocalizes with ICP8 (and the other viral DNA replication proteins) in foci that probably correspond to the true pre-replicative sites and progress to become replication compartments (181). Viral mutants lacking a functional DNA polymerase form similar ICP8-containing foci but PML is not recruited to them suggesting a role for Pol in this process. The PML-containing pre-replicative foci resemble ND10 domains in terms of size and number, but because disruption of ND10 precedes their assembly the precise relationship between the two sites remains uncertain. It is also not known whether PML performs any essential direct or indirect role in HSV-1 DNA synthesis.

10. Concluding remarks

Analysis of the molecular interactions involved in HSV-1 DNA replication is a subject still very much in its infancy. An increasing number of interactions amongst the viral DNA replication proteins has been described, several of which appear to be involved in the assembly of a multienzyme complex or replisome that functions at the replication forks. Further such associations probably remain to be identified, and much work is necessary before the structural basis for these interactions and their roles in regulating viral DNA synthesis are fully understood. Recent studies on the localization of parental genomes and replication compartments in infected cells illustrate that significant interactions between viral components and host proteins are awaiting identification and characterization. Hopefully these investigations will shed further light on mechanisms of HSV-1 DNA synthesis, both at the molecular level, where many aspects of the initiation and elongation modes remain to be eluci-dated, and at the genomic level, where the structures of the replicative intermediates remain poorly understood.

References

1. Fields, B. N., Knipe, D. M., and Howley, P. M. (ed.) (1996) *Virology*, 3rd edition. Lippincott-Raven Publishers, Philadelphia.
2. Roizman, B., Whitley, R. J., and Lopez, C. (ed.) (1993) *The human herpesviruses*. Raven Press, New York.
3. Challberg, M. D. (1996) Herpesvirus DNA replication. In *DNA replication in eukaryotic cells* (ed. M. L. DePamphilis),p. 721. Cold Spring Harbor Laboratory Press.
4. Boehmer, P. E. and Lehman, I. R. (1997) Herpes simplex virus DNA replication. *Annu. Rev. Biochem.*, **66**, 347.
5. Jenkins, F. J. and Roizman, B. (1986) Herpes simplex virus recombinants with noninverting genomes frozen in different arrangements are capable of independent replication. *J. Virol.*, **56**, 494.
6. Efstathiou, S., Minson, A. C., Field, H. J., Anderson, J. R., and Wildy, P. (1986) Detection of herpes simplex virus specific sequences in latently infected mice and humans. *J. Virol.*, **57**, 446.
7. Vlazny, D. A. and Frenkel, N. (1981) Replication of herpes simplex virus DNA: localization of replication signals within defective genomes. *Proc. Natl. Acad. Sci. USA*, **78**, 742.
8. Spaete, R. R. and Frenkel, N. (1982) The herpes simplex virus amplicon: a new eukaryotic defective-virus cloning-amplifying vector. *Cell*, **30**, 295.

9. Stow, N. D. (1982) Localization of an origin of DNA replication within the TR_S/IR_S repeated region of the herpes simplex virus type 1 genome. *EMBO J.*, **1**, 863.

10. Weller, S. K., Spadaro, A., Schaffer, J. E., Murray, A. W., Maxam, A. M., and Schaffer, P. A. (1985) Cloning, sequencing, and functional analysis of ori_L, a herpes simplex virus type 1 origin of DNA synthesis. *Mol. Cell. Biol.*, **5**, 930.

11. Wu, C. A., Nelson, N. J., McGeoch, D. J., and Challberg, M. D. (1988) Identification of herpes simplex virus type 1 genes required for origin-dependent DNA synthesis. *J. Virol.*, **62**, 435.

12. Weir, H. M. and Stow, N. D. (1990) Two binding sites for the herpes simplex virus type 1 UL9 protein are required for efficient activity of the ori_S replication origin. *J. Gen. Virol.*, **71**, 1379.

13. Hernandez, T. R., Dutch, R. E., Lehman, I. R., Gustafsson, C., and Elias, P. (1991) Mutations in a herpes simplex virus type 1 origin that inhibit interaction with origin-binding protein also inhibit DNA replication. *J. Virol.*, **65**, 1649.

14. Stow, N. D. and McMonagle, E. C. (1983). Characterization of the TR_S/IR_S origin of DNA replication of herpes simplex virus type 1. *Virology*, **130**, 427.

15. Stow, N. D., McMonagle, E. C., and Davison, A. J. (1983) Fragments from both termini of the herpes simplex virus type 1 genome contain signals required for the encapsidation of viral DNA. *Nucleic Acids Res.*, **11**, 8205.

16. Fink, D. J., DeLuca, N. A., Goins, W. F., and Glorioso, J. C. (1996) Gene transfer to neurons using herpes simplex virus based vectors. *Annu. Rev. Neurosci.*, **19**, 265.

17. Hardwicke, M. A. and Schaffer, P. A. (1995) Cloning and characterization of herpes simplex virus type 1 ori_L: comparison of replication and protein-DNA complex formation by ori_L and ori_S. *J. Virol.*, **69**, 1377.

18. Polvino-Bodnar, M., Orberg, P. K., and Schaffer, P. A. (1987) Herpes simplex virus type 1 ori_L is not required for virus replication or the establishment of latent infection in mice. *J. Virol.*, **61**, 3528.

19. Igarashi, K., Fawl, R., Roller, R. J., and Roizman, B. (1993) Construction and properties of a recombinant herpes simplex virus-1 lacking both S-component origins of DNA synthesis. *J. Virol.*, **67**, 2123.

20. Weller, S. K. (1991) Genetic analysis of HSV genes requird for genome replication. In *Herpesvirus transcription and its regulation* (ed. E. K. Wagner), p. 105. CRC Press, Boca Raton.

21. McGeoch, D. J., Dalrymple, M. A., Dolan, A., McNab, D., Perry, L. J., Taylor, P., *et al.* (1988) Structures of herpes simplex virus type 1 genes required for replication of virus DNA. *J. Virol.*, **62**, 444.

22. McGeoch, D. J., Dalrymple, M. A., Davison, A. J., Dolan, A., Frame, M. C., McNab, D., *et al.* (1988) The complete DNA sequence of the long unique region in the genome of herpes simplex virus type 1. *J. Gen. Virol.*, **69**, 1531.

23. Heilbronn, R. and zur Hausen, H. (1989) A subset of herpes simplex virus replication genes induces DNA amplification within the host cell genome. *J. Virl.*, **63**, 3683.

24. Stow, N. D. (1992) Herpes simplex virus type 1 origin-dependent DNA replication in insect cells using recombinant baculoviruses. *J. Gen. Virol.*, **73**, 313.

25. Hammarsten, O., Yao, X. D., and Elias, P. (1996) Inhibition of topoisomerase II by ICRF-193 prevents efficient replication of herpes simplex virus type 1. *J. Virol.*, **70**, 4523.

26. Umene, K. and Nishimoto, T. (1996) Replication of herpes simplex virus type 1 DNA is inhibited in a temperature-sensitive mutant of BHK-21 cells lacking RCC1 (regulator of chromosome condensation) and virus DNA remains linear. *J. Gen. Virol.*, **77**, 2261.

27. Poffenberger, K. L. and Roizman, B. (1985) A noninverting genome of a viable herpes simplex virus 1: presence of head-to-tail linkages in packaged genomes and requirements for circularization after infection. *J. Virol.*, **53**, 598.

28. Garber, D. A., Beverly, S. M., and Coen, D. M. (1993) Demonstration of circularization of herpes simplex virus DNA following infection using pulsed-field gel electrophoresis. *Virology*, **197**, 459.

29. Yao, X. D., Matecic, M., and Elias, P. (1997) Direct repeats of the herpes simplex virus *a* sequence promote nonconservative homologous recombination that is not dependent on XPF/ERCC4. *J. Virol.*, **71**, 6842.

30. Kornberg, A. and Baker, T. A. (1992) *DNA replication*, 2nd edition. W.H. Freeman and Co., New York.

31. Zhang, X., Efstathiou, S., and Simmons, A. (1994) Identification of novel herpes simplex virus replicative intermediates by field inversion gel electrophoresis: implications for viral DNA amplification strategies. *Virology*, **202**, 530.

32. Jacob, R. J. and Roizman, B. (1977) Anatomy of herpes simplex virus DNA. VIII. Properties of the replicating DNA. *J. Virol.*, **23**, 394.

33. Jacob, R. J., Morse, L. S., and Roizman, B. (1979) Anatomy of herpes simplex virus DNA. XII. Accumulation of head-to-tail concatemers in nuclei of infected cells and their role in the generation of the four isomeric arrangements of viral DNA. *J. Virol.*, **29**, 448.

34. Jongeneel, C. V. and Bachenheimer, S. L. (1981) Structure of replicating herpes simplex virus DNA. *J. Virol.*, **39**, 656.

35. Severini, A., Morgan, A. R., Tovell, D. R., and Tyrrell, D. L. J. (1994) Study of the structure of replicative intermediates of HSV-1 DNA by pulsed-field gel electrophoresis. *Virology*, **200**, 428.

36. Bataille, D. and Epstein, A. L. (1997) Equimolar generation of the four possible arrangements of adjacent L components in herpes simplex virus type 1 replicative intermediates. *J. Virol.*, **71**, 7736.

37. Slobedman, B., Zhang, X., and Simmons, A. (1999) Herpes simplex virus genome isomerization: origins of adjacent long segments in concatemeric viral DNA. *J. Virol.*, **73**, 810.

38. Honess, R. W., Buchan, A., Halliburton, I. W., and Watson, D. H. (1980) Recombination and linkage between structural and regulatory genes of herpes simplex virus type 1: study of the functional organization of the genome. *J. Virol.*, **34**, 716.

39. Rabkin, S. D. and Hanlon, B. (1990) Herpes simplex virus DNA synthesis at a preformed replication fork *in vitro*. *J. Virol.*, **64**, 4957.

40. Skaliter, R. and Lehman, I. R. (1994) Rolling circle DNA replication *in vitro* by a complex of herpes simplex virus type 1 encoded enzymes. *Proc. Natl. Acad. Sci. USA*, **91**, 10665.

41. Skaliter, R., Makhov, A. M., Griffith, J. D., and Lehman, I. R. (1996) Rolling-circle DNA replication by extracts of herpes simplex virus type 1 infected human cells. *J. Virol.*, **70**, 1132.

42. Martinez, R., Sarisky, R. T., Weber, P. C., and Weller, S. K. (1996) Herpes simplex virus type 1 alkaline exonuclease is required for efficient processing of viral DNA replication intermediates. *J. Virol.*, **70**, 2075.

43. Homa, F. L. and Brown, J. C. (1997) Capsid assembly and DNA packaging in herpes simplex virus. *Rev. Med. Virol.*, **7**, 107.

44. Salmon, B., Cunningham, C., Davison, A. J., Harris, W. J., and Baines, J. D. (1998) The herpes simplex virus type 1 UL17 gene encodes virion tegument proteins that are required for cleavage and packaging of viral DNA. *J. Virol.*, **72**, 3779.

45. Nasseri, M. and Mocarski, E. S. (1988) The cleavage recognition signal is contained within sequences surrounding an a-a junction in herpes simplex virus DNA. *Virology* **167**, 25.

46. Deiss, L. P. and Frenkel, N. (1986) Herpes simplex virus amplicon: cleavage of concatemeric DNA is linked to packaging and involves amplification of the terminally reiterated *a* sequence. *J. Virol.*, **57**, 933.

47. Varmuza, S. L. and Smiley, J. R. (1985) Signals for site-specific cleavage of HSV DNA: maturation involves two separate cleavage events at sites distal to the recognition sequences. *Cell*, **41**, 793.

48. Ruyechan, W. T. and Weir, A. C. (1984) Interaction with nucleic acids and stimulation of the viral DNA polymerase by the herpes simplex virus type 1 major DNA-binding protein. *J. Virol.*, **52**, 727.

49. Boehmer, P. E. and Lehman, I. R. (1993) Physical interaction between the herpes simplex virus-1 origin-binding protein and single-stranded DNA binding protein ICP8. *Proc. Natl. Acad. Sci. USA*, **90**, 8444.

50. Boehmer, P. E. and Lehman, I. R. (1993) Herpes simplex virus type 1 ICP8—helix-destabilizing properties. *J. Virol.*, **67**, 711.

51. Dutch, R. E. and Lehman, I. R. (1993) Renaturation of complementary DNA strands by herpes simplex virus type 1 ICP8. *J. Virol.*, **67**, 6945.

52. Bortner, C., Hernandez T. R., Lehman, I. R., and Griffith, J. (1993) Herpes simplex virus-1 single-strand DNA binding protein (ICP8) will promote homologous pairing and strand transfer. *J. Mol. Biol.*, **231**, 241.

53. Godowski, P. J. and Knipe, D. M. (1986) Transcriptional control of herpesvirus gene expression: gene functions required for positive and negative regulation. *Proc. Natl. Acad. Sci. USA*, **83**, 256.

54. Gao, M. and Knipe, D. M. (1991) Potential role for herpes simplex virus ICP8 DNA replication protein in stimulation of late gene expression. *J. Virol.*, **65**, 2666.

55. Dudas, K. C. and Ruyechan, W. T. (1998) Identification of a region of the herpes simplex virus single-stranded DNA-binding protein involved in cooperative binding. *J. Virol.*, **72**, 257.

56. Wong, S. W. and Schaffer, P. A. (1991) Elements in the transcriptional regulatory region flanking herpes simplex virus type 1 ori$_S$ stimulate origin function. *J. Virol.*, **65**, 2601.

57. Olivo, P. D., Nelson, N. J., and Challberg, M. D. (1988) Herpes simplex virus DNA replication: the UL9 gene encodes an origin-binding protein. *Proc. Natl. Acad. Sci. USA*, **85**, 5414.

58. Koff, A. and Tegtmeyer, P. (1988) Characterization of major recognition sequences for a herpes simplex virus type 1 origin-binding protein. *J. Virol.*, **62**, 4096.

59. Elias, P., Gustafsson, C. M., and Hammarsten, O. (1990) The origin binding protein of herpes simplex virus-1 binds cooperatively to the viral origin of replication ori$_S$. *J. Biol. Chem.*, **265**, 17167.

60. Hazuda, D. J., Perry, H. C., Naylor, A. M., and McClements, W. L.(1991) Characterization of the herpes simplex virus origin binding protein interaction with ori$_S$. *J. Biol. Chem.*, **266**, 24621.

61. Elias, P., Gustafsson, C. M., Hammarsten, O., and Stow, N. D. (1992) Structural elements required for the cooperative binding of the herpes simplex virus origin binding protein to ori$_S$ reside in the N-terminal part of the protein. *J. Biol. Chem.*, **267**, 17424.

62. Martin, D. W., Deb, S. P., Klauer, J. S., and Deb, S. (1991) Analysis of the herpes simplex virus type 1 oriS sequence: mapping of functional domains. *J. Virol.*, **65**, 4359.

63. Dabrowski, C. E., Carmillo, P. J., and Schaffer, P. A. (1994) Cellular protein interactions with herpes simplex virus type 1 oriS. *Mol. Cell. Biol.*, **14**, 2545.

64. Hardwicke, M. A. and Schaffer, P. A. (1997) Differential effects of nerve growth factor and

dexamethasone on herpes simplex virus type 1 oriL- and oriS-dependent DNA replication in PC12 cells. *J. Virol.*, **71**, 3580.

65. Nguyen-Huynh, A. T. and Schaffer, P. A. (1998) Cellular transcription factors enhance herpes simplex virus type 1 oriS-dependent DNA replication. *J. Virol.*, **72**, 3635.

66. Lockshon, D. and Galloway, D. A. (1988) Sequence and structural requirements of a herpes simplex viral DNA replication origin. *Mol. Cell. Biol.*, **8**, 4018.

67. Hubenthal-Vass, J., Starr, L., and Roizman, B. (1987) The herpes simplex virus origins of DNA synthesis in the S component are each contained in a transcribed open reading frame. *J. Virol.*, **61**, 3349.

68. Bruckner, R. C., Crute, J. J., Dodson, M. S., and Lehman, I. R. (1991) The herpes simplex virus-1 origin binding protein—a DNA helicase. *J. Biol. Chem.*, **266**, 2669.

69. Fierer, D. S. and Challberg, M. D. (1992) Purification and characterization of UL9, the herpes simplex virus type 1 origin-binding protein. *J. Virol.*, **66**, 3986.

70. Martinez, R., Shao, L., and Weller, S. K. (1992) The conserved helicase motifs of the herpes simplex virus type 1 origin-binding protein UL9 are important for function. *J. Virol.*, **66**, 6735.

71. Stow, N. D., Hammarsten, O., Arbuckle, M. I., and Elias, P. (1993) Inhibition of herpes simplex virus type 1 DNA replication by mutant forms of the origin-binding protein. *Virology*, **196**, 413.

72. Abbotts, A. P. and Stow, N. D. (1995) The origin-binding domain of the herpes simplex virus type 1 UL9 protein is not required for DNA helicase activity. *J. Gen. Virol.*, **76**, 3125.

73. Weir, H. M., Calder, J. M., and Stow, N. D. (1989) Binding of the herpes simplex virus type 1 UL9 gene product to an origin of viral DNA replication. *Nucleic Acids Res.*, **17**, 1409.

74. Boehmer, P. E., Craigie, M. C., Stow, N. D., and Lehman, I. R. (1994) Association of origin-binding protein and single-strand DNA binding protein, ICP8, during herpes simplex virus type 1 DNA replication *in vivo*. *J. Biol. Chem.*, **269**, 29329.

75. Hazuda, D. J., Perry, H. C., and McClements, W. L. (1992) Cooperative interactions between replication origin-bound molecules of herpes simplex virus origin-binding protein are mediated via the amino terminus of the protein. *J. Biol. Chem.*, **267**, 14309.

76. Stabell, E. C. and Olivo, P. D. (1993) A truncated herpes simplex virus origin-binding protein which contains the carboxyl-terminal origin-binding domain binds to the origin of replication but does not alter its conformation. *Nucleic Acids Res.*, **21**, 5203.

77. Fierer, D. S. and Challberg, M. D. (1995) The stoichiometry of binding of the herpes simplex virus type 1 origin-binding protein, UL9, to ori$_S$. *J. Biol. Chem.*, **270**, 7330.

78. Martin, D. W., Muñoz, R. M., Oliver, D., Subler, M. A., and Deb, S. (1994) Analysis of the DNA-binding domain of the HSV-1 origin-binding protein. *Virology*, **198**, 71.

79. Gustafsson, C. M., Falkenberg, M., Simonsson, S., Valadi, H., and Elias, P. (1995) The DNA ligands influence the interactions between the herpes simplex virus-1 origin-binding protein and the single-strand DNA-binding protein, ICP-8. *J. Biol. Chem.*, **270**, 19028.

80. Simonsson, S., Samuelsson, T., and Elias, P. (1998) The herpes simplex virus origin binding protein: specific recognition of phosphates and methyl groups defines the interacting surface for a monomeric DNA binding domain in the major groove of DNA. *J. Biol. Chem.*, **273**, 24633.

81. Stow, N. D., Brown, G., Cross, A. M., and Abbotts, A. P. (1998) Identification of residues within the herpes simplex virus type 1 origin-binding protein that contribute to sequence-specific DNA binding. *Virology*, **240**, 183.

82. Koff, A., Schwedes, J. F., and Tegtmeyer, P. (1991) Herpes simplex virus origin-binding protein (UL9) loops and distorts the viral replication origin. *J. Virol.*, **65**, 3284.

83. Rabkin, S. D. and Hanlon, B. (1991) Nucleoprotein complex formed between herpes simplex virus UL9 protein and the origin of DNA replication: intermolecular and intramolecular interactions. *Proc. Natl. Acad. Sci. USA*, **88**, 10946.

84. Makhov, A. M., Boehmer, P. E., Lehman, I. R., and Griffith, J. D. (1996) The herpes simplex virus type 1 origin-binding protein carries out origin specific DNA unwinding and forms unwound stem-loop structures. *EMBO J.*, **15**, 1742.

85. Gorbalenya, A. E., Koonin, E. V., Donchenko, A. P., and Blinov, V. M. (1989) Two related superfamilies of putative helicases involved in replication, repair and expression of DNA and RNA genomes. *Nucleic Acids Res.*, **17**, 4713.

86. Boehmer, P. E., Dodson, M. S., and Lehman, I. R. (1993) The herpes simplex virus type 1 origin binding protein—DNA helicase activity. *J. Biol. Chem.*, **268**, 1220.

87. Boehmer, P. E. (1998) The herpes simplex virus type 1 single-strand DNA binding protein, ICP8, increases the processivity of the UL9 protein DNA helicase. *J. Biol. Chem.*, **273**, 2676.

88. Makhov, A. M., Boehmer, P. E., Lehman, I. R., and Griffith, J. D. (1996) Visualization of the unwinding of long DNA chains by the herpes simplex virus type 1 UL9 protein and ICP8. *J. Mol. Biol.*, **258**, 789.

89. Lee, S. S. K. and Lehman, I. R. (1997) Unwinding of the box I element of a herpes simplex virus type 1 origin by a complex of the viral origin binding protein, single-strand DNA binding protein, and single-stranded DNA. *Proc. Natl. Acad. Sci. USA*, **94**, 2838.

90. Malik, A. K., Martinez, R., Muncy, L., Carmichael, E. P., and Weller, S. K. (1992) Genetic analysis of the herpes simplex virus type 1 UL9 gene: isolation of a lacZ insertion mutant and expression in eukaryotic cells. *Virology*, **190**, 702.

91. Perry, H. C., Hazuda, D. J., and McClements, W. L. (1993) The DNA-binding domain of herpes simplex virus type 1 origin-binding protein is a transdominant inhibitor of virus-replication. *Virology*, **193**, 73.

92. Baradaran, K., Dabrowski, C. E., and Schaffer, P. A. (1994) Transcriptional analysis of the region of the herpes simplex virus type 1 genome containing the UL8, UL9, and UL10 genes and identification of a novel delayed-eary gene-product, OBPC. *J. Virol.*, **68**, 4251.

93. Baradaran, K., Hardwicke, M. A., Dabrowski, C. E., and Schaffer, P. A. (1996) Properties of the novel herpes simplex virus type 1 origin-binding protein, OBPC. *J. Virol.*, **70**, 5673.

94. Crute, J. J., Tsurumi, T., Zhu, L., Weller, S. K., Olivo, P. D., Challberg, M. D., *et al.* (1989) Herpes simplex virus-1 helicase-primase—a complex of 3 herpes-encoded gene products. *Proc. Natl. Acad. Sci. USA*, **86**, 2186.

95. Crute, J. J., Mocarski, E. S., and Lehman, I. R. (1988) A DNA helicase induced by herpes simplex virus type 1. *Nucleic Acids Res.*, **16**, 6585.

96. Crute, J. J. and Lehman, I. R. (1991) Herpes simplex virus-1 helicase-primase—physical and catalytic properties. *J. Biol. Chem.*, **266**, 4484.

97. Sherman, G., Gottlieb, J., and Challberg, M. D. (1992) The UL8 subunit of the herpes simplex virus helicase-primase complex is required for efficient primer utilization. *J. Virol.*, **66**, 4884.

98. Tenney, D. J., Sheaffer, A. K., Hurlburt, W. W., Bifano, M., and Hamatake, R. K. (1995) Sequence-dependent primer synthesis by the herpes simplex virus helicase-primase complex. *J. Biol. Chem.*, **270**, 9129.

99. Dodson, M. S., Crute, J. J., Bruckner, R. C., and Lehman, I. R. (1989) Overexpression and assembly of the herpes simplex virus type 1 helicase-primase in insect cells. *J. Biol. Chem.*, **264**, 20835.

100. Klinedinst, D. K. and Challberg, M. D. (1994) Helicase-primase complex of herpes simplex virus type 1. A mutation in the UL52 subunit abolishes primase activity *J. Virol.*, **68**, 3693.

101. Dracheva, S., Koonin, E. V., and Crute, J. J. (1995) Identification of the primase active site of the herpes simplex virus type 1 helicase-primase. *J. Biol. Chem.*, **270**, 14148.

102. Zhu, L. and Weller, S. K. (1992) The 6 conserved helicase motifs of the UL5 gene product, a component of the herpes simplex virus type 1 helicase-primase, are essential for its function. *J. Virol.*, **66**, 469.

103. Graves-Woodward, K. L., Gottlieb, J., Challberg, M. D., and Weller, S. K. (1997) Biochemical analyses of mutations in the HSV-1 helicase-primase that alter ATP hydrolysis, DNA unwinding, and coupling between hydrolysis and unwinding. *J. Biol. Chem.*, **272**, 4623.

104. Graves-Woodward, K. L. and Weller, S. K. (1996) Replacement of gly in helicase motif V alters the single-stranded DNA-dependent ATPase activity of the herpes simplex virus type 1 helicase-primase. *J. Biol. Chem.*, **271**, 13629.

105. Spector, F. C., Liang, L., Giordano, H., Sivaraja, M., and Peterson, M. G. (1998) Inhibition of herpes simplex virus replication by a 2-amino thiazole via interactions with the helicase component of the UL5-UL8-UL52 complex. *J. Virol.*, **72**, 6979.

106. Biswas, N. and Weller, S. K. (1999) A mutation in the C-terminal putative zinc finger motif of UL52 severely affects the biochemical activities of the HSV-1 helicase-primase subcomplex. *J. Biol. Chem.*, **274**, 8068.

107. Barrera, I., Bloom, D., and Challberg, M. (1998) An intertypic herpes simplex virus helicase-primase complex associated with a defect in neurovirulence has reduced primase activity. *J. Virol.*, **72**, 1203.

108. Calder, J. M. and Stow, N. D. (1990) Herpes simplex virus helicase-primase—the UL8 protein is not required for DNA-dependent ATPase and DNA helicase activities. *Nucleic Acids Res.*, **18**, 3573.

109. Dodson, M. S. and Lehman, I. R. (1991) Association of DNA helicase and primase activities with a subassembly of the herpes simplex virus-1 helicase-primase composed of the UL5 and UL52 gene products. *Proc. Natl. Acad. Sci. USA*, **88**, 1105.

110. Parry, M. E., Stow, N. D., and Marsden, H. S. (1993) Purification and properties of the herpes simplex virus type 1 UL8 protein. *J. Gen. Virol.*, **74**, 607.

111. Tenney, D. J., Hurlburt, W. W., Micheletti, P. A., Bifano, M., and Hamatake, R. J. (1994) The UL8 component of the herpes simplex virus helicase-primase complex stimulates primer synthesis by a subassembly of the UL5 and UL52 components. *J. Biol. Chem.*, **269**, 5030.

112. McLean, G. W., Abbotts, A. P., Parry, M. E., Marsden, H. S., and Stow, N. D. (1994) The herpes simplex virus type 1 origin-binding protein interacts specifically with the viral UL8 protein. *J. Gen. Virol.*, **75**, 2699.

113. Calder, J. M., Stow, E. C., and Stow, N. D. (1992) On the cellular localization of the components of the herpes simplex virus type 1 helicase-primase complex and the viral origin-binding protein. *J. Gen. Virol.*, **73**, 531.

114. Barnard, E. C., Brown, G., and Stow, N. D. (1997) Deletion mutants of the herpes simplex virus type 1 UL8 protein: effect on DNA synthesis and ability to interact with and influence the intracellular localization of the UL5 and UL52 proteins. *Virology*, **237**, 97.

115. Falkenberg, M., Bushnell, D. A., Elias, P., and Lehman, I. R. (1997) The UL8 subunit of the heterotrimeric herpes simplex virus type 1 helicase-primase is required for the unwinding of single strand DNA binding protein (ICP8)-coated DNA substrates. *J. Biol. Chem.*, **272**, 22766.

116. Tanguy, Le Gac, N., Villani, G., Hoffmann, J.-S., and Boehmer, P. E. (1996) The UL8 subunit of the herpes simplex virus type 1 DNA helicase-primase optimizes utlization of DNA templates covered by the homologous single-strand DNA-binding protein ICP8. *J. Biol. Chem.*, **271**, 21645.

117. Hamatake, R. K., Bifano, M., Hurlburt, W. W., and Tenney, D. J. (1997) A functional interaction of ICP8, the herpes simplex virus single-stranded DNA-binding protein, and the helicase-primase complex that is dependent on the presence of the UL8 subunit. *J. Gen. Virol.*, **78**, 857.

118. Falkenberg, M., Elias, P., and Lehman, I. R. (1998) The herpes simplex virus type 1 helicase-primase: analysis of helicase activity. *J. Biol. Chem.*, **273**, 32154.

119. Ben-Porat, T., Blakenship, M. L., deMarchi, J. M., and Kaplan, A. (1977) Replication of herpesvirus DNA. III Rate of DNA elongation. *J. Virol.*, **22**, 734.

120. Gallo, M. L., Jackwood, D. H., Murphy, M., Marsden, H. S., and Parris, D. S. (1988) Purification of the herpes simplex virus type 1 65-kilodalton DNA-binding protein: properties of the protein and evidence of its association with the virus-encoded DNA polymerase. *J. Virol.*, **62**, 2874.

121. Crute, J. J. and Lehman, I. R. (1989) Herpes simplex-1 DNA polymerase–identification of an intrinsic 5′–3′ exonuclease with ribonuclease H activity. *J. Biol. Chem.*, **264**, 19266.

122. Gottlieb, J., Marcy, A. I., Coen, D. M., and Challberg, M. D. (1990) The herpes simplex virus type 1 UL42 gene-product: a subunit of DNA polymerase that functions to increase processivity. *J. Virol.*, **64**, 5976.

123. Coen, D. M. (1996) Viral DNA polymerases. In *DNA replication in eukaryotic cells* (ed. M. L. DePamphilis), p. 495. Cold Spring Harbor Laboratory Press.

124. Chartrand, P., Crumpacker, C. S., Schaffer, P. A., and Wilkie, N. M. (1980) Physical and genetic analysis of the herpes simplex virus DNA polymerase locus. *Virology*, **103**, 311.

125. Coen, D. M., Aschman, D. P., Gelep, P. T., Retondo, M. J., Weller, S. K., and Schaffer, P. A. (1984) Fine mapping and molecular cloning of mutations in the herpes simplex virus DNA polymerase locus. *J. Virol.*, **49**, 236.

126. Haffey, M. L., Stevens, J. T., Terry, B. J., Dorsky, D. I., Crumpacker, C. S., Wietstock, S. M., *et al.* (1988) Expression of herpes simplex virus type 1 DNA polymerase in *Saccharomyces cerevisiae* and detection of virus-specific enzyme activity in cell-free lysates. *J. Virol.*, **62**, 4493.

127. Dorsky, D. I., and Crumpacker, C. S. (1988) Expresson of herpes simplex virus type 1 DNA polymerase gene by *in vitro* translation and effects of gene deletions on activity. *J. Virol.*, **62**, 3224.

128. Marcy, A. I., Olivo, P. D., Challberg, M. D., and Coen, D. M. (1990) Enzymatic activities of overexpressed herpes simplex virus DNA polymerase purified from recombinant baculovirus-infected insect cells. *Nucleic Acids Res.*, **18**, 1207.

129. Weisshart, K., Kuo, A. A., Hwang, C. B. C., Kumura, K., and Coen, D. M. (1994) Structural and functional organization of herpes simplex virus DNA polymerase investigated by limited proteolysis. *J. Biol. Chem.*, **269**, 22788.

130. Johnson, P. A., Best, M. G., Friedmann, T., and Parris, D. S. (1991) Isolation of a herpes simplex virus type 1 mutant deleted for the essential UL42 gene and characterization of its null phenotype. *J. Virol.*, **65**, 700.

131. Hernandez, T. R. and Lehman, I. R. (1990) Functional interaction between the herpes simplex-1 DNA polymerase and UL42 protein. *J. Biol. Chem.*, **265**, 11227.

132. Hart, G. J. and Boehme, R. E. (1992) The effect of the UL42 protein on the DNA

polymerase activity of the catalytic subunit of the DNA polymerase encoded by herpes simplex virus type 1. *FEBS Lett.*, **305**, 97.

133. Gottlieb, J. and Challberg, M. D. (1994) Interaction of herpes simplex virus type 1 DNA polymerase and the UL42 accessory protein with a model primer template. *J. Virol.*, **68**, 4937.

134. Weisshart, K., Chow, C. S., and Coen, D. M. (1999) Herpes simplex virus processivity factor UL42 imparts increased DNA-binding specificity to the viral DNA polymerase and decreased dissociation from the primer template without reducing the elongation rate. *J. Virol.*, **73**, 55.

135. Kuriyan, J. and O'Donnell, M. (1993) Sliding clamps of DNA polymerases. *J. Mol. Biol.*, **234**, 915.

136. Tenney, D. J., Hurlburt, W. W., Bifano, M., Stevens, J. T., Micheletti, P. A., Hamatake, R. K., *et al.* (1993) Deletions of the carboxy terminus of herpes simplex virus type 1 UL42 define a conserved amino-terminal functional domain. *J. Virol.*, **67**, 1959.

137. Gao, M., Ditusa, S. F., and Cordingley, M. G. (1993) The C-terminal 3rd of UL42, a HSV-1 DNA replication protein, is dispensable for viral growth. *Virology*, **194**, 647.

138. Digard, P., Chow, C. S., Pirrit, L., and Coen, D. M. (1993) Functional analysis of the herpes simplex virus UL42 protein. *J. Virol.*, **67**, 1159.

139. Monahan, S. J., Barlam, T. F., Crumpacker, C. S., and Parris, D. S. (1993) Two regons of the herpes simplex virus type 1 UL42 protein are required for its functional inteaction with the viral DNA polymerase. *J. Virol.*, **67**, 5922.

140. Chow, C. S. and Coen, D. M. (1995) Mutations that specifically impair the DNA-binding activity of the herpes simplex virus protein UL42. *J. Virol.*, **69**, 6965.

141. Digard, P. and Coen, D. M. (1990) A novel functional domain of an alpha-like DNA polymerase—the binding-site on the herpes simplex virus polymerase for the viral UL42 protein. *J. Biol. Chem.*, **265**, 17393.

142. Haffey, M. L., Novotny, J., Bruccoleri, R. E., Carroll, R. D., Stevens, J. T., and Matthews, J. T. (1990) Structure-function studies of the herpes simplex virus type 1 DNA polymerase. *J. Virol.*, **64**, 5008.

143. Digard, P., Bebrin, W. R., Weisshart, K., and Coen, D. M. (1993) The extreme C-terminus of herpes simplex virus DNA polymerase is crucial for functional interaction with processivity factor UL42 and for viral replication. *J. Virol.*, **67**, 398.

144. Stow, N. D. (1993) Sequences at the C-terminus of the herpes simplex virus type 1 UL30 protein are dispensable for DNA polymerase activity but not for viral origin-dependent DNA replication. *Nucleic Acids Res.*, **21**, 87.

145. Marsden., H. S., Murphy, M., McVey, G. L., MacEachran, K. A., Owsianka, A. M., and Stow, N. D. (1994) Role of the carboxy-terminus of herpes simplex virus type 1 DNA polymerase in its interaction with UL42. *J. Gen. Virol.*, **75**, 3127.

146. Digard, P., Williams, K. P., Hensley, P., Brooks, I. S., Dahl, C. E., and Coen, D. M. (1995) Specific inhibition of herpes simplex virus DNA polymerase by helical peptides corresponding to the subunit interface. *Proc. Natl. Acad. Sci. USA*, **92**, 1456.

147. Owsianka, A. M., Hart, G., Murphy, M., Gottlieb, J., Boehme, R., Challberg, M., *et al.* (1993) Inhibition of herpes simplex virus type 1 DNA polymerase activity by peptides from the UL42 accessory protein is largely nonspecific. *J. Virol.*, **67**, 258.

148. O'Donnell, M. E., Elias, P., and Lehman, I. R. (1987) Processive replication of single-stranded DNA templates by the herpes simplex virus-induced DNA polymerase. J. Biol. Chem., **262**, 4252.

149. Vaughan, P. J., Banks, L. M., Purifoy, D. J. M., and Powell, K. L. (1984) Interactions between herpes simplex virus DNA-binding proteins. *J. Gen. Virol.*, **65**, 2033.

150. Chiou, H. C., Weller, S. K., and Coen, D. M. (1985) Mutations in the herpes simplex virus major DNA-binding protein gene leading to altered sensitivity to DNA polymerase inhibitors. *Virology*, **145**, 213.

151. Marsden, H. S., McLean, G. W., Barnard, E. C., Francis, G. J., MacEachran, K., Murphy, M., *et al.* (1997) The catalytic subunit of the DNA polymerase of herpes simplex virus type 1 interacts specifically with the C-terminus of the UL8 component of the viral helicase-primase complex. *J. Virol.*, **71**, 6390.

152. Monahan, S. J., Grinstead, L. A., Olivieri, W., and Parris, D. S. (1998) Interaction between the herpes simplex virus type 1 origin-binding and DNA polymerase accessory proteins. *Virology*, **241**, 122.

153. Lee, S. S. K., Dong, Q., Wang, T. S. F., and Lehman, I. R. (1995) Interaction of herpes simplex virus-1 origin-binding protein with DNA polymerase α. *Proc. Natl. Acad. Sci. USA*, **92**, 7882.

154. Waga, S. and Stillman, B. (1994) Anatomy of a DNA replication fork revealed by reconstitution of SV40 DNA replication *in vitro*. *Nature*, **369**, 207.

155. Inoue, N. and Pellett, P. E. (1995) Human herpesvirus 6B origin-binding protein: DNA-binding domain and consensus binding sequence. *J. Virol.*, **69**, 4619.

156. van Loon, N., Dykes, C., Deng, H. Y., Dominguez, G., Nicholas, J., and Dewhurst, S. (1997) Identification and analysis of a lytic-phase origin of DNA replication in human herpesvirus 7. *J. Virol.*, **71**, 3279.

157. Yates, J. L. (1996) Epstein–Barr virus DNA replication. In *DNA replication in eukaryotic cells* (ed. M. L. DePamphilis), p. 751. Cold Spring Harbor Laboratory Press.

158. Fixman, E. D., Hayward, G. S., and Hayward, S. D. (1995) Replication of Epstein–Barr virus oriLyt: lack of a dedicated virally encoded origin-binding protein and dependence on Zta in cotransfection assays. *J. Virol.*, **69**, 2998.

159. Sarisky, R. T. and Hayward, G. S. (1996) Evidence that the UL84 gene product of human cytomegalovirus is essential for promoting oriLyt-dependent DNA replication and formation of replication compartments in cotransfection assays. *J. Virol.*, **70**, 7398.

160. Huang, L., Zhu, Y., and Anders, D. G. (1996) The variable 3' ends of a human cytomegalovirus oriLyt transcript (SRT) overlap an essential, conserved replicator element. *J. Virol.*, **70**, 5272.

161. Gao, Z., Krithivas, A., Finan, J. E., Semmes, O. J., Zhou, S., Wang, Y., *et al.* (1998) The Epstein–Barr virus lytic transactivator Zta interacts with the helicase-primase replication proteins. *J. Virol.*, **72**, 8559.

162. Thomas, M. S., Gao, M., Knipe, D. M., and Powell, K. L. (1992) Association between the herpes simplex virus major DNA-binding protein and alkaline nuclease. *J. Virol.*, **66**, 1152.

163. Randall, R. E. and Dinwoodie, N. (1986) Intranuclear localization of herpes simplex virus immediate-early and delayed-early proteins: evidence that ICP4 is associated with progeny virus DNA. *J. Gen. Virol.*, **67**, 2163.

164. Shelton, L. S. G., Albright, A. G., Ruyechan, W. T., and Jenkins, F. J. (1994) Retention of the herpes simplex virus type 1 (HSV-1) UL37 protein on single-stranded DNA columns requires the HSV-1 ICP8 protein. *J. Virol.*, **68**, 521.

165. Shelton, L. S. G., Pensiero, M. N., and Jenkins, F. J. (1990) Identification and characterization of the herpes simplex virus type 1 protein encoded by the UL37 open reading frame. *J. Virol.*, **64**, 6101.

166. Rixon, F. J., Atkinson, M. A., and Hay, J. (1983) Intranuclear distribution of herpes simplex virus type 2 DNA synthesis: examination by light and electron microscopy. *J. Gen. Virol.*, **64**, 2087.

167. de Bruyn Kops, A. and Knipe, D. M. (1988) Formation of DNA replication structures in herpesvirus infected cells requires a viral DNA binding protein. *Cell*, **55**, 857.

168. Liptak, L. M., Uprichard, S. L., and Knipe, D. M. (1996) Functional order of assembly of herpes simplex virus DNA replication proteins into prereplicative site structures. *J. Virol.*, **70**, 1759.

169. Lukonis, C. J. and Weller, S. K. (1996) Characterization of nuclear structures in cells infected with herpes simplex virus type 1 in the absence of viral DNA replication. *J. Virol.*, **70**, 1751.

170. de Bruyn Kops, A. and Knipe, D. M. (1994) Preexisting nuclear architecture defines the intranuclear location of herpesvirus DNA replication structures. *J. Virol.*, **68**, 3512.

171. Maul, G. G., Ishov, A. M., and Everett, R. D. (1996) Nuclear domain-10 as preexisting potential replication start sites of herpes simplex virus type 1. *Virology*, **217**, 67.

172. de Bruyn Kops, A., Uprichard, S. L., Chen, M., and Knipe, D. M. (1998) Comparison of the intranuclear distributions of herpes simplex virus proteins involved in various viral functions. *Virology*, **252**, 162.

173. Wilcock, D. and Lane, D. P. (1991) Localization on p 53, retinoblastoma and host replication proteins at sites of viral replication in herpes-infected cells. *Nature*, **349**, 429.

174. Ebert, S. N., Subramanian, D., Shtrom, S. S., Chung, I. K., Parris, D. S., and Muller, M. T. (1994) Association between the p170 form of human topoisomerase II and progeny viral DNA in cells infected with herpes simplex virus type 1. *J. Virol.*, **68**, 1010.

175. Zhong, L. and Hayward, G. S. (1997) Assembly of complete, functionally active herpes simplex virus DNA replication compartments and recuitment of associated viral and cellular proteins in transient cotransfection assays. *J. Virol.*, **71**, 3146.

176. Lukonis, C. J. and Weller, S. K. (1997) Formation of herpes simplex virus type 1 replication compartments by transfection: requirements and localization to nuclear domain 10. *J. Virol.*, **71**, 2390.

177. Uprichard, S. L. and Knipe, D. M. (1997) Assembly of herpes simplex virus replication proteins at two distinct intranuclear sites. *Virology*, **229**, 113.

178. Lukonis, C. J., Burkham, J., and Weller, S. K. (1997) Herpes simplex virus type 1 prereplicative sites are a heterogeneous population: only a subset are likely to be precursors to replication compartments. *J. Virol.*, **71**, 4771.

179. Ishov, A. M. and Maul, G. G. (1996) The periphery of nuclear domain 10 (ND10) as site of DNA virus deposition. *J. Cell Biol.*, **134**, 815.

180. Everett, R. D. and Maul, G. G. (1996) HSV-1 IE protein Vmw110 causes redistribution of PML. *EMBO J.*, **13**, 5062.

181. Burkham, J., Coen, D. M., and Weller, S. K. (1998) ND10 protein PML is recruited to herpes simplex virus type 1 prereplicative sites and replication compartments in the presence of viral DNA polymerase. *J. Virol.*, **72**, 10100.

4 | Epstein–Barr virus proteins involved in cell immortalization

MARTIN ROWE AND J. EIKE FLOETTMANN

1. Introduction

The discovery in 1964 of herpesvirus particles in tumour cell lines derived from patients with Burkitt's lymphoma (1) was a crucial observation that subsequently led to the so-called Epstein–Barr virus (EBV) being the first virus to be implicated in the pathogenesis of a human cancer. This gammaherpesvirus has since been associated with several other malignant diseases, including nasopharyngeal carcinoma, B cell lymphomas in immunosuppressed patients, certain T cell lymphomas, and Hodgkin's disease (2). Absolute proof of the involvement of EBV in these malignancies was hampered by the unexpected finding that EBV was in fact a ubiquitous virus, being carried as a persistent and largely asymptomatic infection by more than 95% of adults in all populations world-wide (2). However, in support of its potential oncogenic properties, EBV was found to be a powerful transforming agent for normal resting human B lymphocytes *in vitro*, resulting in the regular outgrowth of EBV-immortalized lymphoblastoid cell lines (LCL) (3). Viral gene expression in LCL is restricted to about nine of the approximately 100 genes encoded by EBV and, since LCL are largely non-permissive for virus production, the EBV genes expressed in LCL were defined as latent genes. Six of the latent genes encode the nuclear antigens, EBNA-1,- 2, -3A,-3B, -3C., and -LP, while three encode the latent membrane proteins, LMP-1, -2A, and -2B; the most abundantly expressed latent transcripts in LCL are non-polyadenylated RNA, the EBERs, that do not encode proteins.

Genetic studies have identified EBNA1, EBNA2, EBNA3A, EBNA3C, and LMP1 as being essential for EBV-induced transformation of B cells, while EBNA-LP and LMP2A greatly enhance the transforming efficiency (4–7). Although the process of cellular transformation by EBV is not completely understood, a considerable amount is known about the biological properties and the molecular interactions of the individual latent genes that co-operate to mediate B cell transformation. However, in EBV-associated malignancies, not all of the transforming proteins are regularly expressed (2). For example, in Burkitt's lymphoma, only EBNA1 is regularly detected, while in nasopharyngeal carcinoma, EBNA1 is the only protein invariably expressed but many tumours also express LMP1 and LMP2. A number of reasons may be put forward to explain the absence of apparently critical transforming proteins in EBV-associated tumours. The most difficult to prove is the 'hit-and-run' hypothesis in which transforming genes are postulated to have been expressed early in the oncogenic process, but which are no longer required and are down-regulated in the final tumour. Alternatively, and not mutually exclusive, is the likelihood that other genetic aberrations have replaced the requirement for some of the EBV trans-forming genes, e.g. the defining chromosome translocations of Burkitt's lymphoma which result in deregulated expression of the c-myc cellular gene (8). Lastly, it is possible that different sets of EBV genes may be required for transformation of cell types other than B cells, e.g. the epithelial cells of nasopharyngeal carcinoma.

Whilst the precise role of EBV in the pathogenesis of malignant diseases remains unclear, the effects of EBV infection upon normal primary B cells led to the con-clusion that EBV is the most highly transforming virus known for human cells. In this review, we will focus on the protein–protein interactions of the EBV-encoded pro-teins expressed in LCL, with an emphasis on those proteins known to be essential for transformation of resting B cells.

2. The nuclear proteins expressed in EBV-transformed B cells

2.1 The EBV nuclear antigen-1, EBNA1

In LCL, the viral genome replicates as an episome synchronously with the cellular chromosomes, a property that is absolutely dependent upon EBNA1 expression (9). Through this function alone, therefore, EBNA1 is essential for sustaining EBV-induced transformation of B cells. However, EBNA1 also confers a selective growth advantage to tumour cells (10, 11), and can act as a transcriptional activator of viral (12–14) and cellular genes (15, 16). All of these functions require the specific binding of EBNA1 to multiple copies of partial palindromic DNA sequences (17–19). In addition, EBNA1 was found to act like an 'RGG' RNA binding protein, and thus may play a role in EBV-associated tumours through transcriptional regulation (20).

EBNA1 is a nuclear phosphoprotein that associates with chromosomes during mitosis, a property not shared by the other EBNAs (21–23). The EBNA1 encoded by the prototype B95.8 isolate of EBV is a 641 amino acid protein with an N terminus rich in basic residues, a C terminus rich in acidic residues, and an unusual long stretch of glycine–alanine repeats at amino acids 90–328 (24). The essential structural features of EBNA1 are summarized in Fig. 1. Yates *et al.* (25) were the first to carry out a detailed deletion analysis of EBNA1, but were unable to separate replication from *trans*-activation functions, indicating that the two functions are closely linked. Since then, other groups have managed to map the nuclear localization sequence to amino acids 379–387 and the DNA-binding domain to amino acids 459–598 of EBNA1 (26,

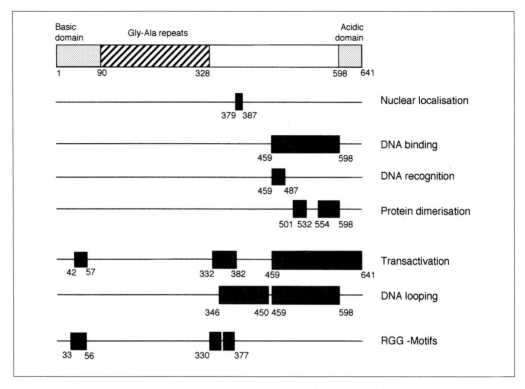

Fig. 1 Schematic summary of the functional domains of EBNA1. The top panel illustrates the features of the primary sequence. The solid black boxes in the lower panels indicate functional domains identified in several studies by analysis of deletion mutants of EBNA1.

27), the latter being further delineated into protein dimerization and DNA recognition sequences (28). Homodimerization of EBNA1 molecules is essential for their ability to bind DNA (28–30). Furthermore, physical interaction can also occur between separate EBNA1 complexes bound to different DNA sequences, resulting in DNA looping (31, 32). The interaction of physically separate EBNA1 complexes has been mapped to three independent regions within the protein: amino acids 54–89, 331–361, and 372–391 (33).

EBNA1 does not display any enzymatic activities that have been identified as important functions for other viral origin-binding proteins involved in DNA replication (34, 35), and neither does it contain a typical *trans*-activation domain. Consequently, the mechanisms by which EBNA1 exerts its functions are unclear. This has prompted the search for cellular proteins interacting with EBNA1 that might effect its functions and, at the time of writing, three such proteins had been identified.

Two groups have reported that EBNA1 binds via its nuclear localization sequence to the nuclear import receptor known variously as Rch1, Karyopherin α1, or importin α (36, 37). Although the binding affinity of Rch1 to EBNA1 does not correlate exactly with the replication activity of EBNA1 (36), it is likely that Rch1 somehow contributes to the functions of EBNA1. Fischer *et al.* (37) speculate that it may be significant that EBNA1 is able to induce expression of the recombination activation protein RAG1 (38) which also binds Rch1. Since EBNA1 is able to loop DNA, it is conceivable that EBNA1 and RAG1 are involved in the formation of DNA complexes and illegitimate recombination as observed in Burkitt's lymphoma. More recently, RPA (hSSB), the replicative single-stranded DNA binding protein of eukaryotic chromosomes, was found to bind to EBNA1 in solution and also to EBNA1 associated with oriP (39), suggesting that RPA plays some role in replication of the EBV episome. A possible mechanism for EBNA1-mediated *trans*-activation has come to light with the observation that EBNA1 interacts via two arginine–glycine repeat regions (amino acids 40–60 and amino acids 325–376) with P32/TAP, a cellular protein that has been implicated in *trans*-activation, splicing, and receptor functions (40). P32/TAP may behave in a manner analogous to the Notch receptor (41) in that it appears to be a plasma membrane receptor which relocates to the nucleus following activation and processing. This is interesting in view of the fact that another EBV-transforming protein, EBNA2, mimics an activated Notch receptor (Section 2.2).

2.2   The EBV nuclear antigen-2, EBNA2

EBNA2 is a nuclear phosphoprotein which localizes to large granules in the nucleus but which does not associate with EBNA1 (23, 42, 43). EBNA2 and EBNA-LP are the first EBV latent proteins to be expressed following EBV infection of B lymphocytes (4, 5), and their importance in cellular transformation was first recognized from studies with the transformation-incompetent virus isolate, P3HR-1 (44, 45). Genetic studies of the P3HR-1 virus identified a deletion in the *Bam*HI Y and H regions of the viral genome that results in a loss of the EBNA2 coding sequence and the last two exons of the EBNA-LP (46–49). Recombination of the P3HR-1 genome, either with transforming viruses in latently infected cells or with plasmids carrying genomic fragments from the *Bam*HI WYH regions, showed that acquisition of the EBNA2 gene is absolutely essential for the reconstitution of the viral immortalization potential. Concomitant acquisition of EBNA-LP was required for optimal transformation, but was not absolutely essential (4, 5, 50).

The importance of EBNA2 in maintaining as well as initiating EBV immortalization, has been demonstrated by Kempkes *et al.* using an EBV recombinant expressing a hormone recptor/EBNA2 chimera whose EBNA2 function is conditional upon the presence of hormone. Using this approach it was shown that EBNA2 function is absolutely essential for proliferation of EBV-transformed B cells (51). Further experiments with LCL immortalized with the conditional EBNA2 recombinant EBV demonstrated that the proliferation-inducing functions of EBNA2 in EBV-transformed B cells can be replaced by a constitutively activated c-myc gene, thus mimicking the situation in Burkitt's lymphoma cells which express only EBNA1, but which also carry a deregulated of c-myc gene as a consequence of a characteristic chromosome translocation (52).

Detection of polymorphism within the EBNA2 gene led to the identification of two major strains of EBV, type-1 and type-2, which show only about 50% amino acid identity between the two proteins (53). This polymorphism has functional consequences, and is largely responsible for the fact that type 1 EBV isolates to transform primary B lymphocytes with greater efficiency than do type-2 isolates (54, 55). However, little is known abut how these polymorphisms affect the various biochemical properties of EBNA2 that in turn effect the observed functional differences between type-1 and type-2 EBV.

Gene transfer experiments led to the identification of EBNA2 as a transcriptional *trans*-activator. Cellular genes activated by EBNA2 include CD21 and CD23 (56–59) and the proto-oncogene c-fgr (60). In addition, EBNA2 *trans*-activates expression of the latent EBV genes (61–64). A map of the functional domains of the EBNA2 is shown in Fig. 2. Unlike EBNA1, EBNA2 does contain a *trans*-activating domain (65–67), but it does not bind directly to specific DNA sequences. Extensive analysis of the EBNA2 response elements in promoter sequences eventually led to the identification of a cellular protein, the ubiquitously-expressed recombination signal binding protein of Kκ (RBP-Jκ protein), that mediates indirect interaction of EBNA2 with DNA (68–71). The core RBP-Jκ binding domain of EBNA2 maps to amino acids 320–330, within which W_{323} and W_{324} are absolutely essential (72). EBNA2 interacts with a transcriptional repressor domain within RBP-Jκ, and thus counteracts RBP-Jκ mediated transcriptional repression (73). The latter observation has led to comparison of EBNA2-mediated signalling and the Notch pathway; like EBNA2, mammalian Notch 1 interacts with RBP-Jκ and thereby partially abolishes RBP-Jκ mediated repression (74, 75). The binding sites of RBP-Jκ to Notch were found to be amino acids 249–251 while amino acid 233 of RBP-Jκ was required for binding of EBNA2 (73, 76), suggesting that

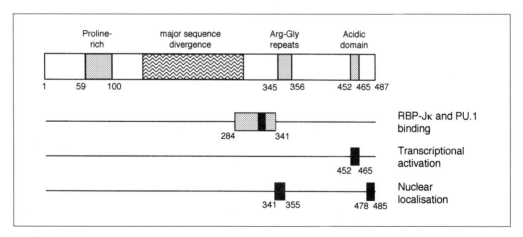

Fig. 2 Schematic summary of the functional domains of EBNA2. The top panel illustrates the features of the primary sequence. The boxes in the lower panels indicate functional domains identified by analysis of deletion mutants of EBNA2. Within the region amino acids 284–341 found to be essential for RBP-Jκ and PU.1 binding, a core sequence for RBP-Jκ binding has been identified at 320–330 (indicated by black within grey box) which contains two critical tryptophan residues.

the mechanisms of abolishing transcriptional repression are evolutionary different. Nevertheless, there is another parallel between the Notch and EBNA2 functions; in *Drosophila* the Hairless gene product negatively regulates Notch/SuH function (41, 77), which is analogous to the regulation of EBNA2/RBP-Jκ function by the EBNA3 family of proteins (Section 2.4).

In addition to RBP-Jκ, EBNA2 also interacts via an overlapping domain with the Ets transcription factors Spi1/PU.1 and Spi-B, which have a binding site within the EBV LMP1 promoter region close to the RBP-Jκ site (68, 78). It is now established that EBNA2 binds cellular transcription factors, such as TFIIB (79) and targets them to EBNA2-responsive promoters by binding to RBP-Jκ and/or PU.1. Other recent work identified hSNF5/Ini21, the human homologue of the yeast transcription factor SNF5, as a binding partner for EBNA2 (80). This finding is further evidence for EBNA2 acting as a transcriptional adapter molecule that enables the formation of RBP-Jκ–EBNA2–polymerase II transcription complexes at EBNA2-responsive target genes. Finally, the Arg–Gly domain of EBNA2 can bind histone HI, which potentially can modulate the interaction of EBNA2 with DNA and/or interfere with the binding of other proteins to the Arg–Gly region which modulate the activity of the acidic *trans*-activating domain of EBNA2 (81, 82).

2.3 The EBV leader protein, EBNA-LP

EBNA-leader protein or EBNA-LP (alternatively known as EBNA5) derives its name from the fact that it is encoded by the leader sequences that form the non-coding 5' end of mRNAs for all of the other EBNAs (83–85). Alternative splicing of the two first exons can create an initiation codon for EBNA-LP translation. The amino terminus of the protein is encoded by a varying number of W1 and W2 exons from the internal repeat region 1 (IR1) of the virus genome. In addition, variation in the number of W1 and W2 exons, encoding 22 and 44 amino acids respectively, results in the protein varying in molecular weight ranging from 20–140 kDa (86, 87). Each of the 44 amino acid repeats is phosphorylated on serine residues with a casein kinase II-like phosphorylation consensus site, which gives rise to a highly phosphorylated protein (23). Phosphorylation of EBNA-LP, possibly by p34 (cdc2) kinase and casein kinase II, was shown to be cell cycle-dependent, with phosphorylation being increased in G2 and maximal at G2/M (88).

As mentioned in Section 2.2, studies with the transformation-defective deletion mutant virus, P3HR1, implicated the C terminus of EBNA-LP in EBV-induced B cell transformation. Both EBNA2 and EBNA-LP are essential for optimal reconstitution of the transforming function of P3HR1 EBV; restoring EBNA2 function alone results in partial reconstitution of transforming ability, while restoring EBNA-LP function alone leaves the virus transformation-deficient (5, 89, 90). EBNA-LP may also be involved in the up-regulation of the expression of an autocrine factor required for LCL outgrowth, since primary B lymphocytes infected with mutant virus strains carrying deletions or nonsense mutations in the EBNA-LP open reading frame can be grown successfully on fibroblast feeder layers (5).

Interest in functional studies on EBNA-LP was heightened following the observation that the C terminal domain of the protein has homology with the conserved region 1 of adenovirus 5 E1a protein, which co-operates with the ras oncogene, and also has similarity with the SV40 T-antigen and the human papillomavirus 16 E7 proteins (91). Subsequently, *in vitro* physical interactions were found between EBNA-LP and the retinoblastoma gene product, pRB, and also p53 (92, 93), suggesting that the protein may play a role analogous to the other viral proteins, such as the E1a protein of adenovirus type-5, in disrupting normal cell cycle control. So far there is little experimental evidence that the interaction of EBNA-LP with pRB and p53 is physiologically relevant (94). However, immunofluorescence studies showed that EBNA-LP and pRB locate to the same nuclear foci that co-localize with heat shock protein 70 (95) and also with nuclear bodies positive for PML, the promyelocytic leukaemia-associated protein (96). Affinity precipitations resulted in the identification of the heat shock proteins HSP72/73 as further potential EBNA-LP binding partners (97, 98). The HSPs have previously been identified as being associated with nuclear proteins involved in gene expression, such as SP1. The site of interaction of EBNA-LP with HSP 72/73 has been mapped to the Y2 exon (97) as well as to the W repeat-encoded domain (98).

Recently, experiments in which EBNA-LP and EBNA2 were transfected into primary B lymphocytes showed that EBNA-LP co-operates with EBNA2 in inducing G_0 to G_1 transition of resting cells re-entering the cell cycle (99). Other groups have demonstrated that EBNA-LP can co-operate with EBNA-2 in up-regulating expression of EBNA2-responsive genes, such as LMP1, through binding as an EBNA-LP/EBNA2 complex to RBP-Jκ binding sites (100, 101). This co-operative function was mapped to the amino terminal W1W2 repeat region of EBNA-LP. The above findings strongly suggest that EBNA-LP plays an important role in EBV-mediated cell growth transformation by modulating EBNA-2 mediated transcriptional activation.

2.4 The EBV nuclear antigen-3 family, EBNA3A, EBNA3B, and EBNA3C

The EBNA3A, EBNA3B, and EBNA3C proteins (alternatively known as EBNA3, EBNA4, and EBNA6 respectively) are transcriptional regulators that localize to the nucleus (23), but they do not demonstrate direct sequence-specific binding to DNA (102). Although the EBNA3A, EBNA3B, and EBNA3C show only a limited similarity in their primary amino acid sequences, and this is largely confined to a short region of the N terminus, there is evidence consistent with the possibility that the genes arose by amplification and divergence from a common ancestor (103–106). The three genes are tandemly arranged on the genome and show a similar organization of the mRNA which, in each case, consist of a short 5′ and a long 3′ exon separated by a short intron (103–105, 107–109). In addition, all three members of the EBNA3 family share certain sequence features, e.g. acidic regions near the middle of the polypeptides and variable numbers of distinct proline-rich repeats in the C terminal half

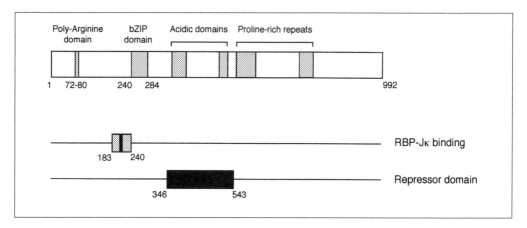

Fig. 3 Schematic summary of the functional domains of EBNA3C. The top panel illustrates the features of the primary sequence. The boxes in the lower panels indicate functional domains identified by analysis of deletion mutants of EBNA3C. Within the region amino acids 183–240, a core sequence for RBP-J binding has been identified by point mutations at 209–212 (indicated by black within grey box). The region from amino acids 182–231 shows the greatest degree of sequence homology with the EBNA3A and EBNA3B proteins (see text).

of the polypeptides that give rise to variation in size of the proteins encoded by different EBV isolates (110, 111). As illustrated in Fig. 3, the EBNA3C protein also contains a region at amino acids 240–284 that is partially homologous to the basic domain/leucine zipper present in the bZip family of transcription factors (112), and there is a poly-arginine sequence at amino acids 72–80 that is a feature of RNA-binding *trans*-activators. The significance of these domains in EBNA3C is not known. Similarly to EBNA2, the EBNA3 genes show divergence in their sequences that correlates with the type-1 or type-2 designation defined by the EBNA2 sequences (110, 113), but the functional significance of this sequence divergence in unclear (114).

If the EBNA3A, 3B, and 3C genes were derived from a common ancestor, then it might be predicted that the proteins perform related but different functions. However, experiments with recombinant viruses carrying stop mutations in either the EBNA3A, 3B, or 3C open reading frames showed that while EBNA3A and 3C are essential for the transformation of normal B lymphocytes, EBNA3B is probably dispensable (115, 116). The EBNA3 family are also distinct with respect to other biological properties. EBNA3C, like the human papillomavirus E7 and adenovirus E1A immortalizing proteins, can co-operate with activated Ha-ras to immortalize and transform primary rodent fibroblasts (117). Gene transfection experiments with human cells, which showed that EBNA3C can induce expression of the CD21 EBV receptor molecule in certain B cell lines, suggested that that EBNA3C may be a transcriptional activator (59), although more recent attempts to confirm transcriptional activation of CD21 have proved unsuccessful (118). Similarly, there has been one unconfirmed report that EBNA3B can up-regulate expression of CD40 and the cytoskeletal protein vimentin, and down-regulate CD77 in the EBV-negative DG75 B lymphoma cell line (119). Intriguingly, it was also reported that co-expression in the

DG75 cell line of all three members of the EBNA3, but not individual expression, correlated with up-regulated expression of the human gene Pleckstrin, which is implicated in modulating certain cell signalling cascades (120).

Although the EBNA3A, 3B, and 3C proteins are distinct with respect to many of their biological properties, they do share two important functions. First, they all behave as repressors of EBNA2-mediated transcriptional activation (121), and secondly, they all bind the RBP-Jκ protein that mediates EBNA2 function (122–126). The regions of EBNA3A, 3B, and 3C that interact with RBP-Jκ have all been mapped to similar locations within their N terminus (123). As shown in Fig. 3, the RBP-Jκ binding site on EBNA3C maps to amino acids 183–240 which spans a region (182–231) that shows limited sequence homology with EBNA3A (172–222) and EBNA 3B (178–228); mutation of a conserved motif within this domain of EBNA3C, at amino acids 209–212, was sufficient to abolish the ability of EBNA3C to repress EBNA2-mediated *trans*-activation of a Jκ reporter gene (127). From that observation, it would be predicted that the binding of RBP-Jκ to EBNA3A and EBNA3B would map to the same conserved regions. However, while this may hold true for EBNA3B (123), where the binding site maps to amino acids 109–311, the situation with EBNA3A is unclear since one group has mapped the binding to the conserved region at amino acids 172–223 (127), but more recent studies by others have mapped sites at amino acids 125–138 and 224–566, which lie outside the conserved domain (123, 128, 129). It may be relevant that EBNA3A binds RBP-Jκ more efficiently *invivo* than *in vitro*, suggesting that some post-translational modification of EBNA3A occurs *in vivo* that affects the interaction (127).

EBNA2, EBNA3B, and EBNA3C appear to interact with overlapping or adjacent regions of the RBP-Jκ protein (73, 126, 127, 130, 131), and it has formally been demonstrated for EBNA2 and EBNA3C that the EBNAs compete with each other for binding to individual RBP-Jκ proteins (131). The binding of EBNA3A to RBP-Jκ appears to be more complex, and may involve more than one binding site (126). B cells express two major isoforms of RBP-Jκ that result from different initiation sites and alternative splicing, and which differ in their N terminal regions (132). It is the smaller isoform, RBP-2N, that preferentially associates with EBNA2, EBNA3A, EBNA3B, and EBNA3C (126, 130, 131). Since the N terminal region of RBP-Jκ may influence the DNA-binding properties, future studies on EBNA/RBP-Jκ interactions ought ideally to include the RBP-2N isoform.

The question remains, how do the EBNA3 proteins repress EBNA2-mediated *trans*-activation? The possibility that EBNA3 proteins bind to RBP-Jκ/EBNA2 complexes and modulate EBNA2 effects has been discounted, at least for EBNA3C, by the demonstration that only one or other of EBNA2 or EBNA3C can bind to a single RBP-Jκ protein, and that separate RBP-Jκ/EBNA2 and RBP-Jκ/EBNA3C complexes are found in EBV-transformed B cells (131). Whilst, the RBP-Jκ binding domains of EBNA3A and 3C are necessary for repression, they are not sufficient (118, 128), which argues against the hypothesis that the EBNA3 proteins exhibit their effects by inhibiting the binding of RBP-Jκ to DNA. In fact, both EBNA3A and EBNA3C are powerful repressors of transcription when tethered to DNA via GAL4-DNA binding domains independently of RBP-Jκ (118, 124, 128, 133); these repressor functions map to amino acids 524–666 of EBNA3A (129), and predominantly to amino acids 346–543 of EBNA3C (118, 124, 133). It is likely, therefore, that RBP-Jκ acts as a bridge to link the EBNA3 proteins to DNA so that they can exert their own repressor functions. At present, it is unclear whether other factors are also involved.

3. Latent membrane proteins expressed in EBV-transformed B cells

3.1 The latent membrane protein-1, LMP1

Of the EBV latent genes, LMP1 was the first to be implicated as having a direct role in oncogenesis following the observation that it transforms immortal rodent fibroblasts, a phenomenon characterized by loss of cell–cell contact inhibition and loss of anchorage dependence *in vitro*, and the acquisition of tumourigenicity in nude mice (134, 135). Subsequent studies with recombinant EBV showed that LMP1 was also essential for EBV-induced transformation of human primary B cells (136). As depicted in Fig. 4A, the LMP1 protein consists of a short cytoplasmic N terminus linked to six hydrophobic transmembrane spanning domains, and a long hydrophilic carboxy tail of about 200 amino acid residues located to the cytoplasm (137, 138). The protein is phosphorylated on serine and threonine residues in the cytoplasmic C terminus (139), and has a short half-life of two to five hours (139, 140). In LCL, LMP1 localizes

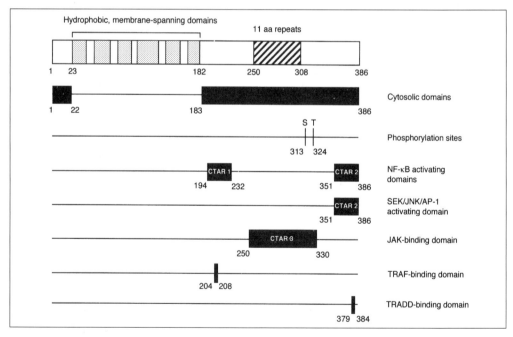

Fig. 4 (A) Schematic summary of the functional domains of LMP1. (B) Summary of the proteins interacting withLMP1, and their relationships to the signalling pathways that are activated by LMP1.

to discrete patches on the plasma membrane, probably due to an association with vimentin-containing intermediate filaments (138, 141), although vimentin itself is not required for LMP1 patch formation (142, 143).

One important function of LMP1 is to enhance cell survival (144, 145). In B cells, this function is mediated at least in part by up-regulation of the anti-apoptotic gene, bcl-2 (144, 146). Although LMP1 appears to up-regulate bcl-2 only in B cells (146), it does enhance cell survival in a number of other cell types, including epithelial cells and T cells (147–149). LMP1 therefore uses additional strategies to enhance cell

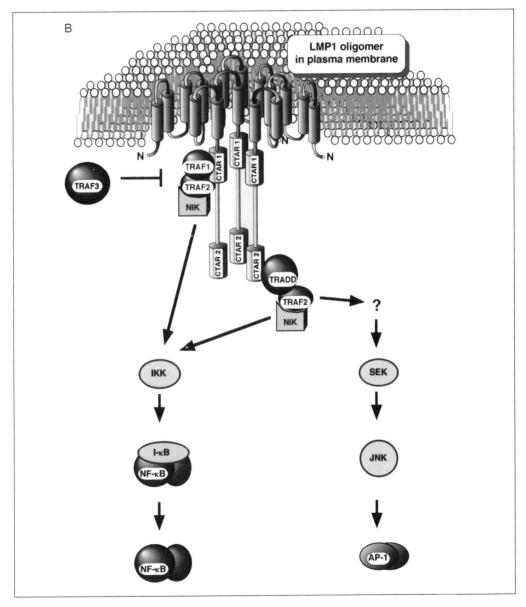

survival., and one principal mechanism involves up-regulation of an anti-apoptotic zinc finger protein, A20 (150, 151). Other possible mechanisms of LMP1-induced anti-apoptotic effects include up-regulation of Mcl-1 (152), down-regulation of c-myc (148, 153), and inhibition of differentiation (154).

Gene transfection experiments demonstrated that LMP1 can also induce a number of cell surface receptor molecules, including the epidermal growth factor receptor (EGFR) in epithelial cells (149) and CD40 in a wider range of cell types (155), suggesting that LMP1 has the potential to modulate proliferative potential through induction of membrane receptor molecules. Furthermore, early studies on the biochemical properties of LMP1, led to speculation that LMP1 itself resembled an activated growth factor receptor (156, 157). Indeed, expression of LMP1 in primary B cells does induce progression through G_1 to the DNA synthesis phase of the cell cycle (158). However, when expressed in cell lines of B cell or epithelial cell origin, LMP1 has growth inhibitory effects (147, 153, 159). These two effects are not necessarily contradictory, since the reduced rate of proliferation of LMP1-expressing lines is a consequence of a partial block at the G_2M phase of the cell cycle (153), and it is likely, therefore, that LMP1 serves to induce DNA synthesis from cells at G_1 following primary infection, and subsequently to modulate continued proliferation at the G_2M checkpoint. The absolute requirement of LMP1 expression for the continued proliferation of EBV-transformed primary B cells has been demonstrated using antisense technology to down-regulate LMP1 in LCL (160) and by using an LCL containing a novel mini-EBV plasmid in which the LMP1 gene can be regulated at will without affecting the expression of other latent EBV genes (161)

A third function of LMP1 is to modulate several components of the antigen processing pathway to render the cells more susceptible to recognition by EBV-specific immune cytotoxic T cells (162, 163). This function represents an important mechanism by which the virus limits its pathogenic potential in healthy infected individuals.

Clearly, LMP1 has many important functions. However, a longstanding question is how does LMP1 effect these functions? The discovery that LMP1 activates the pleiotropic NF-κB transcription factor (150, 164) provided a spur for renewed efforts to define the domains of LMP1 required for signal transduction. Consequently, two functional domains on the cytoplasmic C terminus of LMP1 were identified that were independently able to activate NF-κB; as indicated schematically in Fig. 4A, these domains mapped to amino acids 194–232 and to 351–386, and are designated C terminal activation region-1 and -2 (CTAR1 and CTAR2) respectively (155, 165). The CTAR2 domain also triggers activation of the AP-1 transcription factor via the SEK/JNK kinase cascade (166, 167), and an as yet undetermined region of the C terminus was shown to activate a ras-MAPK-dependent pathway (168). During the preparation of this review, a third functional domain of LMP1 was identified (CTAR3; see Fig 4A) which was shown to interact with Janus kinase 3 (JAK3) and to activate the STAT signalling pathway in B cells (168a).

The question still remains, just how does LMP1 activate the various signalling cascades implicated in its functions? LMP1 has no enzymatic activities itself and, therefore, other protein–protein interactions were long suspected to be important. In this context, the first breakthrough came in 1995 from an important study by Mosialos *et al.* (169). These workers demonstrated that LMP1 associates with two proteins, EB16 and LAP1 (now known as TRAF1 and TRAF3, respectively) belonging to a family of signalling proteins that also associate with the tumour necrosis factor receptor family, which includes, TNFR1, TNFR2, CD40, and CD30. It is now known that LMP1 can associate with several members of the TRAF family of signalling proteins. Thus, TRAF1, TRAF2, TRAF3, and TRAF5 can interact with LMP1 via a PxQxT motif at amino acids 204–208 in the CTAR1 domain of LMP1, and mutation of this core sequence abolishes the NF-κB activating function of CTAR1 (169–173). None of these TRAFs binds directly to CTAR2, although TRAF2 is somehow involved in the generation of signals from this domain (174, 175). Instead, the TNFR-associated death domain protein, TRADD, has been identified as one factor that directly binds to the LMP1 CTAR2 region, and this interaction critically requires tyrosine residues at the extreme C terminus of LMP1 that are contained within a core sequence at amino acids 379–384 that is essential for activation of NF-κB (176, 177). Since TRADD is able to associate TRAF2 (178), it is possible that TRADD is the adapter protein that allows TRAF2 to effect signalling from CTAR2.

The evidence is overwhelming, therefore, that LMP1 utilizes the same signalling pathways as the TNFR family. However, an important requirement for signal transduction by the TNFR family is the formation of oligomers in response to specific binding of extracellular ligands. TNFR family molecules comprises large extra-cellular ligand-binding domains that are linked to the TRAF-binding cytoplasmic domains via a single transmembrane spanning region (179). LMP1 does not conform to this molecular architecture, and there is no evidence that LMP1 interacts with an extracellular ligand. However, studies with chimeric receptor/LMP1 proteins have confirmed that oligomerization of LMP1 is an essential prerequisite for function (177, 180, 181). The available data are consistent with the hypothesis that LMP1 mimics a constitutively active receptor which spontaneously oligomerizes via hydrophobic interactions of the multiple membrane spanning domains. This oligomerization does not necessarily equate with the patching phenomenon seen in many cells with wild-type LMP1 (138, 143), since mutant LMP1 molecules deleted for the cytoplasmic N terminus do not patch but they do retain the ability to activate NF-κB and to mediate other downstream functions (155).

While it is now well-established that LMP1 mimics a constitutively active receptor of the TNFR family, it is by no means clear what is the role of the various signalling proteins that bind LMP1. TRAF3 was initially thought to be the effector of LMP1 signalling (147, 160), but it is now clear that overexpression of wild-type TRAF3 inhibits activation of NF-κB from the CTAR1 domain (170). In contrast, over-expression of wild-type TRAF1 significantly increases LMP1-mediated NF-κB activa-tion, while a TRAF1 mutant deleted for its N terminal zinc finger domain, which is important for the formation of TRAF1/TRAF2 complexes, was inhibitory for CTAR1-mediated activation of NF-κB (170). Transient overexpression of wild-type TRAF2 does not interfere with LMP1-mediated signalling, but is able independently to activate NF-κB and the SEK/JNK pathways (182–185). Overexpression of a RING finger deletion mutant of TRAF2 blocked LMP1-mediated NF-κB activation from CTAR1 (170, 173, 175), and caused a significant but less severe impairment of CTAR2-mediated activation of NF-κB and SEK/JNK (174, 175). To date, equivalent studies with the TRAF5 have not been reported.

How can these observations be reconciled to what is known about the binding of TRAFs and TRADD to each other and to LMP1? Here, one has to be cautious about interpreting data obtained from yeast two-hybrid experiments and from experiments where one or more TRAFs are overexpressed in cells with LMP1. In addition, it has been observed that the affinity of TRAF2 binding to wild-type LMP1 is weaker than the affinity of TRAF2 for a truncated LMP1 lacking the CTAR2 domain (170, 172). This latter observation indicates that the conformation of LMP1 is important for TRAF association, a point which is reinforced by a more recent study showing that wild-type LMP1 function involves some kind of physical interaction between CTAR1 and CTAR2 domains within an oligomeric LMP1 complex (186). Taking all the evidence available to date, one possible scenario (summarized in Fig. 4B) is that, as with other members of the TNFR/CD40 family (182–185), TRAF2 is the crucial effector of both NF-κB activation and JNK/AP-1 activation. TRAF2 can be tethered to LMP1 at CTAR1 via an association with TRAF1, and at CTAR2 via an association with TRADD. TRAF3 probably acts as a negative regulator by competing with TRAF1/TRAF2 for the CTAR1 binding site. It is possible that TRAF2 mediates activation of NF-κB from CTAR1 and from CTAR2 by a common mechanism involving association with an NF-κB-inducing kinase (NIK) which activates an IκB-kinase (IKK) responsible for phosphorylating the NF-κB inhibitory molecule, IκB; this in turn leads to degradation of IκB, and to the release of active NF-κB complexes that can translocate to the nucleus (187–190). TRAF2 appears to be the bifurcation point for activation of NF-κB and of the SEK/JNK/AP-1 pathway by CTAR1 (174), but exactly how LMP1 and TRAF2 effect activation of JNK/AP-1 is not yet clear.

3.2   The latent membrane protein-2, LMP2

There are two LMP2 proteins, LMP2A and LMP2B (alternatively known as TP1 and TP2), which differ only in the length of their N terminus domains. The LMP2A and LMP2B proteins are predicted to include 12 hydrophobic integral membrane sequences with hydrophilic C and N termini locating to the cytosol (191–193). LMP2A has a longer cytosolic N terminus than LMP2B due to the additional 119 amino acids encoded by the unique 5′ exon. LMP2A and LMP2B are located in discrete patches in the plasma membrane in B lymphocytes where they may co-localize with LMP1 (194). Recombinant viruses with stop codons in either LMP2A or LMP2B open reading frames retained the ability to immortalize primary B lymphocytes, suggesting that neither of the LMP2s is required for B lymphocyte transformation (195, 196). Consistent with these observations, LMP2 is also not essential for growth of EBV-transformed B lymphocytes in SCID mice (197). Nevertheless, LMP2A-negative mini-EBV episomes were found to be greatly reduced in their capacity to yield immortalized B cell clones (7), which suggests that while LMP2 is not absolutely required for initiation or maintenance of EBV-induced B cell transformation, it does play a significant role. However, the nature of the involvement of LMP2 in cellular transformation is not known. Another function of LMP2A is to block activation from latent to productive virus infection in B lymphocytes (198), the mechanism of which is related to the ability of LMP2A to block B cell receptor (BCR)-induced tyrosine phosphorylation and calcium mobilization (199).

LMP2A is phosphorylated on tyrosine, serine, and threonine residues (200). The unique 119 amino acids N terminus that is present in LMP2A, but absent in LMP2B, includes multiple tyrosine, serine, and threonine residues. This region also contains an immunoreceptor tyrosine-based activation motif (ITAM), together with several other motifs indicative of binding sites for proteins with src homology 2 (SH2)-containing domains, proteins with SH3-containing domains, and serine and threonine kinases (201, 202). In LCL, LMP2A has been shown to associate with the Syk protein tyrosine kinase (PTK), with Lyn, a src-family PTK, and with the ERK1 and ERK2 isoforms of mitogen-activated protein kinase (MAPK); the binding of all these factors maps to the unique 119 amino acids N terminal region of LMP2A (203–205). Several other proteins are probably recruited to LMP2A, as evidenced by the co-precipitation of at least eight unidentified proteins (204). From the binding motifs present in the N terminus of LMP2A, candidates for other interacting proteins include PI-3 kinase, PLCγ2, Vav, and Csk.

Central to the ability of LMP2A to block BCR-induced signals, and thus entry from latent to virus productive infection, is the LMP2A ITAM motif and its interaction with Syk-PTK. Thus, mutation of the ITAM motif in LMP2A abolished binding of Syk-PTK and restored the signal transduction function of the BCR which also involves sequential activation of the Syk-PTK, and the src family PTKs, and their interactions with ITAMs (202). Thus, it is hypothesized that by sequestering and/or desensitizing Syk and other signalling components that bind both to LMP2A and activated BCR, the constitutively activated LMP2A functions as a negative regulator of BCR signalling. This is consistent with the results of experiments using cells transformed by LMP2A null mutant EBV recombinants which, following BCR cross-linking, were able to trigger the same protein tyrosine kinase cascade as non-infected B lymphocytes, while cells transformed with wild-type virus failed to do so (205).

One of the unanswered questions concerning LMP2 is what is the role of the apparently inactive form of the protein, the N terminus truncated LMP2B? As with LMP1, the multiple membrane-spanning domains of LMP2 serve to allow spontaneous oligomerization of the molecules that is necessary for function. It is possible that the inactive LMP2B may serve to modulate the activity of LMP2A by forming mixed oligomers with reduced activity (201). In effect, therefore, expressed LMP2B could function to allow the cells to switch from latency to virus productive infection.

4. Conclusions

Cellular transformation by EBV is an elaborate process involving the co-operative actions of several viral proteins which take advantage of the cellular signalling machinery for their own ends. It is now apparent that the functions of all the EBNAs are intricately interwoven to provide elaborate mechanisms to regulate the effects of the pivotal EBNA2 protein. Nevertheless, for all that is known about the functions of the individual viral transforming genes, we are still far from fully understanding how EBV transforms cells with such efficiency. Part of the problem with past studies has been the inevitable need to study individual protein functions in isolation from the other transforming proteins, and usually in EBV-negative tumour cell lines that do not fairly reflect the situation in EBV-transformed primary B cells. The availability now of recombinant EBV and mini-EBV technology, in which the expression and/or function of individual latent genes can be regulated in normal LCL, represents a major improvement over previous models for studying EBV latent gene function, and should help further our understanding of the EBV transformation process.

In the context of protein–protein interactions and signal transduction, one neglected area of study has been the mechanisms underlying the phenotypic differences between type-1 and type-2 strains of EBV. Similarly, while considerable energy has been devoted to identifying sequence variations in the LMP1 gene, and the possibility of disease associations, very little is known about the effects of such sequence variations upon signal transduction. Research into LMP1 functions has reached a particularly interesting phase. To the casual observer, there must seem to be no end to the number of biological functions attributed to this protein and to the number of signalling pathways that it utilizes. The challenge now must be to elucidate just how these diverse functions and the multiple signal transduction pathways relate to each other, and to the functions of the other transformation-associated EBV genes.

References

1. Epstein, M. A., Achong, B. G., and Barr, Y. M. (1964) Virus particles in cultured lymphoblasts from Burkitt's lymphoma. *Lancet*, **1**, 702.
2. Rickinson, A. B. and Kieff, E. (1996). Epstein–Barr Virus. In *Fields virology* (ed. B. N. Fields, D. M. Knipe, and P. M. Howley), Vol. 2. pp. 2397. Lippincott-Raven Publishers: Philadelphia.
3. Pope, J. H., Horne, M. K., and Scott, W. (1968) Transformation of foetal human leukocytes *in vitro* by filtrates of a human leukaemic cell line containing herpes-like virus. *Int. J. Cancer*, **3**, 857.
4. Cohen, J., Wang, F., Mannick, J., and Kieff, E. (1989) Epstein–Barr virus nuclear protein 2 is a key determinant of lymphocyte transformation. *Proc. Natl. Acad. Sci. USA*, **86**, 9558.
5. Hammerschmidt, W. and Sugden, B. (1989) Genetic analysis of immortalizing functions of Epstein–Barr virus in human B lymphocytes. *Nature*, **340**, 393.
6. Robertson, E. S. and Kieff, E. D. (1995). Genetic analysis of Epstein–Barr virus in B lymphocytes. *EBV Rep.*, **2**, 73.
7. Breilmeier, M., Mautner, J., Laux, G., and Hammerschmidt, W. (1996) The latent membrane protein 2 gene of Epstein–Barr virus is important for efficient B cell immortalization. *J. Gen. Virol.*, **77**, 2807.
8. Cory, S. (1986). Activation of cellular oncogenes in hemopoietic cells by chromosome translocation. *Adv. Cancer Res.*, **47**, 189.

9. Nonoyama, M. and Pagano, J. S. (1972) Separation of Epstein–Barr virus DNA from large chromosomal DNA in non-virus-producing cells. *Nature*, **333**, 41.

10. Wilson, J. B. and Levine, A. J. (1992) The oncogenic potential of Epstein–Barr virus nuclear antigen 1 in transgenic mice. *Curr. Top. Microbiol. Immunol.*, **182**, 375.

11. Shimizu, N., Tanabe-Tochikura, A., Kuroiwa, Y., and Takada, K. (1994) Isolation of Epstein–Barr (EBV)-negative cell clones from the EBV-positive Burkitt's Lymphoma (BL) line Akata: Malignant phenotypes of BL cells are dependent on EBV. *J. Virol.*, **68**, 6069.

12. Puglielli, M. T., Woisetschlaeger, M., and Speck, S. H. (1996) *oriP* is essential for EBNA gene promoter activity in Epstein–Barr virus-immortalized lymphoblastoid cell lines. *J. Virol.*, **70**, 5758.

13. Sugden, B. and Warren, N. (1989). A promoter of Epstein–Barr virus that can function during latent infection can be transactivated by EBNA-1, a viral protein required for viral DNA replication during latent infection. *J. Virol.*, **63**, 2644.

14. Gahn, T. A. and Sugden, B. (1995) An EBNA-1-dependent enhancer acts from a distance of 10 kilobase pairs to increase expression of the Epstein–Barr virus LMP gene. *J. Virol.*, **69**, 2633.

15. Shiramizu, B., Barriga, F., Neequaye, J., Jafri, A., Dalla-Favera, R., Neri, A., *et al.* (1991) Patterns of chromosomal breakpoint locations in Burkitt's lymphoma: relevance to geography and Epstein–Barr virus association. *Blood*, **77**, 1516.

16. Magrath, I., Jain, V., and Bahitia, K. (1993) The EBNA-1 gene of Epstein–Barr virus has a direct role in the pathogenesis of Burkitt's Lymphoma. *Inserm Colloq.*, **225**, 377.

17. Rawlins, D. R., Milman, G., Hayward, S. D., and Hayward, G. S. (1985). Sequence-specific DNA binding of the Epstein–Barr virus nuclear antigen (EBNA-1) to clustered sites in the plasmid maintenance region. *Cell*, **42**, 859.

18. Kimball, A. S., Milman, G., and Tullious, T. D. (1989) High-resolution footprints of the DNA-binding domain of Epstein–Barr virus nuclear antigen 1. *Mol. Cell. Biol.*, **9**, 2738.

19. Ambinder, R. F., Shah, W. A., Rawlins, D. R., Hayward, G. S., and Hayward, S. D. (1990). Definition of the sequence requirements for the binding of the EBNA-1 protein to its palindromic target sites in Epstein–Barr virus DNA. *J. Virol.*, **64**, 2369.

20. Snudden, D. K., Hearing, J., Smith, P. R., Graesser, F. A., and Griffin, B. E. (1994) EBNA-1, the major nuclear antigen of Epstein–Barr virus, resembles 'RGG' RNA binding proteins. *EMBO J.*, **13**, 4840.

21. Grogan, E. A., Summers, W. P., Dowling, S., Shedd, D., Gradoville, L., and Miller, G. (1983). Two Epstein–Barr virus neoantigens distinguished by gene transfer. *Proc. Natl. Acad. Sci. USA*, **80**, 7650.

22. Ohno, S., Luka, J., Lindahl, T., and Klein, G. (1977) Identification of a purified complement-fixing antigen as the Epstein–Barr virus determined nuclear antigen (EBNA) by its binding to metaphase chromosomes. *Prov. Natl. Acad. Sci. USA*, **74**, 1605.

23. Petti, L., Sample, C., and Kieff, E. (1990) Subnuclear localization and phosphorylation of Epstein–Barr virus latent infection proteins. *Virology*, **176**, 563.

24. Hennessy, K. and Kieff, E. (1983) One of two Epstein–Barr virus nuclear antigens contains a glycine-alanine copolymer domain. *Proc. Natl. Acad. Sci. USA*, **80**, 5665.

25. Yates, J. L. and Camiolo, S. M. (1988) Dissection of DNA replication and enhancer activation function of Epstein–Barr virus nuclear antigan 1. *Cancer Cells*, **6**, 197.

26. Ambinder, R. F., Mullen, M., Chang, Y., Hayward, G. S., and Hayward, S. D. (1991) Functional domains of Epstein–Barr virus nuclear antigen EBNA-1. *J. Virol.*, **65**, 1466.

27. Polvino-Bodnar, M., Kiso, J., and Schaffer, P. (1988) Mutational analysis of Epstein–Barr virus nuclear antigen 1 (EBNA 1). *Nucleic Acids Res.*, **16**, 3415.

28. Chen, M. R., Middeldorp, J. M., and Hayward, S. D. (1993) Separation of the complex DNA binding domain of EBNA-1 into DNA recognition and dimerization subdomains of novel structure. *J. Virol.*, **67**, 4875.

29. Shah, W. A. and Ambinder, R. F. (1992) Binding of EBNA 1 to DNA creates a protease-resistant domain that encompasses the DNA recognition and dimerization functions. *J. Virol.*, **66**, 3355.

30. Jones, C. H., Hayward, S. D., and Rawlins, D. R. (1989) Interaction of the lymphocyte-derived Epstein–Barr virus nuclear antigen EBNA-1 with its DNA binding sites. *J. Virol.*, **63**, 101.

31. Frappier, L. and O'Donnell, M. (1991) Epstein–Barr nuclear antigen 1 mediates a DNA loop within the latent replication origin of Epstein–Barr virus. *Proc. Natl. Acad. Sci. USA*, **88**, 10875.

32. Su, W., Middleton, T., Sugden, B., and Echols, H. (1991) DNA looping between the origin of replication of Epstein–Barr virus and its enhancer site: stabilization of an origin complex with Epstein–Barr nuclear antigen 1. *Proc. Natl. Acad. Sci. USA*, **88**, 10870.

33. Mackey, D., Middelton, T., and Sugden, B. (1995) Multiple regions within EBNA 1 can link DNAs. *J. Virol.*, **69**, 6199.

34. Frappier, L. and O'Donnell, M. (1991) Overproduction, purification, and characcerization of EBNA 1, the origin binding protein of Epstein–Barr virus. *J. Biol. Chem.*, **266**, 7819.

35. Middleton, T. and Sugden, B. (1992) EBNA 1 can link the enhancer element to the initiator element of the Epstein–Barr virus plasmid origin of replication. *J. Virol.*, **66**, 489.

36. Kim, A. L., Maher, M., Hayman, J. B., Ozer, J., Zerby, D., Yates, J. L., *et al.* (1997) An imperfect correlation between DNA replication activity of Epstein–Barr virus nuclear antigen 1 (EBNA1) and binding to the nuclear import receptor, RCH1/importin alpha. *Virology*, **239**, 340.

37. Fischer, N., Kremmer, E., Lautscham, G., Mueller-Lantzsch, N., and Grasser, F. A. (1997) Epstein–Barr virus nuclear antigen 1 forms a complex with the nuclear transporter karyopherin alpha 2. *J. Biol. Chem.*, **272**, 3999.

38. Srinivas, S. K. and Sixbey, J. W. (1995). Epstein–Barr virus induction of recombinase-activating genes RAG1 and RAG2. *J. Virol.*, **69**, 8155.

39. Zhang, D., Frappier, L., Gibbs, E., Hurwitz, J., and O'Donnell, M. (1998) Human RPA (hSSB) interacts with EBNA1, the latent origin binding protein of Epstein–Barr virus. *Nucleic Acids Res.*, **26**, 631.

40. Wang, Y., Finan, J. E., Middeldorp, J. M., and Hayward, S. D. (1997) P32/TAP, a cellular protein that interacts with EBNA-1 of Epstein–Barr virus. *Virology*, **236**, 18.

41. Artevanis-Tsakonas, S., Matsuno, K., and Fortini, M. E. (1995) Notch signalling. *Science*, **268**, 225.

42. Hennessy, K. and Kieff, E. (1985) A second nuclear protein is encoded by Epstein–Barr virus in latent infection. *Science*, **227**, 1238.

43. Grässer, F. A., Haiß, P., Göttel, S., and Mueller-Lantzsch, N. (1991) Biochemical characterization of Epstein–Barr virus nuclear antigen 2A. *J. Virol.*, **65**, 3779.

44. Miller, G., Robinson, J., Heston, L., and Lipman, M. (1974) Differences between laboratory strains of Epstein–Barr virus based on immortalization, abortive infection and interference. *Proc. Natl. Acad. Sci. USA*, **71**, 4006.

45. Menezes, J., Leibold, W., and Klein, G. (1975) Biological differences between Epstein–Barr virus (EBV) strains with regard to lymphocyte transforming ability, superinfection and antigen induction. *Exp. Cell Res.*, **92**, 478.

46. Bornkamm, G. W., Hudewentz, J., Freese, U. K., and Zimber, U. (1982) Deletion of the

nontransforming Epstein–Barr virus strain P3HR-1 causes fusion of the large internal repeat to the DS-L region. *J. Virol.*, **43**, 952.

47. King, W., Dambaugh, T., Heller, M., Dowling, J., and Kieff, E. (1982) Epstein–Barr virus DNA. XII. A variable region of the Epstein–Barr virus is included in the P3HR-1 deletion. *J. Virol.*, **43**, 979.

48. Rabson, M. L., Gradoville, L., Heston, L., and Miller, G. (1982) Non-immortalizing P3J-HR-1 Epstein–Barr virus: a deletion mutant of its transforming parent. *J. Virol.*, **44**, 834.

49. Rowe, D., Heston, L., Metlay, J., and Miller, G. (1985). Identification and expression of a nuclear antigen from the genomic region of the Jijoye strain of Epstein–Barr virus that is missing in its non-immortalizing deletion mutant, P3HR-1. *Proc. Natl. Acad. Sci. USA*, **82**, 7429.

50. Skare, J., Farley, J., Strominger, J. L., Fresen, K. O., Cho, M. S., and zur Hausen, H. (1985) Transformation by Epstein–Barr virus requires DNA sequences in the region of *Bam*HI fragments Y and H. *J. Virol.*, **55**, 286.

51. Kempkes, B., Spitkovsky, D., Jansen-Duerr, P., Ellwart, J. W., Delecluse, H.-J., Rottenberger, C., *et al.* (1995) B-cell proliferation and induction of early G1-regulating proteins by Epstein–Barr virus mutants conditional for EBNA2. *EMBO J.*, **14**, 88.

52. Polack, A., Hortnagel, K., Pajic, A., Christoph, B., Baier, B., Falk, M., *et al.* (1996) C-Myc activation renders proliferation of Epstein–Barr-Virus (EBV)-transformed cells independent of EBV Nuclear Antigen-2 and Latent Membrane-Protein-1. *Proc. Natl. Acad. Sci. USA*, **93**, 10411.

53. Dambaugh, T., Hennessy, K., Chamnankit, L., and Kieff, E. (1984) The U2 region of EBV DNA may encode EBNA2. *Proc. Natl. Acad. Sci. USA*, **81**, 7632.

54. Rickinson, A. B., Young, L. S., and Rowe, M. (1987) Influence of the Epstein–Barr virus nuclear antigen EBNA-2 on the growth of virus-transformed B-cells. *J. Virol.*, **61**, 1310.

55. Cohen, J. I., Wang, F., and Kieff, E. (1991) Epstein–Barr virus nuclear protein 2 mutations define essential domains for transformation and transactivation. *J. Virol.*, **65**, 2545.

56. Wang, F., Gregory, C. D., Rowe, M., Rickinson, A. B., Wang, D., Birkenbach, M., *et al.* (1987) Epstein–Barr virus nuclear protein 2 specifically induces expression of the B cell activation antigen CD23. *Proc. Natl. Acad. Sci. USA*, **84**, 3452.

57. Aman, P., Rowe, M., Kai, C., Finke, J., Rymo, L., Klein, E., *et al.* (1990) Effect of the EBNA-2 gene on the surface antigen phenotype of transfected EBV-negative B-lymphoma lines. *Int. J. Cancer*, **45**, 77.

58. Cordier, M., Calender, A., Billaud, M., Zimber, U., Rousselet, G., Pavlish, O., *et al.* (1990) Stable transfection of Epstein–Barr virus (EBV) nuclear antigen 2 in lymphoma cells containing the EBV P3HR1 genome induces expresson of B-cell activation molecules CD21 and CD23. *J. Virol.*, **64**, 1002.

59. Wang, F., Gregory, C., Sample, C., Rowe, M., Liebowwitz, D., Murray, R., *et al.* (1990) Epstein–Barr virus latent membrane protein (LMP1) and nuclear proteins 2 and 3C are effectors of phenotypic changes in B lymphocytes: EBNA-2 and LMP1 cooperatively induce CD23. *J. Virol.*, **64**, 2309.

60. Knutson, J. C. (1990) The level of c-fgr RNA is increased by EBNA-2, an Epstein–Barr virus gene required for B-cell immortalization. *J. Virol.*, **64**, 2530.

61. Wang, F., Tsang, S., Kurilla, M. G., Cohen, J. I., and Kieff, E. (1990) Epstein–Barr virus nuclear antigen 2 transactivates latent membrane protein LMP1. *J. Virol.*, **64**, 3407.

62. Abbot, S. D., Rowe, M., Cadwallader, K., Ricksten, A., Gordon, J., Wang, F., *et al.* (1990) Epstein–Barr virus nuclear antigen 2 (EBNA2) induces expression of the virus-coded latent membrane protein (LMP). *J. Virol.*, **64**, 2126.

63. Zimber-Strobl, U., Suentzenich, K., Laux, G., Eick, D., Cordier, M., Calender, A., *et al.* (1991) Epstein–Barr virus nuclear antigen 2 transactivates transcription of the terminal protein gene. *J. Virol.*, **65**, 415.

64. Sung, N. S., Kenney, S., Gutsch, D., and Pagano, J. S. (1991) EBNA-2 transactivates a lymphoid-specific enhancer in the *Bam*HI-C promoter of Epstein–Barr virus. *J. Virol.*, **65**, 2164.

65. Cohen, J. I. and Kieff, E. (1991) An Epstein–Barr Virus Nuclear Protein 2 domain essential for transformation is a direct transcriptional activator. *J. Virol.*, **65**, 5880.

66. Ling, P. D., Ryon, J. J., and Hayward, S. D. (1993) EBNA-2 of Herpesvirus Papio diverges significantly from the type A and type B EBNA-2 proteins of Epstein–Barr virus but retains an efficient transactivation domain with a conserved hydrophobic motif. *J. Virol.*, **67**, 2990.

67. Cohen, J. I. (1992) A region from herpes simplex virus VP16 can substitute for a transforming domain of Epstein–Barr virus nuclear protein 2. *Proc. Natl. Acad. Sci. USA*, **89**, 8030.

68. Laux, G., Adam, B., Strobl, L. J., and Moreau-Gachelin, F. (1994) The Spi-/PU.1 and Spi-B ets family transcription factors and the recombination signal binding protein RBP-Jκ interact with the Epstein–Barr virus nuclear antigen 2 responsive cis-elements. *EMBO J.*, **13**, 5624.

69. Zimber-Strobl, U., Strobl, L. J., Meitinger, C., Hinrichs, R., Sakai, T., Furukawa, T., *et al.* (1994) Epstein–Barr virus nuclear antigen 2 exerts its transactivating function through interaction with the recombination signal binding protein RPB-Jκ, the homologue of Drosophila Suppressor of hairless. *EMBO J.*, **13**, 4973.

70. Ling, P. D., Rawlins, D. R., and Hayward, S. D. (1993) The Epstein–Barr virus immortalizing protein EBNA-2 is targeted to DNA by a cellular enhancer-binding protein. *Proc. Natl. Acad. Sci. USA*, **90**, 9237.

71. Henkel, T., Ling, P. D., Hayward, S. D., and Peterson, M. G. (1994) Mediation of Epstein–Barr-virus EBNA2 transactivation by recombination signal-binding protein (κ). *Science*, **265**, 92.

72. Ling, P. D. and Hayward, S. D. (1995). Contribution of conserved amino-acids in mediating the interaction between EBNA2 and CBF1/RBPJκ. *J. Virol.*, **69**, 1944.

73. Hsieh, J. J.-D. and Hayward, S. D. (1995) Masking of the CBF/RBPJκ transcriptional repression domain by Epstein–Barr virus EBNA2. *Science*, **268**, 560.

74. Hsieh, J. J. D., Henkel, T., Salmon, P., Robey, E., Peterson, M. G., and Hayward, S. D. (1996) Truncated mammalian Notch1 activates CBF1/RBPJκ-repressed genes by a mechanism resembling that of Epstein–Barr-Virus EBNA2. *Mol. Cell. Biol.*, **16**, 952.

75. Strobl, L. J., Hofelmayr, H., Stein, C., Marschall, G., Brielmeier, M., Laux, G., *et al.* (1997) Both Epstein–Barr viral nuclear antigen 2 (EBNA2) and activated Notch1 transactivate genes by interacting with the cellular protein RBP-Jκ. *Immunobiology*, **198**, 299.

76. Hsieh, J. J. D., Nofziger, D. E., Weinmaster, G., and Hayward, S. D. (1997) Epstein–Barr virus immortalization: Notch2 interacts with CBF1 and blocks differentiation. *J. Virol.*, **71**, 1938.

77. Brou, C., Logeat, F., Lecourtois, M., Vandekerckhove, J., Kourilsky, P., Schweisguth, F., *et al.* (1994) Inhibition of the DNA-binding activity of Drosophila suppressor of hairless and of its human homolog, KBF1/RBP-J kappa, by direct protein-protein interaction with Drosophila hairless. *Genes Dev.*, **8**, 2491.

78. Johannsen, E., Koh, E., Mosialos, G., Tong, X., Kieff, E., and Grossman, S. R. (1995) Epstein–Barr virus nuclear protein 2 transactivation of the latent membrane protein 1 promoter is mediated by Jκ and PU.1. *J. Virol.*, **69**, 253.

79. Tong, X., Wang, F., Thut, C. J., and Kieff, E. (1995). The Epstein–Barr virus nuclear protein 2 acidic domain can interact with TFIIB, TAF40, and RPA70 but not with TATA-bindng protein. *J. Virol.*, **69**, 585.

80. Wu, D. Y., Kalpana, G. V., Goff, S. P., and Schubach, W. H. (1996) Epstein–Barr-Virus Nuclear-Protein-2 (EBNA2) binds to a component of the human SNF-SWI complex, HSNF5/INI1. *J. Virol.*, **70**, 6020.

81. Grässer, F. A., Sauder, C., Haiss, P., Hille, A., Konig, S., Gottel, S., *et al.* (1993) Immuno-logical detection of proteins associated with the Epstein–barr virus nuclear antigen 2A. *Virology*, **195**, 550.

82. Tong, X., Yalamanchili, R., Harada, S., and Kieff, E. (1994) The EBNA-2 Arginine-Glycine domain is critical but not essential for B lymphocyte growth transformation; the rest of Region 3 lacks essential interactive domains. *J. Virol.*, **68**, 6188.

83. Bodescot, M., Chambraud, B., Farrell, P., and Perricaudet, M. (1984) Spliced RNA from the IR1-U2 region of Epstein–Barr virus: presence of an open reading frame for a repetitive polypeptide. *EMBO J.*, **3**, 1913.

84. Speck, S. H., Pfitzner, A., and Strominger, J. L. (1986) An Epstein–Barr virus transcript from a latently infected, growth-transformed B-cell line encodes a highly repetitive polypeptide. *Proc. Natl. Acad. Sci. USA*, **83**, 9298.

85. Sample, J., Hummel, M., Braun, D., Birkenbach, M., and Kieff, E. (1986) Nucleotide sequences of mRNAs encoding Epstein–Barr virus nuclear proteins: A probable transcriptional initiation site. *Proc. Natl. Acad. Sci. USA*, **83**, 5096.

86. Wang, F., Petti, L., Braun, D., Seung, S., and Kieff, E. (1987) A bicistronic Epstein–Barr virus mRNA encodes two nuclear proteins in latently infected, growth-transformed lymphocytes. *J. Virol.*, **61**, 945.

87. Finke, J., Rowe, M., Kallin, B., Ernberg, I., Rosen, A., Dillner, J., *et al.* (1987) Monoclonal and polyclonal antibodies against Epstein–Barr virus nuclear antigen 5 (EBNA-5) detect multiple protein species in Burkitt's lymphoma and lmphoblastoid cell lines. *J. Virol.*, **61**, 3870.

88. Kitay, M. K. and Rowe, D. T. (1996) Cell cycle stage-specific phosphorylation of the Epstein–Barr virus immortalization protein EBNA-LP. *J. Virol.*, **70**, 7885.

89. Allan, G. J., Inman, G. J., Parker, B. D., Rowe, D. T., and Farrell, P. J. (1992) Cell growth efects of Epstein–Barr virus leader protein. *J. Gen. Virol.*, **73**, 1547.

90. Mannick, J. B., Cohen, J. I., Birkenbach, M., Marchini, A., and Kieff, E. (1991) The Epstein–Barr virus nuclear protein encoded by the leader of the EBNA RNAs is important in B-lymphocyte transformation. *J. Virol.*, **65**, 6826.

91. Huen, D. S., Grand, R. J., and Young, L. S. (1988) A region of the Epstein–Barr virus nuclear antigen leader protein and adenovirus E1A are identical. *Oncogene*, **3**, 729.

92. Jiang, W.-Q., Szekely, L., Wendel-Hansen, V., Ringeritz, N., Klein, G., and Rosen, A. (1991) Co-localization of the retinoblastoma protein and the Epstein–Barr virus encoded nuclear antigen EBNA-5. *Exp. Cell Res.*, **197**, 314.

93. Szekely, L., Selivanova, G., Magnusson, K. P., Klein, G., and Wiman, K. G. (1993) EBNA-5, an Epstein–Barr virus-encoded nuclear antigen, binds to the retinoblastoma and p53 proteins. *Proc. Natl. Acad. Sci. USA*, **90**, 5455.

94. Inman, G. J. and Farrell, P. J. (1995) Epstein–Barr virus EBNA-LP and transcription regulation properties of pRB, p107 and p53 in transfection assays. *J. Gen. Virol.*, **76**, 2141.

95. Szekely, L., Jiang, W. Q., Pokrovskaja, K., Wiman, K. G., Klein, G., and Ringertz, N. (1995) Reversible nucleolar translocation of Epstein–Barr virus-encoded EBNA-5 and hsp70 proteins after exposure to heat shock or cell density congestion. *J. Gen. Virol.*, **76**, 2423.

96. Szekely, L., Pokrovskaya, K., Jiang, W. Q., de The, H., Ringertz, N., and Klein, G. (1996) The Epstein–Barr virus-encoded antigen EBNA-5 accumulates in PML-containing bodies. *J. Virol.*, **70**, 2562.

97. Kitay, M. K. and Rowe, D. T. (1996) Protein-protein interactions between Epstein–Barr virus nuclear antigen-LP and cellular gene products: binding of 70-kilodalton heat shock proteins. *Virology*, **220**, 91.

98. Mannick, J. B., Tong, X., Hemnes, A., and Kieff, E. (1995) The Epstein–Barr virus nuclear antigen leader protein associates with hsp72/hsc73. *J. Virol.*, **69**, 8169.

99. Sinclair, A. J., Palmero, I., Peters, G., and Farrell, P. J. (1994) EBNA-2 and EBNA-LP cooperate to cause G0 to G1 transition during immortalization of resting human B lymphocytes by Epstein–Barr virus. *EMBO J*, **13**, 3321.

100. Nitsche, F., Bell, A., and Rickinson, A. (1997) Epstein–Barr virus leader protein enhances EBNA-2-mediated transactivation of latent membrane protein 1 expression: a role for the W1W2 repeat domain. *J. Virol.*, **71**, 6619.

101. Harada, S. and Kieff, E. (1997) Epstein–Barr virus nuclear protein LP stimulates EBNA-2 acidic domain-mediated transcriptional activation. *J. Virol.*, **71**, 6611.

102. Sample, C. and Parker, B. (1994) Biochemical characterisation of Epstein–Barr virus nuclear antigen 3A and 3C proteins. *Virology*, **205**, 534.

103. Hennessy, K., Wang, F., Woodland-Bushman, E., and Kieff, E. (1986) Definitive identification of a member of the Epstein–Barr virus nuclear protein 3 family. *Proc. Natl. Acad. Sci. USA*, **83**, 5693.

104. Joab, I., Rowe, D., Bodescot, M., Nicholas, J., Farrell, P., and Perricaudet, M. (1987) Mapping of the gene coding for Epstein–Barr virus-determined nuclear antigen EBNA3 and its transient overexpression in a human cell line by using an adenovirus expression vector. *J. Virol.*, **61**, 3340.

105. Petti, L. and Kieff, E. (1988) A sixth Epstein–Barr virus nuclear protein (EBNA3B) is expressed in latently infected growth transformed lymphocytes. *J. Virol.*, **62**, 2173.

106. Ricksten, A., Kallin, B., Alexander, H., Dillner, J., Fahraeus, R., Klein, G., *et al.* (1988) *Bam*HI E region of the Epstein–Barr virus genome encodes three transformation-associated nuclear proteins. *Proc. Natl. Acad. Sci. USA*, **85**, 995.

107. Hennessy, K., Fennewald, S., and Kieff, E. (1985) A third nuclear protein in lymphoblasts immortalized by Epstein–Barr virus. *Proc. Natl. Acad. Sci. USA*, **82**, 5944.

108. Petti, L., Sample, J., Wang, F., and Kieff, E. (1988) A fifth Epstein–Barr virus nuclear protein (EBNA3C) is expressed in latently infected growth-transformed lymphocytes. *J. Virol.*, **62**, 1330.

109. Kerdiles, B., Walls, D., Triki, H., Perricaudet, M., and Joab, I. (1990) cDNA cloning and transient expression of the Epstein–Barr virus-determined nuclear antigen EBNA3B in human cells and identification of novel transcripts from its coding region. *J. Virol.*, **64**, 1812.

110. Sample, J., Young, L., Martin, B., Chatman, T., Kieff, E., Rickinson, A., *et al.* (1990) Epstein–Barr virus types 1 and 2 differ in their EBNA-3A, EBNA-3B, and EBNA-3C genes. *J. Virol.*, **64**, 4084.

111. Falk, K., Gratama, J. W., Rowe, M., Zou, J.-Z., Khanim, F., Young, L. S., *et al.* (1994) The role of repetitive DNA sequences in the size variation of Epstein–Barr virus (EBV) nuclear antigens, and the identification of different EBV isolats using RFLP and PCR analysis. *J. Gen. Virol.*, **76**, 779.

112. Busch, S. J. and Sassone-Corsi, P. (1990) Dimers, leucine zippers and DNA-binding proteins. *Trends Genet.*, **6**, 36.

113. Rowe, M., Young, L. S., Cadwallader, K., Petti, L., Kieff, E., and Rickinson, A. B. (1989) Distinction between Epstein–Barr virus type A (EBNA 2A) and type B (EBNA 2B) isolates extends to the EBNA 3 family of nuclear proteins. *J. Virol.*, **63**, 1031.

114. Tomkinson, B. and Kieff, E. (1992) Second-site homologous recombination in Epstein–Barr virus: insertion of Type 1 EBNA 3 genes in place of Type 2 has no effect on *in vitro* infection. *J. Virol.*, **66**, 780.

115. Tomkinson, B. and Kieff, E. (1992) Use of second-site homologous recombination to demonstrate that Epstein–Barr virus nuclear protein 3B is not important for lymphocyte infection or growth transformation *in vitro*. *J. Virol.*, **66**, 2893.

116. Tomkinson, B., Robertson, E., and Kieff, E. (1993) Epstein–Barr virus nuclear proteins EBNA-3A and EBNA-3C are essential for B-lymphocyte growth transformation. *J. Virol.*, **67**, 2014.

117. Parker, G. A., Crook, T., Bain, M., Sara, E. A., Farrell, P. J., and Allday, M. J. (1996) Epstein–Barr virus nuclear antigen (EBNA)3C is an immoralizing oncoprotein with similar properties to adenovirus E1A and papillomavirus E7. *Oncogene*, **13**, 2541.

118. Radkov, S. A., Bain, M., Farrell, P. J., West, M., Rowe, M., and Allday, M. J. (1997) Epstein–Barr virus EBNA3C represses Cp, the major promoter for EBNA expression, but has no effect on the promoter of the cell gene CD21. *J. Virol.*, **71**, 8552.

119. Silins, S. L. and Sculley, T. B. (1994) Modulation of vimentin, the CD40 activation antigen and Burkitt's Lymphoma antigen (CD77) by the Epstein–Barr Virus Nuclear Antigen EBNA-4. *Virology*, **202**, 16.

120. Kienzle, N., Young, D., Silins, S. L., and Sculley, T. B. (1996) Induction of pleckstrin by the Epstein–Barr virus nuclear antigen 3 family. *Virology*, **224**, 167.

121. Le Roux, A., Kerdiles, B., Walls, D., Dedieu, J. F., and Perricaudet, M. (1994) The Epstein–Barr-virus determined nuclear antigens EBNA-3A, EBNA-3B, and EBNA-3C repress EBNA-2-mediated transactivation of the viral Terminal Protein-1 gene promoter. *Virology*, **205**, 596.

122. Robertson, E. S., Grossman, S., Johannsen, E., Miller, C., Lin, J., Tomkinson, C., *et al.* (1995) Epstein–Barr virus nuclear protein 3C modulates transcription through interaction with the sequence-specific DNA-binding protein J kappa. *J. Virol.*, **69**, 3108.

123. Robertson, E. S., Lin, J., and Kieff, E. (1996) The amino-terminal domains of Epstein–Barr-Virus nuclear proteins 3A, 3B, and 3C interact with RBPJ-kappa. *J. Virol.*, **70**, 3068.

124. Waltzer, L., Perricaudet, M., Sergeant, S., and Manet, E. (1996) Epstein–Barr-virus EBNA3A and EBNA3C proteins both repress RBP-J-Kappa-EBNA2-activated transcription by Inhibiting the binding of RBP-J-Kappa to DNA. *J. Virol.*, **70**, 5909.

125. Marshall, D. and Sample, C. (1995) Epstein–Barr virus nuclear antigen 3C is a transcriptional regulator. *J. Virol.*, **69**, 3624.

126. Krauer, K. G., Kienzle, N., Young, D. B., and Sculley, T. B. (1996) Epstein–Barr nuclear antigen-3 and -4 interact with RBP-2N, a major isoform of RBP-J kappa in B lymphocytes. *Virology*, **226**, 346.

127. Zhao, B., Marshall, D. R., and Sample, C. E. (1996) A conserved domain of the Epstein–Barr virus nuclear antigen 3A and 3C binds to a discrete domain of J-Kappa. *J. Virol.*, **70**, 4228.

128. Cludts, I. and Farrell, P. J. (1998) Multiple functions within the Epstein–Barr virus EBNA-3A protein. *J. Virol.*, **72**, 1862.

129. Bourillot, P. Y., Waltzer, L., Sergeant, A., and Manet, E. (1998) Transcriptional repression by the Epstein–Barr virus EBNA3A protein tethered to DNA does not require RBP-J(kappa). *Gen. Virol.*, **79**, 363.

130. Young, D. B., Krauer, K., Kienzle, N., and Sculley, T. (1997) Both A type and B type Epstein–Barr virus nuclear antigen 6 interact with RBP-2N. *J. Gen. Virol.*, **78**, 1671.

131. Johannsen, E., Miller, C. L., Grossman, S. R., and Kieff, E. (1996) EBNA-2 and EBNA-3C extensively and mutally exclusively associate with RBPJ-Kappa in Epstein–Barr virus-transformed B-lymphocytes. *J. Virol.*, **70**, 4179.

132. Amakawa, R., Jing, W., Ozawa, K., Matsunami, N., Hamaguchi, Y., Matsuda, F., *et al.* (1993) Human Jκ recombination signal binding protein gene (IGKJRB): Comparison with its mouse homologue. *Genomics*, **17**, 306.

133. Bain, M., Watson, R. J., Farrell, P. J., and Allday, M. J. (199) Epstein–Barr virus nuclear antigen 3C is a powerful repressor of transcription when tethered to DNA. *J. Virol.*, **70**, 2481.

134. Wang, D., Liebowitz, D., and Kieff, E. (1985) An EBV membrane protein expressed in immortalized lymphocytes transforms established rodent cells. *J. Virol.*, **43**, 831.

135. Baichwal, V. R and Sugden, B. (1988) Transformation of Balb 3T3 cells by the BNLF-1 gene of Epstein–Barr virus. *Oncogene*, **2**, 461.

136. Kaye, K. M., Izumi, K. M., and Kieff, E. (1993) Epstein–Barr virus latent membrane protein 1 is essential for B-lymphocyte growth transformation. *Proc. Natl. Acad. Sci. USA*, **90**, 9150.

137. Fennewald, S., van Santen, V., and Kieff, E. (1984) Nucleotide sequence of an mRNA transcribed in latent growth transforming infection indicates that it may encode a membrane protein. *J. Virol.*, **51**, 411.

138. Liebowitz, D., Wang, D., and Kieff, E. (1986) Orientation and patching of the latent infection membrane protein encoded by Epstein–Barr virus. *J. Virol.*, **58**, 233.

139. Baichwal, V. R. and Sugden, B. (1987) Posttranslational processing of an Epstein–Barr virus-encoded membrane protein expressed in cells transformed by Epstein–Barr virus. *J. Virol.*, **61**, 866.

140. Moorthy, R. and Thorley-Lawson, D. A. (1990) Processing of the Epstein–Barr virus-encoded latent membrane protein p63/LMP. *J. Virol.*, **64**, 829.

141. Liebowitz, D., Kopan, R., Fuchs, E., Sample, J., and Kieff, E. (1987) An Epstein–Barr virus transforming protein associates with vimentin in lymphocytes. *Mol. Cell Biol.*, **7**, 2299.

142. Mann, K. P. and Thorley-Lawson, D. (1987) Posttranslational processing of the Epstein–Barr virus encoded p63/LMP protein. *J. Virol.*, **61**, 2100.

143. Liebowitz, D. and Kieff, E. (1989) Epstein–Barr virus latent membrane protein: induction of B-cell activation antigens and membrane patch formation does not require vimentin. *J. Virol.*, **63**, 4051.

144. Henderson, S., Rowe, M., Gregory, C., Croom-Carter, D., Wang, F., Longnecker, R., *et al.* (1991) Induction of *bcl-2* expression by Epstein–Barr virus Latent Membrane Protein-1 protects infected B cells from programmed cell death. *Cell*, **65**, 1107.

145. Zimber-Strobl, U., Kempkes, B., Marschall, G., Zeidler, R., van Kooten, C., Banchereau, J., *et al.* (1996) Epstein–Barr virus latent membrane protein (LMP1) is not sufficient to maintain proliferation of B cells but both it and activated CD40 can prolong their survival. *EMBO J.*, **15**, 7070.

146. Rowe, M., Peng-Pilon, M., Huen, D. S., Hardy, R., Croom-Carter, D., Lundgren, E., *et al.* (1994) Upregulation of bcl-2 by the Epstein–Barr virus latent membrane proteinLMP-1: a B-cell-specific response that is delayed relative to NF-κB activation and to induction of cell surface markers. *J. Virol.*, **68**, 5602.

147. Eliopoulos, A. G., Dawson, C. W., Mosialos, G., Floettmann, J. E., Rowe, M., Armitage, R. J., *et al.* (1996) CD40-induced growth inhibition in epithelial cells is mimicked by

Epstein–Barr Virus-encoded LMP1: involvement of TRAF3 as a common mdiator. *Oncogene*, **13**, 2243.

148. Kawanishi, M. (1997) Expresson of Epstein–Barr Virus Latent Membrane Protein 1 protects Jurkat T cells from apoptosis induced by serum deprivation. *Virology*, **228**, 244.

149. Miller, W. E., Earp, H. S., and Raab-Traub, N. (1995) The Epstein–Barr virus latent membrane protein 1 induces expression of the epidermal growth factor receptor. *J. Virol.*, **69**, 4390.

150. Laherty, C. D., Hu, . M., Opipari, A. W., Wang, F., and Dixit, V. M. (1992) Epstein–Barr virus LMP1 gene product induces A20 zinc finger protein expresson by activating nuclear factor κB. *J. Biol. Chem.*, **267**, 24157.

151. Fries, K. L., Miller, W. E., and Raab-Traub, N. (1996) Epstein–Barr-Virus Latent Membrane-Protein-1 blocks p53-mediated apoptosis through the Induction of the A20 gene. *J. Virol.*, **70**, 8653.

152. Wang, S., Rowe, M., and Lundgren, E. (1996) Expression of the Epstein–Barr-virus transforming protein LMP1 causes a rapid and transient stimulation of the Bcl-2 homolog Mcl-1 levels in B-cell lines. *Cancer Res.*, **56**, 4610.

153. Floettmann, J. E., Ward, K., Rickinson, A. B., and Rowe, M. (1996) Cytostatic effect of Epstein–Barr virus latent membrane protein-1 analyzed using tetracycline-regulated expression in B cell lines. *Virology*, **223**, 29.

154. Dawson, C. W., Rickinson, A. B., and Young, L. S. (1990) Epstein–Barr virus latent membrane protein inhibits human epithelial cell differentiation. *Nature*, **344**, 777.

155. Huen, D. S., Henderson, S. A., Croom-Carter, D., and Rowe, M. (1995) The Epstein–Barr virus latent membrane protein-1 (LMP-1) mediates activation of NF-κB and cell surface phenotype via two effector regions in its carboxy-terminal cytoplasmic domain. *Oncogene*, **10**, 549.

156. Martin, J. and Sugden, B. (1991) The latent membrane protein oncoprotein resembles growth factor receptors in the properties of its turnover. *Cell Growth Differentiation*, **2**, 653.

157. Sugden, B. (1989) An intricate route to immortality. *Cell*, **57**, 5.

158. Peng, M. and Lundgren, E. (1992) Transient expression of the Epstein–Barr virus LMP-1 gene in human primary B cells induces cellular activation and DNA synthesis. *Oncogene*, **7**, 1775.

159. Cuomo, L., Ramquist, T., Trivedi, P., Wang, F., Klein, G., and Masucci, M. G. (1992) Expression ofthe Epstein–Barr virus (EBV)-encoded membrane protein LMP1 impairs the *in vitro* growth, clonability and tumorgenicity of an EBV-negative Burkitt Lymphoma line. *Int. J. Cancer*, **51**, 949.

160. Mattia, E., Chichiarelli, S., Hickish, T., Gaeta, A., Mancini, C., Cunningham, D., *et al.* (1997) Inhibition of *in vitro* proliferation of Epstein–Barr Virus infected B cells by an antisense oligodeoxynucleotide targeted against EBV latent membrane protein LMP1. *Oncogene*, **15**, 489.

161. Kilger, E., Kieser, A., Baumann, M., and Hammserschmidt, W. (1998) Epstein–Barr virus-mediated B-cell proliferation is dependent upon latent membrane protein 1, which simulates an activated CD40 receptor. *EMBO J.*, **17**, 1700.

162. Rowe, M., Khanna, R., Jacob, C. A., Argaet, V., Kelly, A., Powis, S., *et al.* (1995) Restoration of endogenous antigen-processing in Burkitt's lymphoma cells by Epstein–Barr-virus latent membrane protein-1—coordinate up-regulation of peptide transporters and HLA-class-I antigen expression. *Eur. J. Immunol.*, **25**, 1374.

163. Rowe, M. (1995) The EBV latent membrane protein-1 (LMP1): A tale of two functions. *EBV Rep.*, 2, 99.

164. Hammarskjold, M.-L. and Simurda, M. C. (1992) Epstein–Barr virus latent membrane protein transactivates the human immunodeficiency virus type 1 long terminal repeat through induction of NF-κB activity. *J. Virol.*, **66**, 6496.

165. Mitchell, T. and Sugden, B. (1995) Stimulation of NF-kappa-B-mediated transcription by mutant derivatives of the latent membrane protein of Epstein–Barr virus. *J. Virol.*, **69**, 2968.

166. Eliopoulos, A. and Young, L. S. (1998) Activation of the c-Jun N-terminal kinase (JNK) pathway by the Epstein–Barr virus-encoded latent membrane protein 1 (LMP1). *Oncogene*, **16**, 1725.

167. Kieser, A., Kilger, E., Gires, O., Ueffing, M., Kolch, W., and Hammerschmidt, W. (1997) Epstein–Barr virus latent membrane protein-1 triggers AP-1 activation via the c-Jun N-terminal kinase cascade. *EMBO J.*, **16**, 6478.

168. Roberts, M. L. and Cooper, N. R. (1998) Activation of a ras-MAPK-dependent pathway by Epstein–Barr virus latent membrane protein 1 is essential for cellular transformation. *Virology*, **240**, 93.

168a Gires, O., Kohlhuber, F., Kilger, E., Baumann, M., Kieser, A., Kaiser, C., *et al.* (1999) Latent membrane protein 1 of Epstein-Barr virus interacts with JAK3 and activates STAT proteins. *EMBO J.*, **18**, 3064.

169. Mosialos, G., Birkenbach, M., Yalamanchili, R., VanArsdale, T., Ware, C., and Kieff, E. (1995) The Epstein–Barr virus transforming protein LMP1 engages signal proteins for the Tumor Necrosis Factor Receptor family. *Cell.*, **80**, 389.

170. Devergne, O., Hatzivassiliou, E., Izumi, K. M., Kaye, K. M., Kleijnen, M. F., Kieff, E., *et al.* (1996) Association of TRAF1, TRAF2, and TRAF3 with an Epstein–Barr virus LMP1 domain important for B-lymphocyte transformation—Role of NF-κB activation. *Mol. Cell. Biol.*, **16**, 7098.

171. Brodeur, S., Cheng, G., Baltimore, D., and Thorley-Lawson, D. A. (1997) Localization of the major NF-κB activating site and the sole TRAF3 binding site of LMP-1 defines two distinct signalling motifs. *J. Biol. Chem.*, **272**, 19777.

172. Sandberg, M., Hammerschmidt, W., and Sugden, B. (1997) Characterization of LMP-1's association with TRAF1, TRAF2, and TRAF3. *J. Virol.*, **71**, 4649.

173. Eliopoulos, A., Stack, M., Dawson, C., Kaye, K., Rowe, M., and Young, L. (1997) Epstein–Barr Virus-encoded LMP1 and CD40 mediate IL6 production in epithelial cells via an NF-κB pathway involving TNF Receptor-associated Factors. *Oncogene*, **14**, 2899.

174. Eliopoulos, A. G., Blake, S. M. S., Floettmann, J. E., Rowe, M., and Young, L. S. (1999) Epstein–Barr virus Latent Membrane Protein-1 activates the JNK pathway through its extreme C-terminus via a mechanism involving TRADD and TRAF2. *J. Virol.*, **73**, 1023.

175. Kaye, K. M., Devergne, O., Harada, J. N., Izumi, K. M., Yalamanchili, R., Kieff, E., *et al.* (1996). Tumor-Necrosis-Factor Receptor-Associated Factor 2 is a mediator of NF-κB activation by Epstein–Barr virus transforming protein. Proc. Natl. Acad. Sci. USA, **93**, 11085.

176. Izumi, K. M. and Kieff, E. D. (1997) The Epstein–Barr virus oncogene product latent membrane protein 1 engages the tumor ncrosis factor receptor-associated death domain protein to mediate B lymphocyte growth transformation and activate NF-κB. *Proc. Natl. Acad. Sci. USA*, **94**, 12592.

177. Floettmann, J. E. and Rowe, M. (1997) Epstein–Barr virus latent membrane protein-1 (LMP1) C-terminus activation region 2 (CTAR2) maps to the far C-terminus and requires oligomerisation for NF-κB activation. *Oncogene*, **15**, 1851.

178. Hsu, H., Shu, H. B., Pan, M. G., and Goeddel, D. V. (1996) TRADD-TRAF2 and TRADD-

FADD interactions define two distinct TNFreceptor 1 signal transduction pathways. *Cell*, **84**, 299.

179. Smith, C. A., Farrah, T., and Goodwin, R. G. (1994) The TNF receptor superfamily of cellular and viral proteins: activation, costimulation, and death. *Cell*, **76**, 959.

180. Gires, O., Zimber-Strobl, U., Gonella, R., Ueffing, M. G. M., Zeidler, R., Pich, D., *et al.* (1997) Latent membrane protein 1 of Epstein–Barr virus mimics a constitutively active receptor molecule. *EMBO J.*, **16**, 6131.

181. Hatzivassiliou, E., Miller, W. E., Raab-Traub, N., Kieff, E., and Mosialos, G. (1998) Fusion of the EBV latent membrane protein-1 (LMP1) transmembrane domains to the CD40 cytoplasmic domain is similar to LMP1 in constitutive activation of epidermal growth factor receptor expression, nuclear factor-kappa B, and stress-activated protein kinase. *J. Immunol.*, **160**, 1116.

182. Song, H. Y., Regnier, C. H., Kirschning, C. J., Goeddel, D. V., and Rothe, M. (1997) Tumor necrosis factor (TNF)-mediated kinase cascades: bifurcationof nuclear factor-kappaB and c-jun N-terminal kinase (JNK/SAPK) pathways at TNF receptor-associated factor 2. *Proc. Natl. Acad. Sci. USA*, **94**, 9792.

183. Reinhard, C., Shamoon, B., Shyamala, V., and Williams, L. T. (1997) Tumor necrosis factor α -induced activation of c-jun N-terminal kinase is mediated by TRAF2. *EMBO J.*, **16**, 1080.

184. Natoli, G., Costanzo, A., Ianni, A., Templeton, D. J., Woodgett, J. R., Balsano, C., *et al.* (1997) Activation of SAPK/JNK by TNF receptor 1 through a noncytotocix TRAF2-dependent pathway. *Science*, **275**, 200.

185. Rothe, M., Sarma, V., Dixit, V. M., and Goeddel, D. V. (1995) TRAF2-Mediated Activation of NF-Kappa-B by TNF Receptor-2 and CD40. *Science*, **2699**, 1424.

186. Floettmann, J. E., Eliopoulos, A. G., Jones, M., Young, L. S., and Rowe, M. (1998) Epstein–Barr virus Latent Membrane Protein-1 (LMP1) signalling is distinct from CD40 and involves physical cooperation of its two C-terminus functional regions. *Oncogene*. **17**, 2383.

187. Maniatis, T. (1997) Catalysis by a multiprotein IkappaB kinase complex. *Science*, **278**, 818.

188. Lee, F. S., Hagler, J., Chen, Z. J., and Maniatis, T. (1997) Activation of the IkappaB alpha kinase complex by MEKK1, a kinase of the JNK pathway. *Cell*, **88**, 213.

189. DiDonato, J. A., Hayakawa, M., Rothwarf, D. M., Zandi, E., and Karin, M. (1997) A cytokine-responsive IkappaB kinase that activates the transcription factor NF-kappaB. *Nature*, **388**, 548.

190. Malinin, N. L., Boldin, M. P., Kovalenko, A. V., and Wallach, D. (1997) MAP3K-related kinase involved in NF-κB induction by TNF, CD95 and IL-1. *Nature*, **385**, 540.

191. Laux, G., Perricaudet, M., and Farrell, P. J. (1988) A spliced Epstein–Barr virus gene expressed in immortalized lymphocytes is created by circularization of the linear viral genome. *EMBO J.*, **7**, 769.

192. Laux, G., Economou, A., and Farrell, P. J. (1989) The terminal protein gene 2 of Epstein–Barr virus is transcribed from a bidirectional latent promoter region. *J. Gen. Virol.*, **70**, 3079.

193. Sample, J., Liebowitz, D., and Kieff, E. (1989) Two related Epstein–Barr virus membrane proteins are encoded by separate genes. *J. Virol.*, **63**, 933.

194. Longnecker, R. and Kieff, E. (1990) A second Epstein–Barr virus membrane protein (LMP2) is expressed in latent infection and colocalizes with LMP1. *J. Virol.*, **64**, 2319.

195. Longnecker, R., Miller, C. L., Tomkinson, B., Miao, X.-Q., and Kieff, E. (1993) Deletion of

DNA encoding the first five transmembrane domains of Epstein–Barr virus latent membrane proteins 2A and 2B. *J. Virol.*, **67**, 5068.

196. Longnecker, R., Miller, C., Maio, X.-Q., Marchini, A., and Kieff, E. (1992) The only domain that distinguishes Epstein–Barr virus membrane protein 2A (LMP2A) from LMP2B is dispensable for lymphocyte infection and growth transformation *in vitro*. *J. Virol.*, **66**, 6461.

197. Rochford, R., Miller, C. L., Cannon, M. J., Izumi, K. M., Kieff, E., and Longnecker, R. (1997) *In vivo* growth of Epstein–Barr virus transformed B cells with mutations in latent membrane protein 2 (LMP2). *Arch. Virol.*, **142**, 707.

198. Miller, C. L., Lee, J. H., and Kieff, E. (1994) An integral membrane-protein (LMP2) blocks reactivation of Epstein–Barr-Virus from latency following surface-immunoglobulin cross-linking. *Proc. Natl. Acad. Sci. USA*, **91**, 772.

199. Miller, C. L., Longnecker, R., and Kieff, E. (1993) Epstein–Barr virus latent membrane protein 2A blocks calcium mobilization in B-lymphocytes. *J. Virol.*, **67**, 3087.

200. Longnecker, R., Drucker, B., Roberts, T. M., and Kieff, E. (1991). An Epstein–Barr virus protein associated with cell growth transformation interacts with a tyrosine kinase. *J. Virol.*, **65**, 3681.

201. Frueling, S., Caldwell, R., and Longknecker, R. (1997) LMP2 function in EBV latency. *EBV Rep.*, **4**, 151.

202. Frueling, S. and Longknecker, R. (1997) The immunoreceptor tyrosine-based activation motif of Epstein–Barr virus LMP2A is essential for blocking BCR-mediated signal transduction. *Virology*, **235**, 241.

203. Panousis, C. G. and Rowe, D. T. (1997) Epstein–Barr virus latent membrane protein 2 associates with and is a substrate for mitogen-activated protein kinase. *J. Virol.*, **71**, 4752.

204. Burkhardt, A. L., Bolen, J. B., Kieff, E., and Longnecker, R. (1992) An Epstein–Barr virus transformation-associated membrane protein interacts with src family tyrosine kinases. *J. Virol.*, **66**, 5161.

205. Miller, C. L., Burkhardt, A. L., Lee, J. H., Stealey, B., Longnecker, R., Bolen, J. B., *et al.* (1995) Integral membrane protein 2 of Epstein–Barr virus regulates reactivation from latency through dominant negative effects on protein-tyrosine kinases. *Immunity*, **2**, 155.

5 | Molecular piracy by KSHV: a strategy of live and let live

SUKHANYA JAYACHANDRA, YUAN CHANG, AND
PATRICK S. MOORE

1. Introduction

Viruses have developed elaborate strategies to subvert the host cell metabolism and evade host immune surveillance in order to ensure their survival. One strategy makes use of molecular piracy in which homologues to cellular proteins involved in critical pathways are encoded by the virus. Molecular piracy is used most effectively by members of the herpesvirus and poxvirus families. For example, a common characteristic of the herpesvisuses is that they encode not only viral structural proteins but also enzymes involved in nucleic acid metabolism that play important roles in the synthesis of genomic viral DNA (1).

However, some herpesvirus have an even greater degree of molecular piracy. Epstein–Barr virus (EBV) and equine herpesvirus-2 (EHV-2) encode homologues to the IL-10 cytokines, and herpesvirus saimiri (HVS) encodes an IL-17-like cytokine (2–4). Herpesviruses also frequently encode receptors for cytokines and cell cycle regulatory proteins (5). Similarly, the poxviruses encode chemokine homologues, and anti-apoptotic factors (6–9). Molecular piracy of these cellular genes appears primarily to serve as a mechanism by which viruses can abrogate cellular defense pathways (10, 11).

2. KSHV as a model for molecular piracy

Among the herpesviruses, Kaposi's sarcoma-associated herpes virus (KSHV) has perhaps made the most extensive use of molecular piracy of host cell genes. Functional studies suggest that these pirated genes may help the virus to evade immune responses, prevent cell cycle shutdown, and interrupt activation of apoptotic pathways. The consequences of the viral defense strategies may well be cell transformation and tumourigenesis under certain conditions.

KSHV was discovered by Chang and co-workers in 1994, using representational difference analysis (12). KSHV, also known as human herpesvirus-8 (HHV-8) represents the first human *Rhadinovirus* or *gamma-2* herpesvirus and belongs to the subfamily *Gammaherpesvirinae* of the family *Herpesviridae* (10, 13). This virus was originally isolated from Kaposi's sarcoma (KS), a malignancy frequently occurring in AIDS patients (12, 14). Subsequently, epidemiological and serological studies have shown KSHV to be associated with all clinical forms of KS, body cavity-based lymphomas (BCBLs) of B cell origin, and a subset of Castleman's disease, a hyperplastic lymphadenopathy (for review see refs 15–17).

Although KSHV is detected in KS lesions, the virus is difficult to culture directly from KS tumours (18). KSHV can be maintained in cell lines derived from BCBLs and induced into lytic replication using phorbol esters, such as 12-*O*-tetradecanoyl phorbl-13-acetate (TPA), or sodium butyrate (19–21).

Determination of the complete KSHV genome sequence (13) shows that it retains sequence similarity to other gammaherpesviruses including, herpesvirus saimiri (HVS), murine gammaherpesvirus 68 (MHV68), and Epstein–Barr Virus (EBV). The ~ 165 kb KSHV viral genome contains over 80 open reading frames (ORF) arranged in a long unique region which is flanked by multiple 801 bp terminal repeat units of high G + C content (13, 22).

The long unique region contains blocks of conserved genes, that are found in most herpesviruses, interspersed with blocks of non-homologous genes that are specific for KSHV and related rhadinoviruses (Fig. 1) (2, 13, 22, 23). KSHV ORFs with homologies to ORFs from HVS (the prototype rhadinovirus) are given the corresponding HVS gene number, whereas ORFs unique to KSHV are prefixed with the letter 'K' (Fig. 1). KSHV-encoded proteins with recognizable homology to cellular proteins

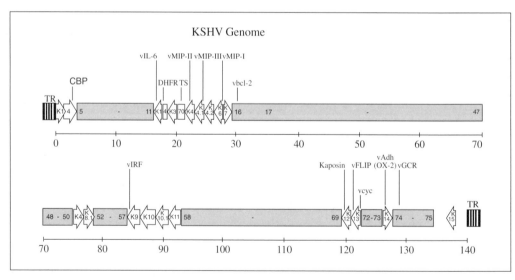

Fig. 1 Schematic representation of the KSHV genome organization. The unique 'K' open reading frames (ORFs) are drawn as arrows to indicate the coding direction of the genes. The shaded boxes are blocks of the genome conserved with Herpesvirus saimiri. Also shown are viral homologues of cellular genes: complement-binding protein (CBP), dihydrofolate reductase (DHFR), thymidylate synthase (TS), viral bcl-2 (vbcl-2), and viral cyclin (vcyc) genes. The high G:C terminal repeats (TR) flank the long unique region of the genome.

include a complement binding protein(ORF 4), an IL-6-like cytokine (ORF K2), three chemokines (ORF K4, ORF K4.1, and ORF K6, a Bcl-2 (ORF 16) anti-apoptotic factor, an interferon regulatory factor (ORF K9), a D-type cyclin (ORF 72), a FLICE inhibitory protein (ORF K13), a cell adhesion-like molecule (ORF K14) and a G protein-coupled receptor (ORF 74). In addition, KSHV has a number of genes such as ORF K12 (encodes the highly expressed transcript, kaposin), and ORF K1, a transmembrane protein that interacts with immunoreceptor kinases (24), which are likely to play a role in viral tumourigenesis.

3. Immunomodulatory proteins encoded by KSHV

3.1 Viral interleukin-6 (vIL-6)

One of the prominent features of KSHV is its possession of functionally active and secreted virus-encoded cytokines (25). While other gammaherpesviruses, including HVS and EBV, do not encode an IL-6-like homologue, the recently discovered rhesus monkey rhadinovirus (RRV) (26) encodes an IL-6-like protein highly similar to the corresponding KSHV IL-6 (vIL-6).

Human IL-6 (huIL-6) belongs to a class of related cytokines that includes oncostatin M, leukaemia inhibitory factor, and ciliary neurotrophic factor (27). huIL-6 signalling is initiated by binding to a high-affinity subunit receptor called IL-6Rα or gp80 which associates with a signal transducer protein gp130 responsible for conducting IL-6 cytokine signalling across the plasma membrane (Fig. 2). The gp130 signal transducer is used by many members of the IL-6 cytokine family for transmembrane signalling in a variety of cell types (28), with receptor specificity for each cytokine being provided by the presence of specific high-affinity co-receptors that activate gp130 on cytokine binding.

huIL-6 activation of gp130 leads to activation of two major pathways (29). The Jak-STAT signal cascades via phosphorylation of Jak1, Jak2, Tyk2, STAT1, and STAT3 and the MAP-kinase pathway via activation of a phosphotyrosine phosphatase (SHP-2) (29) (Fig. 2). Homo- and heterodimerization of cytoplasmic STAT1 and STAT3 after phosphorylation leads to nuclear translocation of these activated transcriptional factors, which in turn bind to and activate specific promoter elements in huIL-6 responsive genes (30, 31). Also, MAP-kinase pathway activation simultaneously occurs in some cell types although it is unknown whether or not this activity is concordant or antagonistic to Jak-STAT signal pathway activation (31). It has been suggested that STAT3 activation leads to differentiation, growth arrest, and anti-apoptosis whereas the MAP-kinase pathways leads to cell proliferation (29).

The biological effects of human IL-6 (huIL-6) varies considerably depending on a number of factors including cell type. For example, huIL-6 acts in an autocrine or paracrine pathway for maintenance of many myeloma B cell lines (32–34). There is considerable evidence that for some B cell lymphoproliferative disorders such as multiple myeloma and Castleman's disease, huIL-6 is required to prevent apoptosis (34) and stimulates B cell division through phosphorylation of the retinoblastoma

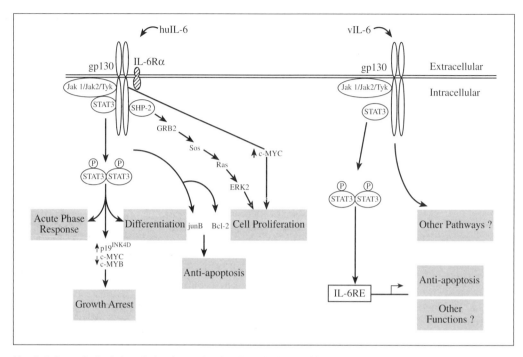

Fig. 2 Schematic depiction of signal transduction through the gp130 receptor complex. The huIL-6 and the vIL-6 signalling pathways that may regulate cell growth, proliferation, and differentiation. Binding of the huIL-6 to the gp130/IL-6Rα receptor complex leads to activation of two major intracellular signalling pathways: the JAK (Janus kinase)/STAT (signal transducer and activator of transcription) and the MAP (mitogen activated protein) kinase pathways. Whereas huIL-6 absolutely requires the IL-6Rα subunit, vIL-6 appears to directly activate the gp130 receptor inducing STAT3 homodimerization and probably other signal transduction pathways.

protein (pRb) (35, 36). huIL-6 can inhibit cell cycle arrest induced by interferon treatment, despite overlapping STAT protein activation by γ-interferon (IFN-γ) and huIL-6 (37). In contrast, some cell lines, including some EBV-transformed B lymphocyte and the macrophage cell lines either differentiate and stop developing or undergo apoptosis in response to prolonged huIL-6 treatment (38).

The vIL-6 encoded by ORF K2 consists of a 204 amino acid protein that is structurally and functionally conserved with huIL-6 (25, 39, 40). vIL-6 contains a hydrophobic secretory signal sequence of 19 amino acids and displays 24.8% amino acid identity and 62.2% similarity to the human protein. It also contains four critically positioned cysteine residues that are characteristic of IL-6 (25, 39). Like huIL-6, vIL-6 is active in bioassays which measure apoptotic cell death in IL-6-dependent mouse myeloma cells (25, 40). However, there are differences between vIL-6 and huIL-6, vIL-6 has a highly conserved central core motif with huIL-6, but differs from huIL-6 in regions that interact with the IL-6Rα (25, 39–41). BAF/130 cells (a transfected mouse pro-B cell line, BAF-B03) that stably express gp130 alone are stimulated by vIL-6 to phosphorylate STAT1 and STAT3, whereas the huIL-6 will only activate this pathway in the presence of soluble IL-6Rα (41). Direct activation of gp 130 could lead to a broader tissue tropism for the viral cytokine and independence of vIL-6 signalling from the IL-6Rα receptor has been confirmed through antibody neuralization assays of BAF/130 cells and human myeloma cell lines (41, 42). A bacterially-expressed recombinant vIL-6 has been found to initiate IL-6 signalling although it has lower specific activity for the human receptor complex than huIL-6 (42).

Transcriptional mapping studies indicate that the ORF K2 gene encoding vIL-6 is a class II gene expressed during presumed latency in BCBL cell lines but is inducible to high level expression with phorbol ester treatment (25, 43). Expression of vIL-6 appears to be restricted to KSHV-infected lymphoid cells and is seen in CD, BCBLs, and cell lines derived from BCBLs. vIL-6 expression is not detected in the spindle cells of KS lesions (25). In addition to vIL-6, BCBLs also express high levels of huIL-6 (44, 45) which may be an autocrine factor for BCBL cells in culture. While vIL-6 is minimally expressed in KS lesions, huIL-6 signalling can induce vascular endothelial growth factor (46) expression leading to the possibility that circulating viral cytokines may play a role in maintenance and growth of the KS tumour.

It is likely that vIL-6 contributes to the B cell expansion of hyperplastic KSHV-infected Castleman's disease lymph nodes (47). Most HIV positive patients, but only approximately half of all HIV-negative Castleman's disease cases, are associated with KSHV infection (48, 49). Among those patients with KSHV infection, vIL-6 expression is readily detected in scattered KSHV-infected B cells localized to the marginal zones of germinal centres (47). These vIL-6 expressing cells appear to represent only a fraction of the infected cell population of the marginal zone as measured by co-immunostaining for the latency associated nuclear antigen (LANA) (C. Parravicini and Y. Chang, unpublished observation). The remaining KSHV-free Castleman's disease patients may have an underlying disorder in huIL-6 expression leading to a pathologically indistinguishable B cell hyperplasia.

Prior to the discovery of KSHV, cell lines established from KS lesions were commonly found to be autocrine-dependent on several IL-6 family cytokines (50, 51) leading to speculation that these cytokines directly play a role in the genesis of the tumour. However, *in vitro* established 'KS' cell lines do not contain KSHV (52, 53), and the origin and usefulness of these non-transformed spindle cells as a model for KS has been questioned.

3.2 Viral macrophage inflammatory proteins (vMIPs)

The KSHV genes ORF K6, K4, and K4.1 encode three secreted chemokines (chemo-attractant cytokines) respectively designated vMIP-I (or vMIP-1α), vMIP-II (or vMIP-1β), and vMIP-III (or BCK) (13, 22, 25, 40). Sequencing studies of the human molluscum contagiosum poxvirus (MCV), which also exhibits extensive genetic piracy, has identified a protein ORF MC148R (8) and murine CMV ORF HJ1 (54) with homology to the macrophage inflammatory protein (MIP) family of β-chemokines.

vMIP-I and vMIP-II are highly homologous to each other and have approximately 40% amino acid identity with human β-chemokines, such as RANTES and MIP-1α. These chemokines are distinguished from α-chemokines by the presence of a dicysteine (X-C-C-X) motif instead of the α-chemokine C-X-C motif (55). vMIP-I and -II probably arose from gene duplication within the virus after piracy of the original host cell gene but have retained conserved regions of the β-chemokines (25). In contrast, vMIP-III has far lower homology to known cellular chemokines and may belong to a new chemokine class (13, 22).

Both α- and β-chemokines bind to seven transmembrane spanning G protein-coupled receptors (GCR), and serve as chemoattractants for various classes of leucocytes (56). Chemokines and their receptors may have overlapping specificities for chemotaxis and progenitor cell proliferation (57, 58). The CCR5 receptor which is capable of binding several CC chemokines is present on a variety of leucocyte cell types whereas eosinophil chemotaxis is primarily inducible through activation of the CCR3 receptor by eotaxin. The α-chemokine receptors bind C-X-C chemokines such as IL-8, and the KSHV GCR (ORF 74, see below) is most closely related to the α-chemokine receptor family.

Unlike many of the other cellular homologues encoded by KSHV, no clear role for the vMIP chemokines has been described in the viral life cycle. Since cellular chemokines are involved in generation of inflammatory responses, the virus chemokine-like proteins may block signalling from native ligands as a means to limit antiviral inflammatory responses. While early studies suggested that vMIP-II might be a broad-spectrum chemoattractant receptor antagonist (59), subsequent studies have shown that vMIP-II can activate Ca^{2+} flux via CCR3 and to a lesser extent CCR5 binding and activate eosinophil chemotaxis (60). These studies were performed using a synthetic peptide which includes a portion of the secretory peptide which might artificially switch the natural antagonistic effect of the viral chemokine to an agonist (61). Unpublished studies suggest that the naturally cleaved vMIP-II retains agonistic properties (61).

Recently, vMIP-III has been demonstrated to be expressed in KS lesions and to bind to CCR4 (J. Stine, personal communication). Since Th2 cells preferentially express CCR4, activation of this co-receptor in $CD4^+$ T cells has been hypothesized to contribute to a shift from a Th1 cell-mediated immune response to a Th2 antibody-dependent immune response which is likely to be less effective in controlling KSHV infection. An alternative hypothesis is that the viral chemokines may serve to attract susceptible non-infected cells to sites of lytic virus production. However, uninfected target cells are likely to be plentiful given the low percentage of circulating PBMC that are infected among persons with demonstrated KSHV infection (62). An intriguing finding is that both vMIP-I and vMIP-II potently induce angiogenesis in chicken chorioallantoic membrane assays (60). Although vMIP-I and vMIP-II, like vIL-6, are only marginally expressed in KS lesions, the angiogenic activity of these viral chemokines may contribute to paracrine stimulation of KS tumours or to angiogenesis occurring in Castleman's disease tumours (60). While other human β-chemokines generally do not induce angiogenic responses in this assay, some α-chemokines such as IL-8 are potent angiogenic factors.

Considerable scientific excitement was generated by the discovery that HIV strains utilize specific chemokine GCRs as co-receptors with CD4 for cell entry, and that chemokine ligands have potential to act as specific HIV inhibitors (63–65). T cell lymphotrophic HIV strains tend to use the SDF-1 α-chemokine receptor, CXCR4, as the HIV co-receptor whereas macrophage-trophic strains tend to use the β-chemokine receptors, CCR5, and to a lesser extent CCR3, for cell entry (for review see ref. 66). Initial co-transfection studies using vMIP-I suggested that this chemokine can inhibit macrophage-trophic HIV strain entry into cells by competing for CCR5 (25). A more rigorous test of the ability of vMIP-I and vMIP-II to inhibit HIV entry has been performed using exogenously administered proteins (59, 60). vMIP-II in particular appears to have a broad spectrum of chemokine receptor binding and HIV inhibition. While K_d for vMIP-II is generally far higher than corresponding cellular chemokines on various receptors, vMIP-II is able to bind to CCR3 with a reasonably low K_d that may be therapeutically achievable (59, 60). This is of particular interest since CCR3 may be the predominant receptor for microglial HIV infection. Specific blocking of HIV at this receptor could have useful therapeutic implications for AIDS-related dementias and two preliminary studies do indeed suggest that AIDS patients with KS are less likely to develop neurologic complications of HIV than AIDS patients without KS, even after controlling for $CD4^+$ counts and other measures of underlying immunosuppression (67).

3.3 Viral G protein-coupled receptor (vGCR)

KSHV ORF 74 encodes a 342 amino acid G protein-coupled receptor (GCR) (13, 68). Other Rhadinoviruses namely, HVS (ORF 74 or ECRF3) (2, 5) and MHV68 (ORF 74) (69) also encode GCR homologues with high homology to human IL-8 receptor proteins (Table 1) (70).

The cellular GCRs are a family of cell surface receptors that span the membrane seven times and bind the chemoattractant chemokines (71). These transmembrane receptors are key modulators of such diverse physiological processes such as neuro-transmission, cellular metabolism, secretion, cellular differentiation, and growth as well as inflammatory and immune responses. The GCRs are generally class restricted, binding to one or multiple chemokines within a given family.

The KSHV-encoded GCR is structurally similar to the cellular GCR consisting of seven hydrophobic domains representing the characteristic transmembrane domains. The viral GCR (vGCR) has glycosylation sites in the N terminal extracellular loop, proline residues in the potential fifth, sixth, and seventh transmembrane regions and two cysteine residues in the putative second, third, and fourth extracellular loops (68, 71, 72). vGCR mRNA transcripts are detected in a small portion of KS lesions and more abundantly in BCBL cells upon TPA stimulation suggesting that vGCR is not required for maintenance of KS tumours but may play a supportive role in patho-genesis *in vivo* (68, 72).

vGCR like its cellular counterparts, the CXCR1, CXCR2, and CXCR3 receptors binds huIL-8 with high affinity (70). However, in contrast to the cellular chemokine receptors the vGCR signalling is constitutive and agonist independent. Transient expression of vGCR in COS-1 cells results in increased accumulation of phospho-

Table 1 Cellular cytokine, cytokine receptor proteins, and cell regulatory proteins encoded by gammaherpesviruses

KSHV[a]	HVS[b]	MHV68[c]	EHV2[d]	RRV[e]
CBP (ORF 4)	CBP (ORF 4)	CBP (ORF 4)	–	–
vIL-6 (ORF K2)	–	–	–	vIL-6
–	–	–	IL-10	–
–	vIL-17 (S13)	–	–	–
vMIP-1 (ORF K4)	–	–	–	–
vMIP-2 (ORF K6)	–	–	–	–
vMIP-3 (ORF K4.1)	–	–	–	–
vBcl-2 (ORF 16)	Bcl-2 (ORF 16)	Bcl-2 (ORF M11)	–	–
vIRF (ORF K9)	–	–	–	–
vFLIP (ORF K13)	vFLIP (ORF 71)	–	vFLIP (E8)	–
v-Cyclin (ORF 72)	Cyclin-D (ORF 72)	Cyclin-D (ORF 72)	–	–
vGCR (ORF 74)	GCR (ORF 74)	GCR (ORF 74)	GCR (ORF 74) GCR(E1) GCR(E6)	–
vAdh/Ox-2 (ORF K14)	–	–	–	–

[a] KSHV: Kaposi's sarcoma herpesvirus (13, 73).
[b] HVS: Herpesvirus saimiri (2).
[c] Murine herpesvirus-68 (69).
[d] EHV-2: Equine hepesvirus-2 (4).
[e] RRV: Rhesus monkey herpesvirus (26).
'–' denote genes absent or unidentified thus far.

inositide-specific phospholipid C, suggesting the activation of the phosphoinositide–inositoltrisphophate protein kinase C pathway. Transient expression of vGCR in 293T cells activates the JNK/SAPK and p38/MAK kinases implicating recruitment of the protein kinase pathways similar to the cellular inflammatory cytokines (70, 73).

In addition to receptor function, vGCR has oncogenic and angiogenic activities. Stable vGCR expression causes NIH3T3 cells to proliferate *in vitro* and when injected into nude mice these stable vGCR expressing cells cause vascular tumour formation (73). Supernatants of vGCR-expressing NIH3T3 cells can induce microtubular formation when applied to human umblical vein endothelial cells (HUVECs) and the angiogenic response is mediated through the vascular endothelial growth factor (VEGF) (73). The cellular VEGF has been shown to induce endothelial cell proliferation, enhance vascular permeability, and stimulate KS spindle cells (74). Also, anti-VEGF blocks VEGF activity and vGCR responses (73, 74). Moreover, VEGF is transcriptionally regulated by the AP-1 responsive promoters (70, 73).

Thus, vGCR may modulate cell differentiation and angiogenesis through the involvement of the p38/HOG-1, JNK/MAPK kinase pathways and the autocrine factor VEGF (70, 73). Expression of the VEGF receptor in AIDS-KS lesions could indicate that vGCR aids in proliferation of the KS tumours through a paracrine signalling cascade involving VEGF induction.

3.4 Viral interferon regulatory factor (vIRF)

KSHV ORF K9 encodes a 449 amino acid named vIRF with 13% amino acid identity to IFN-consensus sequence binding protein (ICSBP), a member of the cellular inteferon regulatory factor (IRF) family (13, 75). The IRF family of transcription factors both positively and negatively regulate interferon (IFN) signal transduction by interacting with ISRE elements present in the promoters of interferon-inducible genes (Fig. 3) (76–78). Up to nine members of the IRF transcriptional family have been characterized thus far, many of which exert varying degrees of activity on

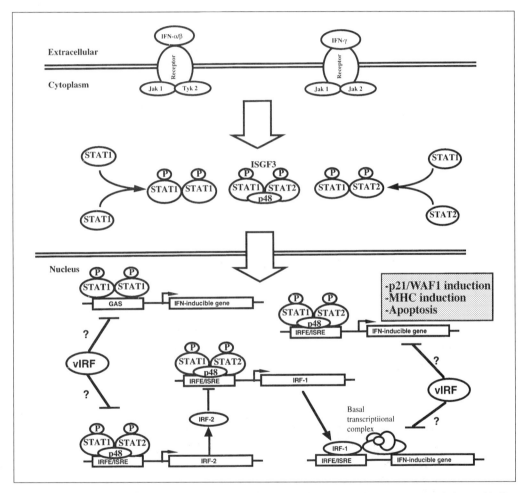

Fig. 3 vIRF inhibits IFN-α/β and IFN-γ signalling pathways. Signalling through the IFN receptors following binding of IFN-α/β or IFN-γ leads to recruitment of STAT homodimers, heterodimers, and the ISGF3 (STAT1 + STAT2 + p48) complexes. Activated STATs are translocated to the nucleus where, by themselves or in combination, they bind IFN stimulated response elements (ISREs) and IFN regulatory factor elements (IRFEs) to effect transcription. vIRF may act either by directly binding to the IFN response elements or by interacting with the basal transcription machinery.

biological processes such as interferon signalling, cell cycle regulation, and apoptosis (79, 80).

A characteristic feature of the IRF family members is the presence of an N-terminal DNA binding domain with five tryptophan residues (25, 81). Both IRF-1 and IRF-2 have high homology to each other in their amino terminal DNA binding regions. The C terminal portions of IRF-1 has *trans*-activation activity whereas the corresponding regions in other members of this transcriptional family, namely ICSBP may have transcriptional repressor activity (82). The functional significance of the high homology between these proteins is their ability to bind overlapping sequences in the promoters of IFN-inducible genes (83, 84).

IRFs also function as regulators of cell growth and differentiation. IRF-1 is involved in cell cycle regulation functioning as a tumour suppressor protein. IRF-1 knock-out cells are resistant to radiation-induced cell cycle arrest/apoptosis (85) and can be transformed by activation of c-H-ras (80). In contrast, IRF-2 is an antagonist to IRF-1 which binds to the same DNA elements as IRF-1 and functions as a transcriptional repressor (76, 84, 86, 87). In some cell lines overexpression of IRF-1 can lead to cell cycle arrest and an antiviral state (88) while the overexpression of IRF-2 in NIH3T3 cells leads to tumourigenicity that can be reversed by co-overexpression of IRF-1 (89). Both IRF-1 and IRF-2 mRNA are constitutively expressed at low levels, however, IRF-1 protein levels are highest in growth arrested cells while the protein levels of IRF-2 remain relatively constant throughout the cell cycle. Also, IRF-1 is induced in response to virus and IFNs. This results in induction of IRF-1 levels relative to IRF-2 suggesting a delicate balance between these proteins that is critical for cell cycle regulation.

KSHV-encoded vIRF, with its homology to IRF family members, was examined for its potential role in inhibiting the interferon response (Fig. 3). In transient expression assays, vIRF inhibits signalling through the ISRE reporter element by type-1 interferons (90). vIRF also inhibits IRF-1 induced transcription (91) suggesting that the block in interferon signalling occurs downstream of IRF-1. None the less, vIRF does not appear to directly bind and compete with IRF proteins at the DNA response elements in interferon-responsive promoters (91, 92). This is not unexpected because of the poor sequence conservation between the DNA-binding domains of IRF members and the corresponding region of vIRF.

Like IRF-2 (89), vIRF not only inhibits interferon signalling but also has *in vitro* oncogenic activity. Stable expression of vIRF transforms NIH3T3 cells which lose contact inhibition and form colonies in soft agar and foci forming assays (90, 93). The cells become fully transformed and form tumours when injected into nude mice (90). Consistent with these results, vIRF suppressed IFN-β mediated induction of the cyclin-dependent kinase inhibitor p21 (also known as WAF1 or CIP1) (90, 94). In response to extracellular or intracellular DNA damaging signals, p21 binds the cyclin E–CDK2 complex and arrest cells in G1 phase of the cell cycle and is induced by IFN treatment (94–96). Recent evidence suggests that vIRF can also function as a transcriptional activator using as GAL4 fusion proteins (Offerman, personal communication) and that vIRF may play a role in vIL-6 cytokine expression (93).

vIRF mRNA is detected in BCBL cells induced by TPA, but is not detected in uninduced cells (43). However, BCBL cells expressing antisense vIRF do not show growth inhibition and preliminary evidence suggests that vIRF is expressed *in vivo* in Castlemans's disease tissue, but not KS or BCBL lesions (Parravicini, personal communication). Thus, the contribution of vIRF to cell transformation in naturally occurring tumours is unclear.

In addition to ORF K9, the KSHV genome has three additional putative IRF like ORFs, however, whether these ORF code for functional proteins is yet to be determined (13, 22).

4. KSHV proteins affecting the cell cycle and apoptosis

In order to replicate, viruses require the cellular transcription and translation machinery. Viral proteins exploit the host cell transcription mechanisms by inducing quiescent cells to enter the S-phase of the cell cycle to promote preferential replication of viral DNA. At the same time, as part of the host defense mechanism, cellular proteins regulating cell cycle are also being recruited to control programmed cell death pathways that induce apoptosis. The apoptotic pathway is a physiological process characterized by certain morphological features including changes in plasma membrane symmetry and attachment, condensation of the cytoplasm and nucleus, and internucleosomal cleavage of DNA. Viruses can inhibit apoptosis induced during lytic replication, preventing premature death of the host cells. Many DNA viruses (e.g. SV40 and adenoviruses) control apoptotic pathway activation by encoding anti-apoptotic proteins that interact directly or indirectly with components of cell death pathways.

4.1 Viral cyclin (v-Cyc)

Several of the DNA tumour viruses encode oncoproteins which directly target and inactivate the pRB tumour suppressor pathway (17). These oncoproteins include simian virus 40 T-antigen, adenovirus E1A, and papillomavirus E7 which all bind pRB through conserved LXCXE sequences (97). Inhibition of pRB results in the release of the E2F family of transcription factors which in turn allows the cell to traverse the G1/S cell cycle checkpoint. Although KSHV does not encode a protein that has been found to directly inactivate pRB thus far, it does possess a gene, ORF 72 (v-CYC) whose product can induce phosphorylation of pRB through interaction with cellular cyclin-dependent kinases (CDKs).

Cyclins are a family of proteins that regulate the cell cycle and DNA replication by binding to and stimulating specific CDKs to phosphorylate proteins involved in cell cycle regulation (for review ref. 98). Subtypes of cyclins are defined by where they act in the cell cycle. The D-type cyclins, for which KSHV v-CYC (ORF 72) has strongest homology, are transiently expressed in the G1 phase of the cell cycle. They preferentially bind CDK6 or CDK4 to phosphorylate pRB which results in release of the cell from pRB enforced G1 block (see ref. 99) for a review). Therefore, KSHV encodes a protein which has the potential to disrupt the pRB pathway similar to pRB-binding oncoproteins of other DNA tumour viruses.

Many neoplasms, whether caused by viruses or genetic anomalies, demonstrate disruption of the pRB tumour suppressor pathway suggesting that cell growth dysregulation at the G1 checkpoint is essential to tumour formation (99). It is clear now that dysregulation of the pRB pathway can occur at several critical points. At the most direct level, a genetic mutation can occur in the RB gene itself as is seen in human retinoblastomas or wild-type pRB can be inactivated by being bound to viral oncoproteins. At another level, anomalous expression of regulators of pRB can cause the same effect. Overexpression of cellular cyclin D1, a proto-oncogene, caused by chromosomal translocation has been described in mantle cell lymphomas and parathyroid tumours (100). Additionally mutations in CDK inhibitors (e.g. p16, p21, p27) can also disrupt normal pRB function (101).

Biochemical characterisation of KSHV v-CYC demonstrates that it has the ability to associate with and activate CDK6. The v-CYC-CDK6 complex has kinase activity that phosphorylates pRB at authentic sites (102). The kinase activity associated with v-CYC appears to have a broader substrate specificity as compared to cellular cyclin D in that it can also result in the phosphorylation of histone H1 (103, 104). Functional studies indicate that expression of v-CYC overcomes pRB-induced growth arrest: when v-CYC is co-transfected with wild-type pRB into Saos-2 cells, which have homozygous deletions at the pRB locus, it prevents expression of the senescent phenotype which results from transfection of wild-type pRB alone (102). Ojala and colleagues found that overexpression of v-CYC together with CDK6 but not CDK4 in several human cell lines leads to rapid apoptosis independent of serum concentrations as would be expected if v-CYC abrogated the G1/S checkpoint by its effect on pRB (105). Additionally, in contrast to cellular cyclins that are responsive to CDK inhibitors, v-CYC appears to be resistant to the CDKI activities of p16, p21, and p27 (106). Overexpression of c-CYC is likely to cause dysregulated cell cycling however, single gene transfection experiments suggest that v-CYC alone cannot effect a transformed phenotype *in vitro* and, in fact, drives the cell into apoptosis (105, and unpublished observations, R. Sarid, C. Boshoff, Y. Chang, P. S. Moore).

4.2   Viral Bcl-2 (vBcl-2)

The Bcl-2 family of proteins regulate the process of apoptosis in cells either by preventing or promoting cell death. These cellular proteins homodimerize and heterodimerize with other members of the family through conserved BH1 and BH2 domains. Cellular Bcl-2, the prototypical anti-apoptotic protein, heterodimerizes and acts in a reciprocally antagonistic manner with Bax, the prototypical pro-apoptotic protein (107). This dimerization, although mechanistically enigmatic, is felt to be functionally critical. Some evidence also suggest that the conserved BH3 site in this family of proteins may serve as a site for caspase cleavage resulting in a product that accelerates cell death via amplification of the caspase cascade (108). Bcl-2 is implicated in the pathogenesis of a variety of human neoplasms including follicular lymphomas in which a t(14;18) chromosomal translocation results in overexpression of the gene (109, 110). Rather than effecting transformation through increased cell proliferation, the Bcl-2 proto-oncogene is believed to enhance cell survival when overexpressed in neoplastic cells.

KSHV as well as a number of other DNA tumour viruses including EBV, herpesvirus saimiri, and adenovirus encode Bcl-2 homologues (111, 112). It is felt that these apoptosis-inhibiting genes may allow the virus to escape apoptotic host responses during productive and/or persistent infection. Although the KSHV vBcl-2 (ORF 16) has only 16% amino acid identity to its cellular counterpart, its BH1 and BH2 domains are well conserved (113). In contrast, the BH3 domain is poorly conserved (108). An epitope-tagged version of vBcl-2 shows punctate cytoplasmic and perinuclear distribution suggestive of membrane localization similar to that described for cellular Bcl-2 (113). vBcl-2 was found to heterodimerize with human Bcl-2 using a yeast two-hybrid system (113) but not by co-immunoprecipitation (114). Despite this difference which is likely to reflect sensitivities of techniques used, both groups found that vBcl-2 was able to prevent Bax-mediated apoptosis in a variety of systems including yeast, human fibroblasts, and Sindbis virus-infected cells (113, 114). In addition, partial reversal of apoptotis was observed in v-CYC-CDK6 transfected cells by co-expression of vBcl-2 (105). It has also been proposed that since vBcl-2 has a poorly preserved BH3 domain, it may be resistant to cleavage by recombinant caspases or by apoptotic cell extracts (108).

4.3 Viral FLIP (vFLIP)

KSHV also contains a viral FLICE-like inhibitory protein (vFLIP) encoded by ORF K13 which may inhibit caspase-induced apoptosis (115). vFLIP proteins are encoded by other gammaherpesviruses namely HVS (ORF 71), EHV-2 (ORF E8), and BHV-4 (E1.1 or BORF E2), and the poxvirus MCV (ORF MC159) (115, 116). vFLIPs and the more recently discovered c-FLIP, (117) act as dominant-negative inhibitors of FLICE (Capsase 8) activation. The crucial players in this signal transduction cascade are the caspase family of aspartate–cysteine proteases that are homologous to the interleukin 1β-converting enzymes (ICE) (reviewed in ref. 118).

The caspase cascade is triggered by the specific ligation of cell surface receptors that contain the death domains (DD), namely Fas (also known as CD95 or Apo1), tumour necrosis receptor factor (TNFR), and TRAMP. The DD is a stretch of 60–80 amino aids present in the cytoplasmic tails of the death signalling receptors that serves as an interaction domain for cytoplasmic adaptor proteins that also have DD. Activation of the receptors by binding with their cognate ligands or with cross-linking antibodies on the cell surface results in recruitment of adaptor proteins such as Fas associated death domain (FADD) or TNF receptor associated death domain (TRADD), to the cytoplasmic DD of the receptor. This in turn results in binding of Fas Ligand induced ICE (FLICE) or Caspase 8 via their amino terminal death-effector domains (DED) to the activated cytoplasmic portions of the receptor/FADD or receptor/TRADD complexes leading to the activation the caspase cascade. FLIPs (viral and cellular) have DED domains but are unable to fully activate FLICE and thus act as dominant-negative inhibitors of Fas pathway activation.

Sequence comparisons between the various FLIPs show that the KSHV vFLIP like the other gammaherpesvirus homologues (HVS, EHV-2, and BHV-4) has two DEDs in its carboxy terminus. The DEDs of KSHV vFLIP are similar to the DEDs found in the amino terminus of FLICE, suggesting that vFLIP could bind FADD and thus prevent caspase cascade activation.

Functional studies with KSHV encoded vFLIP are yet to be performed, however, co-immunoprecipitation experiments with the closely related EHV-2 (ORF E8) and HVS (ORF 71) vFLIP suggest that these homologues are potent inhibitors of TNF and FAS-mediated apoptosis (115). Inhibition occurs through protein–protein inter-actions of the DED motifs of vFLIPs with the DED domains of either FADD or the prodomain of FLICE (6, 115, 116). This interaction does not interfere with the ability of FADD to bind the DD of FAS or TNF. The EHV-2 and HVS-encoded vFLIP block cell death triggered by the TRAMP and TRAIL-R receptors. Interestingly, the EHV-2 vFLIP did not block apoptosis induced by staurosporine or growth factor with-drawal, indicating the involvement of different apoptotic pathways (115).

The vFLIP (ORF K13) transcript is encoded as part of a larger spliced latent tran-script that includes v-CYC (ORF 72) and the latent nuclear antigen LANA (ORF 73) (119, 120)]. *In situ* hybridization demonstrates vFLIP mRNA expression in spindle cells of KS tumours (121–123). However, immunohistochemical data suggests that the protein may not be expressed in KS lesions, suggesting post-transcriptional regulation of the vFLIP latent transcripts (Parravicini, manuscript in preparation). The functional significance of KSHV to encode vFLIP could be to aid in virus survival, persistence, and latency by inhibiting the Fas-mediated apoptosis and evading host immune mechanisms.

5. Novel ORFs encoded by KSHV

In this section, we have concentrated on KSHV genes with identifiable homology to cellular regulatory genes. Functional studies are currently underway to decipher the additional unique ORFs encoded by KSHV. These proteins do not appear to be homologous to any known cellular genes, however their importance in tunouri-genesis and signal transduction is evident from functional studies.

5.1 ORF K1

Present at the right end of the KSHV genome is an open reading frame K1. ORF K1 is positionally in the same place as the saimiri transforming protein (STP) of HVS, however, its transcriptional orientation, amino acid sequence, and functions are dif-ferent from the latter (124). This ORF encodes a 289 amino acid type-1 membrane glycoprotein with a potential N-terminal signal peptide, an extracellular 12 cysteine-rich domain, a transmembrane domain, and a cytoplasmic tail with the consensus sequence for the *i*mmunoreceptor *t*yrosine-based *a*ctivation *m*otif (ITAM) (24). The presence of a functional ITAM motif within the cytoplasmic tail of this glycoprotein is significant in transducing cellular lymphocyte activation signals. The ITAM motifs (conserved Tyr-X-X-Leu/Ile in a 26 amino acid sequence) are present within the cytoplasmic domains of various membrane bound B cell, T cell, and immunoglobulin Fc receptors that are expressed on all cells of the immune system. The activation of ITAM-containing receptors results in a rapid and transient phosphorylation of the ITAMs' tyrosine residues that initiate a cascade of biochemical events that lead to cell proliferation, differentiation, and acquisition of unique effector functions. In addition to its ITAM signalling, ORF K1 is involved in cellular transformation. Expression of ORF K1 in rat fibroblasts causes foci formation (124). Also, a HVS chimeric virus that contains ORF K1 in place of the HVS STP gene causes *in vivo* lymphoma induction in the common marmosets (124), providing further evidence of ORF K1 transforming potential. These two functions of transformation and intracellular signalling suggest a significant role for ORF K1 in tumourigenesis.

5.2 Kaposin

KSHV ORF K12 encodes a 0.7 kb mRNA transcript unique to KSHV and called kaposin. The 0.7 kb transcript translates to a 66 amino acid hydrophobic peptide that is abundantly expressed in latently-infected BCBL cells and in all stages of KS tumours (125, 126). Immunohistochemical analysis of BCBLs shows expression of kaposin primarily in the cytoplasm (127). *In vitro* expression of kaposin in rat fibroblasts leads to foci formation and injection of these cells into nude mice results in highly vascular tumours (127). However, the precise mechanism of transformation by kaposin is yet to be determined.

Functional studies have not been published on other KSHV proteins with homology to cellular proteins. This includes the complement-binding protein (ORF 4) and the extensive repertoire of DNA synthesis enzymes such as the dihydrofolate reductase (ORF 2), thymidylate synthase (ORF 70), DNA polymerase (ORD 9), thymidine kinase (ORF 21), and ribonucleotide reductase (ORF 60 and ORF 61).

6. Conclusions

KSHV contains a number of genes captured by the virus during its evolution. Genetic and biochemical evidence to date suggests that the products of these viral genes are constitutively expressed in KS lesions (e.g. ORF 72) while others may have tissue-specific expression patterns and are inducible during the viral lytic cycle (e.g. IL-6, vIRF). Functional studies of these proteins are at an early stage, but similarity to common features (such as pRB inhibition and antagonism of apoptosis) of other tumour viruses is evident. Further characterization of the proteins produced by KSHV will delineate their critical roles in perturbing cellular responses. These viral products are likely to be important in the pathogenesis of KSHV-related diseases, but also as important reagents for the study of virus–host cell interaction.

Acknowledgements

We thank Dr Julie Osborne, Dr Kenneth Low, and Dr Carlo Parravicini for careful reading and helpful comments on the manuscript.

References

1. Roizman, B. and Sears, A. E. (1996) Herpes simplex viruses and their replication. In *Virology* (ed. B. N. Fields, D. M. Knipe, and P. M. Howley), Vol. 2, pp. 2231–97. Lippincott-Raven Publishers, Philadelphia.

2. Albrecht, J.-C., Nicholas, J., Biller, D., Cameron, K. R., Biesinger, B., Newman, C., *et al.* (1992) Primary structure of the *Herpesvirus saimiri* genome. *J. Virol.*, **66**, 5047.

3. Kieff, E. (1996) Epstein–Barr Virus and its replication. In *Virology* (ed. B. N. Fields, D. M. Knipe, and P. M. Howley), Vol. 2, pp. 2343–96. Lippincott-Raven Publishers, Philadelphia.

4. Telford, E. A., Watson, M. S., Aird, H. C., Perry, J. and Davison, A. J. (1995) The DNA sequence of equine herpesvirus 2. *J. Mol. Biol.*, **249**, 520.

5. Ahuja, S. K. and Murphy, P. M. (1993) Molecular piracy of mammalian interleukin-8 receptor type B by herpesvirus saimiri. *J. Biol. Chem.*, **268**, 20691.

6. Bertin, J., Armstrong, R. C., Ottilie, S., Martin, D. A., Wang, Y., Banks, S., *et al.* (1997) Death effector domain-containing herpesvirus and poxvirus proteins inhibit both Fas- and TNFR1-induced apoptosis. *Proc. Natl. Acad. Sci. USA*, **94**, 1172.

7. Massung, R. F., Liu, L. I., Qi, J., Knight, J. C., Yuran, T. E., Kerlavage, A. R., *et al.* (1994) Analysis of the complete genome of smallpox variola major virus strain Bangladesh-1975. *Virology*, **201**, 215.

8. Senkevich, T. G., Bugert, J. J., Sisler, J. R., Koonin, E. V., Darai, G., and Moss, B. (1996) Genome sequence of a human tumoriegenic poxvirus: Prediction of specific host response-evasion genes. *Science*, **273**, 813.

9. Turner, P. C., and Moyer, R. W. (1998) Control of apoptosis by poxviruses. *Semin. Virol.*, **8**, 453.

10. Moore, P. S., Gao, S. J., Dominguez, G., Cesarman, E., Lungu, O., Knowles, D. M., *et al.* (1996) Primary characterization of a herpesvirus agent associated with Kaposi's sarcomae [published erratum appears in *J. Virol.* 1996 Dec; 70(12): 9083]. *J. Virol.*, **70**, 549.

11. O'Brien, V. (1998) Viruses and apoptosis. *J. Gen. Virol.*, **19**, 1833.

12. Chang, Y., Cesarman, E., Pessin, M. S., Lee, F., Culpepper, J., Knowles, D. M., *et al.* (1994) Identification of herpesvirus-like DNA DNA sequences in AIDS-associated Kaposi's sarcoma. Science, **265**, 1865.

13. Russo, J. J., Bohenzky, R. A., Chien, M. C., Chen, J., Yan, M., Maddalena, D., *et al.* (19960 Nucleotide sequence of the Kaposi sarcoma-associated herpesvirus (HHV8). *Proc. Natl. Acad. Sci. USA*, 93, 14862.

14. Tappero, J. W., Conant, M. A., Wolfe, S. F., and Berger, T. G. (1993) Kaposi's sarcoma: epidemiology, pathogenesis, histology, Clin. specrum, staging criteria and therapy. . *Am. Acad. Detmatol.*, **28**, 371.

15. Boshoff, C. and Weiss, R. A. (1998) Kaposi's sarcoma-associated herpesvirus. *Adv. Cancer Res.*, **75**, 57.

16. Olsen, S. J. and Moore, P. S. (1998) Kaposi's sarcoma -associated herpesvirus (KSHV/HHV8) and the etiology of KS. In *Herpesviruses and immunity* (ed. H. Friedman, P. Medveczky and M. Bendinelli), pp. 115–147. Plenum Publishing, New York.

17. Schulz, T. F., Chang, Y., and Moore, P. S. (1998) Kaposi's sarcoma-associated herpesvirus (KSHV/HHV8). In *Human tumor viruses* (ed. D. J. McCance). ASM press, Washington D.C.

18. Foreman, K. E., Friborg, J. J., Kong, W. P., Woffendin, C., Polverini, P. J., Nickoloff, B. J., *et al.* (1997) Propagation of a human herpesvirus from AIDS-associated Kaposi's sarcoma. *N. Eng. J. Med.,* **336**, 163.

19. Miller, G., Heston, L., Grogan, E., Gradoville, L., Rigsby, M., Sun, R., *et al.* (1997) Selective switch between latency and lytic replication of Kaposi's sarcoma herpesvirus and Epstein–Barr virus in dually infected body cavity lymphoma cells. *J. Virol.,* **71**, 314.

20. Miller, G., Rigsby, M. O., Heston, L., Grogan, E., Sun, R., Metroka, C., *et al.* (1996) Antibodies to butyrate-inducible antigens of Kaposi's sarcoma-associated herpesvirus in patients with HIV-1 infection. *N. Eng. J. Med.,* **334**, 1292.

21. Renne, R., Zhong, W., Herndier, B., McGrat, M., Abbey, N., Kedes, D., *et al.* (1996) Lytic growth of Kaposi's sarcoma-associated herpesvirus (human herpesvirus 8) in culture. *Nature Med.,* **2**, 342.

22. Niepel, F., Albrecht, J. C., and Fleckenstein, B. (1997) Cell-homologous genes in the Kaposi's sarcoma-associated rhadinovirus human herpesvisus 8: determinants of its pathogenicity? *J. Virol.,* **71**, 4187.

23. Moore, P. S. and Chang, Y. (1998) Antiviral activity of tumor-suppressor pathways: clues from molecular piracy by KSHV. *Trends Genet.,* **14**, 144.

24. Lee, H., Guo, J., Li, M., Choi, J.-K., DeMaria, M., Rosenzweig, M., *et al.* (1998) Identification of an immunoreceptor tyrosine-based activation motif (ITAM) of theK1 transforming protein of Kaposi's sarcoma-associated herpesvirus. *Mol. Cell Biol.,* **18**, 5219.

25. Moore, P. S., Boshoff, C., Weiss, R. A., and Chang, Y. (1996) Molecular mimicry of human cytokine and cytokine response pathway genes by KSHV. *Science,* **274**, 1739.

26. Desrosiers, R. C., Sasseville, V. G., Czajak, S. C., Zhang, X., Mansfield, K. G., Kaur, A., *et al.* (1997) A herpesvirus of rhesus monkeys related to the human Kaposi sarcoma-associated herpesvirus. *J. Virol.,* **71**, 9764.

27. Kishimoto, T., Akira, S., Narazaki, M., and Taga, T. (1995) Interleukin-6 family of cytokines and gp130. *Blood,* **86**, 1243.

28. Gearing, D. P., Comeau, M. R., Friend, D. J., Gimpel, S. D., Thut, C. J., McGourty, J., *et al.* (1992) The IL-6 signal transducer, gp130: an oncostatin M receptor and affinity converter for the LIF receptor. *Science,* **255**, 1434.

29. Hirano, T., Nakajima, K., and Hibi, M. (1997) Signalling mechanisms through gp130: a model of the cytokine system. *Cytokine Growth Factor Rev.,* **8**, 241.

30. Stahl, N., Boulton, T. G., Farruggella, T., Ip, N. Y., Davis, S., Witthuhn, B. A., et al. (1994) Association and activation of Jak-Tyk kinases by CNTF-LIF-OSM-IL-6 beta receptor components. Science, **263**, 92.

31. Stahl, N., Farruggella, T. J., Boulton, T. G., Zhong, Z., Darnell, J. J., and Yancopoulos, G. D. (1995) Choice of STATs and other substrates specified bymodular tyrosine-based motifs in cytokine receptors. *Science,* **267**, 1349.

32. Burger, R., Wendler, J., Antoni, K., Helm, G., Kalden, J. R., and Gramatzki, M. (1994) Interleukin-6 production in B-cell neoplasias and Castleman's disease: evidence for an additional paracrine loop. *Ann. Hematol.,* **69**, 25.

33. Klein, B. and Bataille, R. (1992) Cytokine network in human multiple myeloma. *Hematol. Oncol. Clin. North Am.,* **6**, 273.

34. Lichtenstein, A., Tu, Y., Fady, C., Vescio, R., and Berenspon, J. (1995) Interleukin-6 inhibits apoptotis of malignant plasma cells. *Cell. Immunol.,* **162**, 248.

35. Juge, M. N., Harousseau, J. L., Amiot, M., and Bataille, R. (1997) The retinoblastoma susceptibility gene RB-1 in multiple myeloma. *Leuk. Lymph.*, **24**, 229.

36. Urashima, M., Ogata, A., Chauhan, D., Vidriales, M. B., Teh, G., Hoshi, Y., *et al.* (1996) Interleukin-6 promotes multiple myeloma cell growth via phosphorylation of retinoblastoma protein. *Blood*, **88**, 2219.

37. Urashima, M., Teoh, G., Chauhan, D., Hoshi, Y., Ogata, A., Treon, S. P., *et al.* (1997) Interleukin-6 overcomes p21WAF1 upregulation and G1 growth arrest induced by dexamethasone and interferon-famma in multiple myeloma cells. *Blood*, **90**, 279.

38. Dolcetti, R. and Boiocchi, M. (1996) Cellular and molecular bases of B-cell clonal expansions. *Clin. Exp. Rheumatl. Suppl.*, **14**, S21.

39. Niepel, F., Albrecht, J. C., Ensser, A., Huang, Y. Q., Li, J. J., Friedman, K. A., *et al.* (1997) Human herpesvirus 8 encodes a homologue of interleukin-6. *J. Virol.*, **71**, 839.

40. Nicholas, J., Ruvolo, V. R., Burns, W. H., Sandford, G., Wan, X., Ciufo, D., *et al.* (1997) Kaposi's sarcoma-associated human herpesvirus-8 encodes homologues of macrophage inflammatory protein-1 and interleukin-6. *Nature Med.*, **3**, 287.

41. Molden, J., Chang, Y., You, Y., Moore, P. S., and Goldsmith, M. A. (1997) A Kaposi's sarcoma-associated herpesvirus-encoded cytokine homologue (vIL-6) activates signalling through the shared gp130 receptor subunit. *J. Biol. Chem.*, **272**, 19625.

42. Burger, R., Neipel, F., Fleckenstein, B., Savino, R., Ciliberto, F., Kalden, J. R.,, *et al.* (1998) Human herpesvirus type 8 interleukin-6 homologue is functionally active on human myeloma cells. *Blood*, **91**, 1858.

43. Sarid, R., Flore, O., Bohenzky, R. A., Chang, Y., and Moore, P. S. (1998) Transcription mapping of the Kaposi's sarcoma-associated herpesvirus (human herpesvirus 8) genome in a body cavity-based lymphoma cell line (BC-1). *J. Virol.*, **72**, 1005.

44. Asou, H., Said, J. W., Yang, R., Munker, R., Park, D. J., Kamada, N., *et al.* (1998) Mechanisms of growth control of Kaposi's Sarcoma-associated herpes virus-associated primary effusion lymphoma cells. *Blood*, **91**, 2475.

45. Komanduri, K. V., Luce, J. A., McGrath, M. S., Herndier, B. G., and Ng, V. L. (1996) The natural history and molecular heterogeneity of HIV-associated primary malignant lymphomatous effusions. *J. AIDS Hum. Retrovirol.*, **13**, 215.

46. Cohen, T., Nahari, D., Cerem, L. W., Neufeld, G., and Levi, B. Z. (1996) Interleukin 6 induces the expression of vascular endothelial growth factor. *J. Biol. Chem.*, **271**, 736.

47. Parravicini, C., Corbellino, M., Paulli, M., Magrini, U., Lazzarino, M., Moore, P. S., *et al.* (1997) Expression of a virus-derived cytokine, KSHV vIL-6, in HIV seronegative Castleman's disease. *Am. J. Pathol.*, **151**, 1517.

48. Corbellino, M., Poirel, L., Aubin, J. T., Paulli, M., Magrini, U., Bestetti, G., *et al.* (1996) The role of human herpesvirus 8 and Epstein–Barr virus in the pathogenesis of giant lymph node hyperplasia (Castleman's disease). *Clin. Infect. Dis.*, **22**, 1120.

49. Soulier, J., Grollet, L., Oskenhendler, E., Cacoub, P., Cazals-Hatem, D., Babinet, P., *et al.* (1995) Kaposi's sarcoma-associated herpesvirus-like DNA sequences in multicentric Castleman's disease. *Blood*, **86**, 1276.

50. Amaral, M. C., Miles, S., Kumar, G., and Nel. A. E. (1993) Oncostatin-M stimulates tyrosine protein phosphorylation in parallel with the activation of p42MAPK/ERK-2 in Kaposi's cells. Evidence that this pathway is important in Kaposi cell growth. *J. Clin. Invest.*, **92**, 848.

51. Miles, S. A., Rezai, A. R., Salazar-Gonzalez, J. F., Vander Meyden, M., Stevens, R. H., Logan, D. M., *et al.* (1990). AIDS Kaposi sarcoma-derived cells produce and respond to interleukin 6. *Prov. Natl. Acad. Sci. USA*, **87**, 4068.

52. Flamand, L., Zeman, R. A., Bryant, J. L., Lunardi, I. Y., and Gallo, R. C. (1996) Absence of human herpesvirus 8 DNA sequences in neoplastic Kaposi's sarcoma cell lines. *J. AIDS Hum. Retrovirol.*, **13**, 194.

53. Lebbé, C., de Crémoux, P., Rybojad, M., Costa da Cunha, C., Morel, P., and Calvo, F. (1995) Kaposi's sarcoma and new herpesvius. *Lancet.*, **345**, 1180.

54. MacDonald, M. R., Li, X. Y., and Virgin, H. T. (1997) Late expression of a beta chemokine homologue by murine cytomegalovirus. *J. Virol.*, **71**, 1671.

55. Murphy, P. M. (1994) The molecular biology of leukocyte chemoattractant receptors. *Annu. Rev. Immunol.*, **12**, 593.

56. Ahuja, S. K., Ozcelik, Milatovitch, A., Francke, U., and Murphy, P. M. (1992) Molecular evolution of the human interleukin-8 receptor gene cluster. *Nature Genet.*, 2, 31.

57. Broxmeyer, H. E., Sherry, B., Cooper, S., Lu, L., Maze, R., Beckmann, M. P., *et al.* (1993) Comparative analysis of the human macrophage inflammatory protein family of cytokines (chemokines) on proliferation of human myelid progenitor cells. Interacting effects involving suppression, synergistic suppression, and blocking of suppression. *J. Immunol.*, **150**, 3448.

58. Roth, S. J., Carr, M. W., and Springer, T. A. (1995) C-C chemokines, but not the C-X-C chemokines interleukin-8 and interferon-gamma inducible protein-10, stimulate trans-endothelial chemotaxis of T lymphocytes. *Eur. J. Immunol.*, **25**, 3482.

59. Kledal, T. N., Rosenkilde, M. M., Coulin, F., Simmons, G., Johnsen, A. H., Alouani, S., *et al.* (1997) A broad-spectrum chemokine antagonist encoded by Kaposi's sarcoma-associated herpesvirus. *Science*, **277**, 1656.

60. Boshoff, C., Endo, Y., Collins, P. D., Takeuchi, Y., Reeves, J. D., Schweickart, V. L., *et al.* (1997) Angiogenic and HIV inhibitory functions of KSHV-encoded chemokines. *Science*, **278**, 290.

61. Dittmer, D. and Kedes, D. H. (1998) Do viral cemokins modulate Kaposi's sarcoma? *Bioessays*, **20**, 367.

62. Whitby, D., Howard, M. R., Tenant-Flowers, M., Brink, N. S., Copas, A., Boshoff, C., *et al.* (1995) Detection of Kaposi's sarcoma-associated herpesvirus (KSHV) in peripheral blood of HIV-infected individuals predicts progression to Kaposi's sarcoma. *Lancet*, **364**, 799.

63. Cocchi, F., DeVico, A. L., Garzino, D. A., Arya, S. K., Gallo, R. C., and Lusso, P. (1995) Identification of RANTES, MIP-1 alpha, and MIP-1 beta as the major HIV-suppressive factors produced by CD8[+] T cells. *Science*, **270**, 1811.

64. Dragic, T., Litwin, V., Allaway, G. P., Martin, S. R., Huang, Y., Nagashima, K. A., *et al.* (1996) HIV-1 entry into CD4[+] cells is mediated by the chemokine receptor CC-CKR-5. *Nature*, **381**, 667.

65. Wu, L., Gerard, N. P., Wyatt, R., Choe, ., Parolin, C., Ruffing, N., *et al.* (1996) CD4-induced interaction of primary HIV-1 gp120 glycoproteins with the chemokine receptor CCR-5 [see comments]. *Nature*, **384**, 179.

66. Garzino, D. A., Devico, A. L., and Gallo, R. C. (1998) Chemokine receptors and chemokines in HIV infection. *J. Clin. Immunol.*, **18**, 243.

67. Liestael, K., Goplen, A. K., Dunlop, O., Bruun, J. N., and Maehlen, J. (1998) Kaposi's sarcoma and protection from HIV dementia. *Science*, **280**, 361.

68. Cesarman, E., Nador, R. G., Bai, F., Bhenzky, R. A., Russo, J. J., Moore, P. S., *et al.* (1996) Kaposi's sarcoma-associated herpesvirus contains G protein-coupled recptor and cyclin D homologues which are expressed in Kaposi's sarcoma and malignant lymphoma. *J. Virol.*, **70**, 8218.

69. Virgin, H. T., Latreille, P., Wamsley, P., Hallsworth, K., Weck, K. E., Dal, C. A., *et al.* (1997)

Complete sequence and genomic analysis of murine gammaherpesvirus 68. *J. Virol.*, **71**, 5894.

70. Arvanitaskis, L., Geras, R. E., Varma, A., Gershengorn, M. C., and Cesarman, E. (1997) Human herpesvirus KSHV encodes a constitutively active G-protein-coupled receptor linked to cell proliferation. *Nature*, **385**, 347.

71. Stadler, M., Chelbi, A. M., Koken, M. H., Venturini, L., Lee, C., Saib, A., *et al.* (1995) Transcriptional induction of the PML growth suppressor gene by interferons is mediated through an ISRE and a GAS element. *Oncogene*, **11**, 2565.

72. Guo, H. G., Browning, P., Nicholas, J., Hayward, G. S., Tschachler, E., Jiang, Y. W., *et al.* (1997) Characterization of a chemokine receptor-related gene in human herpesvirus 8 and its expression in Kaposi's sarcoma. *Virology*, **228**, 371.

73. Bais, C., Santomasso, B., Coso, O., Arvanitakis, L., Raaka, E. G., Gutkind, J. S., *et al.* (1998) G-protein-coupled receptor of Kaposi's sarcoma-associated herpesvirus is a viral oncogene and angiogenesis activator. *Nature*, **391**, 86. [Published erratum appears in *Nature*, **392**, 210].

74. Masood, R., Lunardi, I. Y., Jean, L. F., Murphy, J. R., Waters, C., Gallo, R. C., *et al.* (1994) Inhibition of AIDS-associated Kaposi's sarcoma cell growth by DAB389-interluekin 6. *AIDS Res. Hum. Retroviruses.*, **10**, 969.

75. Moore, P. S., Kingsley, L. A., Holmberg, S. D., Spira, T., Gupta, P., Hoover, D. R., *et al.* (1996) Kaposi's sarcoma-associated hepesvirus infection prior to onset of Kaposi's sarcoma. *Aids*, **10**, 175.

76. Harada, H., Fujita, T., Miyamoto, M., Kimura, Y., Maruyama, M., Furia, A., *et al.* (1989) Structurally similar but functionally distinct factors, IRF-1 and IRF-2, bind to the same regulatory elements of IFN and IFN-inducible genes. *Cell*, **58**, 729.

77. Miyamoto, M., Fujita, T., Kimura, Y., Maruyama, M., Harada, H., Sudo, Y., *et al.* (1988) Regulated expression of a gene encoding a nuclear factor, IRF-1, that specifically binds to IFn-beta gene regulatory elements. *Cell*, **54**, 903.

78. Tagashira, Y. (1982) Specific suppression of the local GVH reaction by F1 spleen cells. *Acta Med. Okayama*, **36**, 253.

79. Nguyen, H., Hiscott, J., and Pitha, P. M. (19970 The growing family of interferon regulatory factors. *Cytokine Growth Factor Rev.*, **8**, 293.

80. Taniguchi, T., Harada, H., and Lamphier, M. (1995) Regulation of the interferon system and cell growth by the IRF transcription factors. *J. Cancer Res. Clin. Oncol.*, **121**, 516.

81. Veals, S. A., Santa, M. T., and Levy, D. E. (1993) Two domains of ISGF3 gamma that mediate protein-DNA and protein–protein interactions during transcription factor assembly contribute to DNA-binding specificity. *Mol. Cell. Biol.*, **13**, 196.

82. Weisz, A., Marx, P., Sharf, R., Appella, E., Driggers, P. H., Ozato, K., *et al.* (1992) Human interferon consensus sequence binding protein is a negative regulator of enhancer elements common to interferon-inducible genes. *J. Biol. Chem.*, **267**, 25589.

83. Harada, H., Willison, K., Sakakibara, J., Miyamoto, M., Fujita, T., and Taniguchi, T. (1990) Absence of the type I IFN system in EC cells: transcriptional activator (IRF-1) and repressor (IRF-2) genes are developmentally regulated. Cell, **63**, 303.

84. Tanaka, N., Kawakami, T., and Taniguchi, T. (1993) Recognition DNA sequences of interferon regulatory factor 1 (IRF-1) and IRF-2, regulators of cell growth and the interferon system. *Mol. Cell. Biol.*, **13**, 4531.

85. Tanaka, N., Ishihara, M., Lamphier, M. S., Nozawa, H., Matsuyama, T., Mak, T. W., et al. (1996) Cooperation of the tumour suppressors IRF-1 and p53 in response to DNA damage. *Nature*, **382**, 816.

86. Tanaka, N. and Taniguchi, T. (1992) Cytokine gene regulation: regulatory cis-elements and DNA binding factors involved in the interferon system. *Adv. Immunol.*, **52**, 263.

87. Yamamoto, H., Lamphier, M. S., Fujita, T., Taniguchi, T., and Harada, H. (1994) The oncogenic transcription factor IRF-2 possesses a transcriptional repression and a latent activation domain. *Oncogene*, **9**, 1423.

88. Pine, R. (1992) Constitutive expression of an ISGF2/IRF1 transgene leads to interferon-independent activation of interferon-inducible genes and resistance to virus infection. *J. Virol.*, **66**, 4470.

89. Harada,H., Kitagawa, M., Tanaka, N., Yamamoto, H., Harada, K., Ishihara, M., *et al.* (1993) Anti-oncogenic and oncogenic potentials of interferon regulatory factors-1 and -2. *Science*, **259**, 971.

90. Gao, S.-J., Boshoff, C., Jayachandra, S., Weiss, R. A., Chang, Y., and Moore, P. S. (1997) KSHV ORF K9 (vIRF) is an oncogene that inhibits the interferon signalling pathway. *Oncogene*, **15**, 1979.

91. Zimring, J. C., Goodbourn, S., and Offermann, M. K. (1998) HHV8 envodes an IRF homologue that represses IRF-1 mediated transcription. *J. Virol.*, **72**, 701.

92. Taniguchi, T. (1995) IRF-1 and IRF-2 as regulators of the interferon system and cell growth. *Indian J. Biochem. Biophys.*, **32**, 235.

93. Li, M., Lee, H., Guo, J., Neipel, F., Fleckenstein, B., Ozato, K., *et al.* (1998) Kaposi's sarcoma-associated herpesvirus viral interferon regulatory factor. *J. Virol.*, **72**, 5433.

94. Chin, Y. E., Kitagawa, M., Su, W. C., You, Z. H. Iwamoto, Y., and Fu, X. Y. (1996) Cell growth arrest and induction of cyclin-dependent kinase inhibitor p21 WAF1/CIP1 mediated by STAT1. *Science*, **272**, 719.

95. Hobeika, A. C., Subramaniam, P. S., and Johnson, H. M. (1997) IFNa induces the expression of the cyclin-dependent kinase inhibitor p21 in human prostate cancer cells. Oncogene, **14**, 1165.

96. Sangfelt, O., Erickson, S., Einhorn, S., and Grander, D. (1997) Induction of Cip/Kip and Ink4 cyclin dependent kinase inhibitors by interferon-alpha in hematopoietic cell lines. *Oncogene*, **14**, 415.

97. Dyson, N., Howley, P. M., Munger, K., and Harlow, E. (1989) The human papilloma virus-16 E7 oncoprotein is able to bind to the retinoblastoma gene product. *Science*, **243**, 934.

98. Pines, J. (1995) Cyclins and cyclin-dependent kinases: themes and variations. *Adv. Cancer Res.*, **66**, 181.

99. Sherr, C. J. (1995) D-type cyclins. *Trends Biochem. Sci.*, **20**, 187.

100. Motokura, T. and Arnold, A. (1993) PRAD1/Cyclin 1 proto-oncogenes: genomic organization, 59 DNA sequence of a tumor-specific rearrangement breakpoint. *Gene Chromosomes Cancer*, **7**, 89.

101. Sherr, C. J. (1996) Cancer cell cycles. *Science*, **274**, 1672.

102. Chang, Y., Moore, P. S., Talbot, S. J., Boshoff, C. H., Zarkowska, T., Godden-Kent, D., *et al.* (1996) Cyclin encoded by KS herpesvirus. *Nature*, **382**, 410.

103. Godden-Kent, D., Talbot, S. J., Boshoff, C., Chang, Y., Moore, P. S., Weiss, R. A., *et al.* (1997) The cyclin encoded by Kaposi's sarcoma associated herpesvirus (KSHV) stimulates cdk6 to phosphorylate the retinoblastomas protein and Histone H1. *J. Virol.*, **71**, 4193.

104. Li, M., Lee, H., yoon, D. W., Albrecht, J. C., Fleckenstein, B., Neipel, F., *et al.* (1997) Kaposi's sarcoma-associated herpesvirus encodes a functional cyclin. *J. Virol.*, **71**, 1984.

105. Ojala, P. M., Tiainen, M., Moore, P. S., and Makela, T. P. (1998) KSHV cyclin-induced

apoptosis is CDK6-dependent. *The First Annual Meeting on Kaposi's Sarcoma Associated Herpesvirus (KSHV) and Related Agents*, Santa Cruz, California, USA.

106. Swanton, C., Mann, D. J., Fleckenstein, B., Neipel, F., Peters, G., and Jones, N. (199) Herpes viral cyclin/Cdk6 complexzes evade inhibition by CDK inhibitor proteins. *Nature*, **390**, 184.

107. Oltvai, Z. N., Milliman, C. L., and Korsmeyer, S. J. (1993) Bcl-2 heterodimerizes in vivo with a conserved homologue, Bax, that accelerates programmed cell death. *Cell*, **74**, 609.

108. Hardwick, J. M., Bellows, D., Chau, N., Burns, W., Doseff, A., and Lazebnik, Y. (1998) Viral Bcl-2 homologues employ multiple mechanisms to escape regulation that governs cellular Bcl-2 homologues. *The First Annual Meeting on Kaposi's Sarcoma Associated Herpesvirus (KSHV) and Related Agents*, Santa Cruz, California, USA.

109. Cleary, M. L., Smith, S. D., and Sklar, J. (1986) Cloning and structural analysis of cDNAs for bcl-2 and a hybrid bcl-2/immunoglobulin transcript resulting from the t(14:18) translocation. *Cell*, **47**, 19.

110. Tsujimoto, Y. and Croce, C. M. (1986) Analysis of the structure, transcripts, and protein products of bcl-2, the gene involved in human follicular lmphoma. *Proc. Natl. Acad. Sci. USA*, **83**, 5214.

111. Henderson, S., Huen, D., Rowe, M., Dawson, C., Johnson, G., and Rickinson, A. (1993) Epstein–Barr virus-coded BHRF1 protein, a viral homologue of Bcl-2, protects human B cells from programmed cell deat. *Proc. Natl. Acad. Sci. USA*, **90**, 8479.

112. White, E. (1993) Regulation of apoptosis by the transforming genes of the DNA rumor virus adenovirus. *Proc. Soc. Exp. Biol. Med.*, **204**, 30.

113. Sarid, R., Sato, T., Bohenzky, R. A., Russo, J. J., and Chang, Y. (1997) Kaposi's sarcoma-associated herpesvirus encodes a functional bcl-2 homologue. *Nature Med.*, **3**, 293.

114. Cheng, E. H., Nicholas, J., Bellows, G. S., Hayward, G. S., Guo, H. G., Reitz, M. S., *et al.* (1997) A Bcl-2 homologue encoded by Kaposi sarcoma-associated virus, human herpesvirus 8, inhibits apoptosis but does not heterodimerize with Bax or Bak. *Proc. Natl. Acad. Sci. USA*, **94**, 690.

115. Thome, M., Schneider, P., Hofman, K., Fickenshcer, H., Meinl, E., Neipel, F., *et al.* (1997) Viral FLICE-inhibitory proteins (FLIPs) prevent apoptosis induced by death receptors. *Nature*, **386**, 517.

116. Hu, S., Vincenz, C., Buller, M., and Dixit, V. M. (1997) A novel family of viral death effector domain-containing molecules that inhibit both CD-95- and tumor necrosis factor receptor-1-induced apoptosis. *J. Biol. Chem.*, **272**, 9621.

117. Irmler, M., Thome, M., Hahne, M., Schneider, P., Hofman, K., Steiner, V., *et al.* (1997) Inhibition of death receptor signals by cellular FLIP. *Nature*, **388**, 190.

118. Cohen, G. M. (1997) Caspases: the executioners of apoptosis. *Biochem. J.*

119. Dittmer, D., Lagunoff, M., Renne, R., Staskus, K., Haase, A., and Ganem, D. (1998) A cluster of latently expressed genes in Kaposi's sarcoma-associated herpesvirus. *J. Virol.*, **72**, 8309.

120 Sarid, R., Weizorek, J. S., Moore, P. S. and Chang, Y. (1999) Characterization and cell cycle regulation of the major kaposi's sarcoma-associated herpesvirus (human-herpesvirus 8) latent genes and their promoter. *J. Virol.*; **73**, 1438.

121 Kellam, P. (1998) Molecular identification of novel viruses. *Trends Microbiol.*, **6**, 160.

122. Rainbow, L., Platt, G. M., Simpson, G. R., Sarid, R., Gao, S. J., Stoiber, ., *et al.* (1997) The 222- to 234-kilodalton latent nuclear protein (LNA) of Kaposi's sarcoma-associated

herpesvirus (human herpesvirus 8) is encoded by orf73 and is a component of the latency-associated nuclear antigen. *J. Virol.*, **71**, 5915.

123. Stürzl, M., Blasig, C., Schreier, A., Neipel, F., Hohenadl, C., Cornali, E., *et al.* (1997) Expression of HHV-8 latency-associated T0.7 RNA in spindle cells and endothelial cells of AIDS-associated, classical and African Kaposi's sarcoma. *Int. J. Cancer*, **72**, 68.

124. Lee, H., Veazey, R., Williams, K., Li, M., Guo, J., Neipel, F., *et al.* (1998) Deregulation of cell growth by the K1 gene of Kaposi's sarcoma-associated herpesvirus. *Nature Med.*, **4**, 435.

124. Zhong, W. and Ganem, D. (1997) Characterization of ribonucleoprotein complexes containing an abundant polyadenylated nuclear RNA encoded by Kaposi's sarcoma-associated herpesvirus (human herpesvirus 8). *J. Virol.*, **71**, 1207.

126. Zhong, W., Wang, H., Herndier, B., and Ganem, D. (1996) Restricted expression of Kaposi sarcoma-associated herpesvirus (human herpesvirus 8) gnes in Kaposi sarcoma. *Proc. Natl. Acad. Sci. USA*, **93**, 6641.

127. Muralidhar, S., Pumfery, A. M., Hassani, M., Sadaie, M. R., Azumim N., Kishishita, M., *et al.* (1998) Identification of kaposin (open reading frame K12) as a human herpesvirus 8 (Kaposi's sarcoma-associated herpesvirus) transforming gene. *J. Virol.*, **72**, 4980.

6 Cellular sites and mechanisms of human cytomegalovirus latency

JOHN SINCLAIR

1. Introduction

Human cytomegalovirus (HCMV) is a member of the herpesvirus family whose seroprevalence generally depends on socio-economic status but can vary from 50–90% depending on the population. Like all herpesviruses, after primary infection, HCMV maintains a latent infection throughout the lifetime of the infected host. Normally, primary infection of immunocompetent individuals with HCMV rarely causes disease but problems arise when infection or reactivation occurs in immunosupressed individuals, particularly transplant patients and patients with AIDS (1, 2).

During productive infection in the primary human fibroblast, a cell type used extensively *in vitro* for culture of HCMV, the first viral gene products to be expressed in a temporal cascade of viral gene expression have been termed the immediate early (IE) genes (Fig. 1). The most abundantly expressed viral genes at this IE time are transcribed from the major IE locus which encodes the IE1 and IE2 gene products (3). These viral gene products are generated by differential splicing of a primary transcript (Fig. 2) and play a pivotal role in controlling viral and cellular gene expression to optimize the cellular environment for the production of viral DNA (3–9). After IE

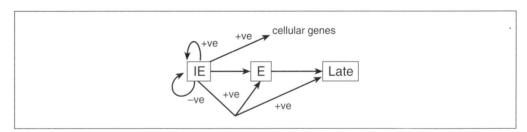

Fig. 1 Temporal cascade of HCMV gene expression. Viral gene expression begins with transcription of immediate early (IE) genes which positively (+) regulate both viral early (E) and late gene expression and also positively effect cellular gene expression. IE genes are also able to autoregulate expression from their own promoters both positively (+) and negatively (–).

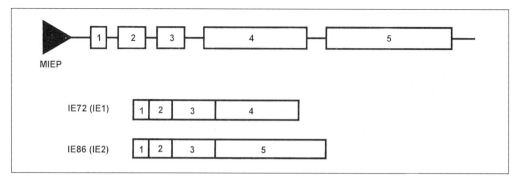

Fig. 2 The HCMV major immediate early transcription unit. The primary transcript from the major immediate early region of the viral genome is expressed from the major IE promoter (MIEP) and differentially spiced to generate two major protein species at immediate early times, IE1 and IE2.

expression, transcription of viral early(E) genes and then late gene products occurs. Early genes generally encode functions associated with viral DNA replication and late genes encode virus structural proteins. Both E and late gene expression are dependent on IE gene expression (3).

During active HCMV infection and during recovery from illness, virus can be detected in many cell types and excreted in the urine, saliva, and breast milk confirming the involvement of a number of tissues as sites of productive infection and viral transmission during active infection (10, 11). However, whether all these cells represent true sites of carriage of HCMV in normal healthy individuals is not known. Similarly it is unclear whether any identified sites of carriage of HCMV are true sites of reversible non-productive infection (latency) or sites of low-level productive infection (persistence). These questions, crucial to our understanding of HCMV pathogenesis, have long been difficult to address for a number of reasons. Clearly, it is difficult to obtain tissue samples etc, for analysis from healthy individuals. Also, in contrast to herpes simplex virus (HSV) and Epstein–Barr virus (EBV), which both have good latent model systems, there is no easy model system for HCMV latency. Consequently, our knowledge of the sites of carriage of HCMV in the healthy seropositive, or the extent of virus gene expression in those sites, is limited. One site of carriage of the virus in the healthy normal which has, however, been extensively studied, is the peripheral blood and the ease of obtaining samples as well as the advent of sensitive techniques to detect viral DNA and RNA is now allowing these questions to be addressed.

2. Cells of the myeloid lineage are an important site of carriage of HCMV

The ability of healthy human HCMV seropositive blood donors to transmit HCMV infection to recipients, and the fact that transmission by blood can be reduced by using leucocyte depleted blood products, has long suggested that one site of carriage of this virus is in the peripheral blood compartment (12–15). However, it is difficult, if not impossible, to isolate HCMV directly from the blood of healthy donors (16). Consequently, while the peripheral blood of healthy seropositive individuals does not carry infectious virus, it must carry virus in a form that is reactivatable.

Using extremely sensitive detection techniques such as polymerase chain reaction (PCR) a number of laboratories have demonstrated the presence of HCMV DNA and/or RNA in the peripheral blood of healthy HCMV seropositive individuals (17–20). Such analyses suggest that the frequency of cells that carry the HCMV genome is extremely low (probably less than 1 in 10 000 peripheral blood mononuclear cells) and it is perhaps for this reason that other laboratories have been unable to detect HCMV DNA in the blood of healthy carriers (21). Using sorted peripheral blood cell populations, our laboratory has observed that monocytes are a major site of carriage of HCMV DNA in healthy carriers with little viral DNA detectable in T or B cells or in polymorphonuclear cells (19, 20, 22). However, the extent of viral gene expression in healthy carriers and particularly whether or not cells in the peripheral blood that carry viral DNA are latently or persistently infected is slightly more contentious.

In at least one study of autopsy tissue, extensive HCMV IE1 protein expression has been detected in many tissues (23) although it should be stressed that what effect post-mortem changes may have had on viral gene expression is unknown. While some analyses using *in situ* hybridization have demonstrated the presence of RNA from the major IE region of HCMV in mononuclear cells of healthy carriers (17), other analyses have failed to detect HCMV IE RNA in seropositive donor tissue even though biopsies taken after transplant of these tissues into seronegative recipients showed high levels of HCMV IE RNA (24). Such conflicting results may reflect limitations in specificity and/or sensitivity of the detection method. For this reason, our laboratory has looked directly for HCMV lytic gene expression in monocytes of healthy seropositive carriers by the use of highly sensitive reverse transcription followed by PCR (RT-PCR). This type of analysis routinely fails to detect viral IE gene expression even though they clearly carry HCMV DNA (22, 25) and is consistent with the inability to culture virus from these peripheral blood monocytes.

Consequently, at least in monocytic cells, it would appear that HCMV is carried in a true latent state with little or no accompanying lytic gene expression. Confirmation that monocytic cells, and in particular a subpopulation of monocytes with specific dendritic markers, do, indeed, carry reactivatable virus has recently been presented by Soderborg-Nauclear *et al.* (26, 27). These investigators have recently shown that infectious HCMV can be co-cultured from monocytes of healthy seropositive individuals, but only after long-term culture and differentiation. This correlation between cellular differentiation and productive infection is a fundamental characteristic of HCMV persistence and is discussed extensively below.

3. Viral gene expression associated with latency

While it is now generally accepted that myeloid cells do indeed carry latent viral genomes in the absence of infectious virus or viral lytic gene expression, attention has been drawn to the possibility that these cells may express other viral gene products required for maintenance of latency. These may, perhaps, be functionally equivalent to HSV latency associated transcripts (LATs) or EBV EBNAs. Recently, Kondo and Mocarski have analysed HCMV transcripts expressed upon infection of granulocyte–macrophage precursor cells which can be infected *in vitro* with HCMV but which carry viral genomes in the absence of virus production or viral lytic gene expression in long-term culture (28). They have detected novel spliced and unspliced RNA transcripts mapping to both strands of the HCMV major IE region in these cells and have also been able to detect similar endogenous transcripts in the same granulocyte—macrophage precursor cells from healthy seropositive individuals (29). A number of these appear to be driven from a novel promoter slightly upstream of the major IE promoter/enhancer and start at a putative latent start site within the major IE promoter/enhancer. Expression of the supposed open reading frames (ORFs) present in these RNAs as bacterial fusion proteins has allowed the generation of antibodies specific for some of these putative polypeptides. Analysis of infected granulocyte–macrophage precursor cells has shown that some of these small ORFs can be detected in these granulocyte–macrophage precursor cells at the protein level in the absence of lytic infection although some were also detected in lytically infected cells.

Consequently, it does appear that carriage of HCMV in peripheral blood cells can be independent of infectious virus or detectable levels of viral lytic gene expression but other, as yet unidentified transcripts perhaps associated with viral latency, may be present in these cells.

4. Permissiveness of peripheral blood cells for HCMV infection requires differentiation

In the hope that some insight may be obtained as to the type of cell in the peripheral blood that carries and disseminates HCMV, a number of laboratories have analysed the permissiveness of different types of peripheral blood cells for HCMV infection, *in vitro*. There have been many reports detailing cell types that can apparently be infected with HCMV, *in vitro*. However, very often the frequency of infection is extremely low and viral gene expression is often restricted to IE events (30). Generally, it is accepted that *in vitro* infection of peripheral blood cells with clinical isolates of HCMV occurs at extremely low frequency (31). In contrast, it has been well documented that long-term culture of peripheral blood monocytes towards a macrophage phenotype results in a cell population that is permissive for HCMV (32, 33). Consistent with this, is the observation that *in vivo* differentiated alveolar macrophages are fully permissive for HCMV infection (34). Consequently, there is now good evidence for a link between the state of differentiation of myeloid cells and their permissiveness for HCMV infection. Such a link between productive infection and differentiation also exists with EBV (35).

Interestingly, this association between differentiation and exogenous infection, *in vitro*, is also reflected in the spectrum of endogenous viral gene expression in healthy seropositive individuals. For instance, differentiation of peripheral blood monocytes of healthy carriers to macrophages, *in vitro*, results in the induction of endogenous HCMV IE expression from monocytes (25). Similarly, the ability to isolate virus from monocytes of healthy seropositive individuals after long-term culture and differentiation *in vitro* (26, 27) confirms the link between differentiation and HCMV gene expression in the myeloid lineage. These analyses, then, point to a situation in which HCMV is carried in monocytes of healthy individuals in the absence of an appreciable level of viral lytic gene expression and formally rules out the possibility that HCMV is carried in the blood as a chronic low-level infection. One interpretation of this scenario is that monocytes act to carry and disseminate virus silently in the peripheral blood and the lack of viral gene expression in these cells facilitates immune evasion. In contrast to this restriction of HCMV gene expression in monocytes, terminal differentiation of these cells to tissue macrophages results in signals permitting reactivation of lytic gene expression and subsequent productive infection.

5. Bone marrow—a reservoir of virus in the healthy carrier

The data accrued so far clearly points to a lack of infectious virus in the peripheral blood of healthy seropositive individuals and, as monocytes are only transiently present in the circulation (36), the question arises as to how monocytes acquire virus in the first instance. One possibility is that monocytes acquire HCMV earlier in their lineage—perhaps in bone marrow. As with all lymphoid and myeloid cells, monocytes arise from pleuripotent CD34$^+$ stem cells present in bone marrow. These CD34$^+$ stem cells are believed to be the progenitors of all blood cell lineages and capable of self-renewal (36). Stem cells differentiate along the myeloid lineage to monoblasts then promonocytes in the bone marrow and then enter the blood stream where they lose CD34 cell surface antigen and develop into monocytes (37). This development depends upon responses to haemopoietic growth factors, particularly granulocyte–macrophage colony stimulating factor (Gm-CSF) and interleukin 3 (IL-3), which stimulate cell division and differentiation (38).

In order to ask whether such bone marrow progenitor cells may act as a reservoir for HCMV, a number of laboratories (28, 31, 39–42) have asked whether these cells can be infected with HCMV *in vitro* and, if so, to what extent do they support virus production. These types of experiments consistently show that CD34$^+$ bone marrow progenitors cells can, indeed, be infected with clinical isolates of HCMV but that, except for one report (43), little or no viral lytic gene expression occurs within these cells until they have been cultured, long-term, in the presence of haemopoetic growth factors (28, 39, 41, 42). Interestingly, the infected bone marrow progenitors can maintain the viral genome in the absence of detectable IE gene expression for two to three weeks. After this time induction of viral IE gene expression and productive infection is concomitant with the appearance of monocytic colonies in these longterm bone marrow cultures (28, 39, 41, 42). Once again, these observations reinforce the relationship between permissiveness for HCMV production infection and the differentiation-state of the cell. Interestingly, it is possible to detect endogenous HCMV DNA in colonies of monocytic cells grown from GM-CSF stimulated CD34$^+$ bone marrow progenitors by PCR (42) and PCR analysis has also detected HCMV DNA in CD34$^+$ cells purified from bone marrow by cell sorting (22). As with monocytes (25), CD34$^+$ bone marrow progenitors which carry HCMV DNA show no evidence of viral IE gene expression. This clearly implies that endogenous HCMV DNA is present in CD34$^+$ bone marrow progenitors in normal virus carriers and that it is maintained in the absence of a productive infection (22). Consequently, it would appear that the latent HCMV genome may be passed from monocyte progenitor cells in the bone marrow to daughter cells but upon terminal differentiation of these cells to macrophages, a cell type arises which then allows virus reactivation.

While experimental observations to date are, generally, all consistent with a scenario in which CD34$^+$ bone marrow progenitors carry latent virus during their terminal differentiation to macrophages, it would appear that not all cells of the myeloid lineage carry viral DNA. Polymorphonuclear cells from healthy sero-positive individuals appear not to contain HCMV DNA (20) yet their precursors, CD34$^+$ bone marrow progenitors, as well as sub-lineages such as monocytes, do. This necessarily argues for a mechanism that differentially maintains carriage of the latent viral genome in certain myeloid lineages but not others. To date, we are uncertain as to what that mechanism might be but it is possible that latent viral gene products, if they exist, may include such functions.

While the exact mechanism of carriage of the viral genome in bone marrow is unclear, bone marrow does appear to act as a reservoir for HCMV. However, the question as to how do these progenitor cells, themselves, originally acquire virus also arises. It is possible that these CD34$^+$ stem cells continually acquire HCMV by a chronic low-grade infection of the bone marrow. For instance, bone marrow stomal cells are fully permissive for HCMV infection (43). Alternatively, HCMV may infect these CD34$^+$ stem cells progenitors directly and then be maintained in the self-renewing progenitor cell population. At present, no data exist to distinguish between these two possibilities but if replication and partition of the HCMV genome does occur in self-renewing CD34$^+$ stem cells, the latent viral genome could be disseminated by the peripheral blood in myeloid cells in the absence of viral gene expression. This would help avoid cell-mediated immune surveillance (44, 45).

6. Conditionally permissive cell lines as models for factors controlling latency and reactivation

Myeloid progenitors early in their lineage are unable to be productively infected with HCMV *in vitro* and do not express viral IE genes from endogenous viral genomes. In the case of non-permissiveness of early myeloid cells, this lack of productive infection is not simply due to the inability of virus to enter target cells. We and others

(46, 47) have observed that clinical isolates of HCMV do enter early bone marrow progenitor cells and monocytes at least as determined by detection of viral pp65, a viral lower matrix protein which enters infected cells with the virion (47) and acts as a marker for viral entry (48). However, there is evidence that some laboratory strains of HCMV enter cells less efficiently than clinical isolates (48). The reasons for apparent differences in the ability of different viral isolates to bind/enter target cells, particularly with respect to laboratory isolates, is unclear at present. However, it is possible that it may be a result of the known large genomic deletions that have recently been observed in these laboratory strains (49).

While it is believed that the additional coding capacity found in clinical HCMV isolates may encode functions associated with increased pathogenesis (49), it is possible they may also include functions associated with virus binding or internalization in cells of specific lineages. In contrast to laboratory strains of virus, clinical isolates of HCMV are able to enter cells at all stages of the myeloid lineage and the lack of productive infection of early myeloid cells appears to be due to lack of viral lytic gene expression. Many experiments have addressed these observations and have concluded that this lack of viral lytic gene expression is probably due to a specific block in expression of viral IE genes which then prevents later classes of viral gene expression. Unfortunately, there is no model system of HCMV latent infection that allows easy analysis of the mechanisms by which early myeloid cells regulate viral IE gene expression. However, a number of cell lines do exist which are conditionally permissive for HCMV IE gene expression and productive infection and they have been used to try to identify factors that regulate IE expression in the hope that they may shed some light on the control of viral latency (50).

Two cell types that have been used extensively are the human teratocarcinoma cell line NTera2D1 (T2) (51–54) and the monocytic cell line, THP1 (55). Both these cell types are non-permissive for HCMV infection due to a block in viral IE expression. However, differentiation of T2 cells to a neuronal phenotype with retinoic acid (RA) or of THP1 to a macrophage phenotype with phorbol esters (PMA) lifts this block in IE expression resulting in productive infection (51–55). Fortunately, transfection assays of reporter plasmids that are driven by viral IE promoter/regulatory regions recapitulate the lack of IE gene expression observed in infection assays. That is, transfection of undifferentiated cells results in low levels of reporter gene expression and this is substantially increased upon differentiation (54–57). These cell types, therefore, represent systems in which permissiveness for viral IE expression and productive infection is differentiation-dependent. This is somewhat similar to that observed for peripheral blood myeloid cells *in vitro* and a number of laboratories have used these cell systems to try to identify differentiation-specific regulators of IE gene expression on the basis that they may help define factors that play a role in viral latency.

Such analyses have concentrated on defining regions of the promoter/regulatory region of the HCMV major IE gene transcription unit which regulate the major IE gene expression in a differentiation-dependent manner. The major IE transcription unit is under the control of the major IE promoter (MIEP) (Fig. 3). This region is ex-

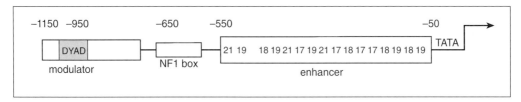

Fig. 3 The HCMV major immediate early promoter/regulatory region. The MIEP comprises a TATA box as well as an enhancer which contains 17, 18, 19, and 21 bp repeat motifs. Upstream from the enhancer lies a domain which contains a number of consensus binding motifs for the cellular factor NF1 (the NF1 box) and another region, the modulator, which contains an imperfect dyad symmetry.

tremely complex and contains one of the strongest known transcriptional enhancers (made up of an array of 17, 18, 19, and 21 bp repeat elements) and a far upstream element which has been termed the modulator and which includes an imperfect dyad symmetry (56–60). The modulator region has been shown to be highly sensitive to DNase I and the extent of hypersensitivity changes upon differentiation of cells to a permissive phenotype for HCMV infection (52). It has, therefore, been argued that changes in chromatin structure around the MIEP may play an important role in determining the permissiveness of cells for HCMV IE expression and hence productive infection and this results from changes in binding of specific proteins to this region of DNA (48).

We and others have found that the lack of expression from the MIEP in non-permissive T2 and THP1 cells is due, at least in part, to repression mediated by DNA sequences 5′ to the promoter (52, 54, 56, 57, 61, 62). Indeed, sequences have been identified within the modulator region which bind differentiation-specific cellular factors (54, 56, 57, 61–63). Interestingly, these cellular factors often decrease when T2 or THP1 cells are differentiated to a permissive phenotype and deletion of the binding sites for these factors from promoter/reporter constructs results in increased promoter activity in transient transfection assays in normally non-permissive cells (54, 61–64). Consequently, these factors have characteristics consistent with a role in differentiation-specific negative regulation of the MIEP. Analysis of the DNA sequences present in the MIEP which are responsible for its differentiation-specific repression in undifferentiated T2 and THP1 cells show consensus binding sites for a number of known cellular transcription factors. These include YY1 (65) and a member of the ets family of transcription factors, ERF (64). Interestingly, identical factors also bind to the 21 bp repeat elements of the HCMV major IE enhancer (Fig. 4) which is another site that has been identified as mediating differentiation-specific repression of the MIEP in non-permissive T2 and THP1 cells (61, 65). Consistent with these factors playing a role in differentiation-specific repression of the MIEP, YY1 and ERF are differentially expressed in T2 and THP1 cells. YY1 appears to be controlled post-translationally, in that steady-state levels of YY1 RNA are constant through differentiation whereas YY1 protein is substantially reduced in differentiated T2 and THP1 cells. In contrast, differentiation of both T2 and THP1 cells leads to decreases in steady-state levels of ERF RNA (64).

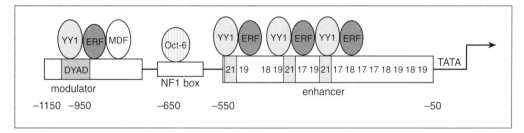

Fig. 4 Cellular transcription factors bind to negative regulatory elements in the MIEP. The MIEP has been shown to bind a number of already identified cellular transcription factors which include YY1, ERF, and Oct-6. At least one other so far uncharacterized factor, MDF, has also been shown to bind the MIEP in a differentiation-dependent manner. The regions of the promoter binding these factors and these cellular factors, themselves, have been implicated in differentiation-specific repression of the MIEP.

Work from our laboratory and others have confirmed that YY1 (65, 66) and ERF (64) do, indeed, bind to the modulator and the 21 bp repeat elements in the MIEP *in vitro* (Fig. 4) and can mediate repression of the MIEP in transient co-transfection experiments (64, 65). The mechanism by which these differentiation-specific cellular factors repress the MIEP is unclear. However, preliminary experiments suggest that YY1 may prevent recruitment of the general transcription factor TFIIB to the transcription pre-initiation complex on the MIEP (67) and may also recruit transcriptional inhibitors such as the histone deacetylase HDAC to this promoter (68). As yet, we do not know how ERF mediates its transcriptional repression effects. We have also identified one additional site within the so-called NF1 cluster of the MIEP (Fig. 4) that binds Oct-6, a member of the Octamer family of cellular transcription factors (64). Oct-6 is able to mediate repression of the MIEP in transient co-transfection assays in T2 cells and Oct-6 expression is also regulated differentiation-specifically (64). If these putative cellular repressors of the MIEP are also differentially expressed in peripheral blood myeloid cell lineages, it is an attractive possibility that they may be responsible for the observed lack of IE gene expression in peripheral blood monocytes *in vivo* and help maintain latency in these cell types. Interestingly, YY1 has also been implicated in the repression of human papillomavirus (69) and, more recently, lytic gene expression of EBV (70).

While it is clear that such differentiation-specific negative regulators of IE gene expression are likely to play an important role in regulating viral IE expression in undifferentiated myeloid cells, it is likely, too, that other cellular factors may also be involved. Other factors that mediate repression of the MIEP have yet to be identified (63) and it is also likely that differentiation-specific positive regulators of the MIEP, not found in undifferentiated cells, will also be required for maximal MIEP activity (60). For instance, elegant experiments by Meier and Stinski (71) using recombinant HCMV in which the modulator had been deleted showed that this virus still showed differentiation-specific IE gene expression in THP1 cells. This strongly argues that multiple sites within the MIEP mediate differentiation-specific negative regulation.

Such promoter analysis, then, does point to the control of the MIEP by differentiation-specific cellular transcription factors as a mechanism which could restrict viral lytic gene expression and maintain latency in the peripheral blood monocyte, *in vivo*. Clearly, blocks in viral DNA replication could occur at other stages in the viral life cycle, perhaps after IE expression but before DNA replication. Indeed there is evidence that some cultured cells are able to express the major IE1 protein after infection but are unable to replicate viral DNA (72). However, a block in expression of HCMV IE proteins, which are heavily expressed during lytic infection, would certainly be more consistent with the mechanism of maintenance of latency of other herpesviruses such as HSV and EBV which occurs at least in part by controlling expression of their respective IE proteins.

7. Many cell types are infected during viraemia—does this reflect other sites of latency?

HCMV is associated with a number of clinical syndromes. These include congenital deafness and mental retardation in the new-born, mononucleosis, and often fatal diseases with lung and gut involvement, particularly in the immunocompromised (73). How the cell type infected during HCMV-mediated disease reflects the pathophysiology of infection and what link these cell types have to the cell type and extent of viral gene expression observed during asymptomatic infection is unclear but is important for a full understanding of virus reactivation and viral pathogenesis.

A number of studies have addressed the extent of viral gene expression in cells during viraemia, at which time HCMV can be isolated from peripheral blood cells and plasma (74). Such analyses, using co-culture with fully permissive human fibroblasts, antigen detection, *in situ* hybridization, or PCR have shown the presence of HCMV DNA and RNA and viral antigens in many cell types (11, 75, 76). Clearly, endothelial, epithelial, fibroblast, and vascular smooth muscle cells can all be productively infected during HCMV disease (11). In the blood, the spectrum of cells that are productively infected is more contentious. Using viral antigen detection, macrophages have clearly been shown to support productive infection *in vivo* (76) whereas monocytes show little evidence of productive infection. In contrast, PMNLs have been shown to contain viral antigens but this is believed to be due to phagocytosis (77). Some studies, however, using sensitive RT-PCR assays have apparently detected IE, early, and late viral RNAs (78, 79) although other studies have failed to detect evidence, at the RNA level, for productive infection in PMNLs (80). Consequently, it is likely that PMNLs do contain viral transcripts to some extent but whether or not this is a result of bona fide infection or limited expression of genes from phagocytosed virus, from sites of active infection, is open to question. Clearly, any viral gene expression observed in PMNLs is likely to be a result of phagocytosis or *de novo* infection and unlikely to be due to reactivation of endogenous virus as we have been unable to detect endogenous HCMV genome in PMNLs of healthy carriers (20).

Even though there is controversy regarding the extent of viral gene expression in leucocytes during active infection, it is clear that many other cell types are productively infected during HCMV disease. Monocyte/macrophage cells are able to transmit virus to cultured endothelial cells *in vitro*, consistent with a view that cells of the myeloid lineage act to disseminate reactivated virus to a variety of tissues during active infection (81, 82). But the question as to whether endothelial, epithelial, and vascular smooth muscle cells, themselves, contain latent virus which becomes reactivated or are infected *de novo* is so far unresolved and represents an important area of cytomegalovirus research, especially as carriage of virus in cells such as vascular smooth muscle cells has been suggested as being involved in the pathogenesis of atherosclerosis (83).

8. Conclusions

A working definition of viral latency is the reversible non-productive infection of cells and this is distinct from viral persistence which involves continual low level production of virus or irreversible non-productive (abortive) infection of cells (84). In the case of HCMV, it is becoming clear that the mechanism of carriage of virus in the healthy individual may be cell-specific. At least in the blood, a speculative model of HCMV latency/persistence in the healthy seropositive carrier would be that virus is maintained in monocytes in a truly latent state with the absence of viral lytic gene expression helping to evade the host immune response (Fig. 5). This results from the transcriptional milieu of the monocyte preventing viral IE expression and would correlate well with the known inability to productively infect monocytes *in vitro* or to culture virus from peripheral blood monocytes of healthy seropositive carriers. However, as monocytes become terminally differentiated to macrophages the block in viral IE expression is lifted and the viral genome may become fully reactivated such that these sites may be looked upon as being persistently infected. In this situation, however, the host's cytotoxic T cell response acts efficiently to clear these productively infected cells unless the host is immunocompromised at which time virus disseminates, resulting in disease.

Unfortunately our understanding of the carriage of HCMV in bone marrow is somewhat less clear. Virus can clearly enter bone marrow progenitors *in vitro* and persist without lytic gene expression over long-term culture. Endogenous HCMV DNA can also be detected in bone marrow progenitors by PCR. These observations would be consistent with the bone marrow acting as a reservoir of HCMV after primary infection which then seeds latent virus into the peripheral blood *via* the monocyte. Differentiation of these cells to tissue macrophages then results in local reactivation. Such a mechanism has been shown for visna virus which has been shown to infect monocyte progenitors in bone marrow without viral gene expression. Differentiation then results in the expression of the integrated visna virus proviral genome and productive infection (85). This model of HCMV carriage and latency/persistence of HCMV in the healthy carrier, though still speculative, is consistent with many of the observations that have been made to date.

Our understanding of HCMV latency and reactivation has clearly been hampered by the lack of a good model system. However, sensitive detection techniques that

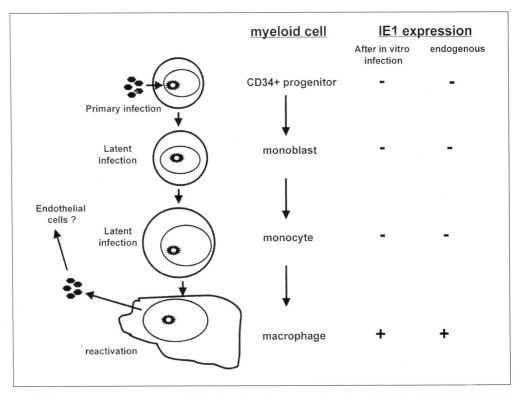

Fig. 5 Scheme for the maintenance of HCMV in cells of the myeloid lineage. CD34$^+$ bone marrow progenitors become infected with HCMV and latent infection is maintained in the myeloid lineage until cells become differentiated to macrophages at which time virus IE gene expression is reactivated, resulting in virus production. Released virus is then free to infect other cell types such as endothelial cells. Permissiveness of cells for IE1 expression after exogenous infection or expression of IE1 from endogenous virus is also linked to the differentiation state of the cell.

have allowed the analysis of sites of carriage of virus as well as studies of specific cell types, which can be infected with HCMV *in vitro*, have begun to give insights into a mechanism by which HCMV is carried in the healthy host. The difficulty in obtaining samples for analysis from healthy carriers has generally restricted these analyses to the peripheral blood. It cannot be overstated that a comprehensive analysis of other sites, besides the peripheral blood, which may act as true sites of latency/persistence of HCMV in the healthy host is an important goal in HCMV research. However, a substantial body of work, to date, clearly points to the important role that myeloid cells play in the maintenance of HCMV in the normal host and to the role of cellular differentiation in the control of viral gene expression.

Acknowledgements

We would like to thank many members of our laboratory, past and present, whose work has contributed to the studies described here. We also apologize to colleagues in the field whose work could not be sited due to space limitations. This work was supported by the Medical Research Council and the Wellcome Trust.

References

1. Rubin, R. H. (1990) Impact of cytomegalovirus infection on organ transplant recipients. *Rev. Infect. Dis.*, **12**, S754.
2. Drew, W. (1988) Cytomegalovirus infection in patients with AIDS. *J. Infect. Dis.*, **158**, 449.
3. Stenberg, R. M. (1996) The human cytomegalovirus major immediate early-gene. *Intervirology*, **39**, 343.
4. Colberg-Poley, A. M., Stantomenna, L. D., Harlow, P. P., Benfield, P., and Tenney, D. J. (1992) Human cytomegalovirus US3 and UL36-38 immediate early proteins regulate gene expression. *J. Virol.*, **66**, 95.
5. Hagemeier, C., Walker, S. M., Sissons, J. G. P., and Sinclair, J. H. (1992) The 72k IE1 and 80K IE2 proteins of human cytomegalovirus independently transactivate the c-fos, c-myc and hsp 70 promoters via basal promoter elements. *J. Gen. Virol.*, **73**, 2385.
6. Malone, C. L., Vesole, D. H., and Stinski, M. F. (1990) Transactivation of a human cytomegalovirus early promoter by gene products from the immediate early gene IE2 and augmentation by IE1: Mutational analysis of the viral proteins. *J. Virol.*, **64**, 1498.
7. Hermiston, T. W., Malone, C. L., Witte, P. R., and Stinski, M. F. (1987) Identification and characterisation of the HCMV immediate early-region 2 gene that stimulates gene expression from an inducible promoter. *J. Virol.*, **61**, 3214.
8. Stenberg, R. M., Fortney, J., Barlow, S. W., Magrane, B. P., Nelson, J. A., and Ghazal, P. (1990) Promoter specific trans activation and repression by the HCMV IE proteins involves common and unique domains. *J. Virol.*, **64**, 1556.
9. Spector, D. (1996) Activation and regulation of HCMV early genes. *Intervirology*, **39**, 361.
10. Stagno, S., Reynolds, D. W., Pass, R. F., and Alford, C. A. (1980) Breast milk and the risk of cytomegalovirus disease. *N. Engl. J. Med.*, **302**, 1073.
11. Sinzger, C., Grefte, A., Plachter, B., Gouw, A. S. H., The, T. H., and Jahn, G. (1995) Fibroblast, epithelial cells, endothelial cells and smooth muscle cells are major targets of HCMV infection in lung and gastrointestinal tract. *J. Gen. Virol.*, **76**, 741.
12. Tolpin, M. D., Stewart, J. A., Warren, D., Mojica, B. A., Collins, M. A., Doveikis, S. A., *et al.* (1985) Transfusion transmission of cytomegalovirus confirmed by restriction endonuclease analysis. *Pediatr.*, **107**, 953.
13. Adler, S. P. (1983) Transfusion-associated cytomegalovirus infections. *Rev. Infect. Dis.*, **5**, 977.
14. Yeager, A. S., Grumet, F. C., Hafleigh, E. B., Arvin, A. M., Bradley, J. S., and Prober, C. G. (1981) Prevention of transfusion-acquired cytomegalovirus infections in newborn infants. *J. Pediatr.*, **98**, 281.
15. De Graan-Hentzen Y. C. E., Gratama, J. W., Mudde, G. C., Verdonck, L. F., Houbiers, J. G. A., Brand, A., *et al.* (1989) Prevention of primary cytomegalovirus infection in patients with hematologic malignancies by intensive white cell depletion of blood products. *Transfusion*, **29**, 757.
16. Jordan, M. C. (1983) Latent infection and the elusive cytomegalovirus. *Rev. Infect. Dis.*, **5**, 205.
17. Schrier, R. D., Nelson, J. A., and Oldstone, M. B. A. (1985) Detection of human cytomegalovirus in peripheral blood lymphocytes in a natural infection. *Science*, **230**, 1048.

18. Stannier, P., Taylor, D. L., Kichen, A. D., Wales, N., Tryhorn, Y., and Tyms, A. S. (1989) Persistence of cytomegalovirus in mononuclear cells in peripheral blood from blood donors. *Br. Med. J.*, **299**, 897.

19. Taylor-Wiedeman, J. A., Sissons, J. G. P., Borysiewicz, L. K., and Sinclair, J. H., (1991) Monocytes are major site of persistence of human cytomegalovirus in peripheral blood mononuclear cells. *J. Gen. Virol.*, **72**, 1059.

20. Taylor-Wiedeman, J. A., Hayhurst, G. P., Sissons, J. G. P., and Sinclair, J. H. (1993) Polymorphonuclear cells are not sites of persistence of human cytomegalovirus in healthy individuals. *J. Gen. Virol.*, **74**, 265.

21. Urushibira, N., Kwon, K. W., Takahashi, T. A., and Sekiguchi, S. (1995) HCMV DNA is not detectable with nested double polymerase chain reaction in healthy blood donors. *Vox Sang.*, **68**, 9.

22. Mendelson, M., Monard, M., Sissons, J. G. P., and Sinclair, J. (1996) Detection of endogenous HCMV in CD34$^+$ bone marrow progenitors. *J. Gen. Virol.*, **77**, 3099.

23. Toorkey, C. B. and Carrigan, D. R. (1989) Immunohistochemical detection of an immediate early antigen of human cytomegalovirus in normal tissues. *J. Infect. Dis.*, **160**, 741.

24. Gnann, J. W., Ahlmén, J., Svalander, C., Olding, L., Oldstone, M. B. A., and Nelson, J. A. (1988) Inflammatory cells in transplanted kidneys are infected by human cytomegalovirus. *Am. J. Pathol.*, **132**, 239.

25. Taylor-Wiedeman, J. A., Sissons, J. G. P., and Sinclair, J. H. (1994) Induction of endogenous human cytomegalovirus gene expression after differentiation of monocytes from healthy carriers. *J. Virol.*, **68**, 1597.

26. Soderberg-Nauclear, C., Fish, K., and Nelson, J. A. (1997) Reactivation of latent human cytomegalovirus by allogeneic stimulation of blood cells from healthy donors. *Cell*, **91**, 119.

27. Soderberg-Nauclear, C., Fish, K., and Nelson, J. A. (1997) Interferon gamma and tumour necrosis factor alpha specifically induces formation of cytomegalovirus permissive monocyte-derived macrophages that are refractory to the antiviral activity of these cytokines. *J. Clin. Invest.*, **12**, 3154.

28. Kondo, K. and Mocarski, E. S. (1995) Cytomegalovirus latency and latency specific transcription in haematopoetic progenitors. *Scan. J. Infect. Dis.*, **S99**, 63.

29. Kondo, K., Kaneshima, H., and Mocarski, E. S. (1994) Human cytomegalovirus infection of granulocyte macrophage progenitors. *Proc. Natl. Acad. Sci. USA*, **91**, 11879.

30. Rice, G. P. A., Schrier, R. D., and Oldstone, M. B. A. (1984) Cytomegalovirus infects human lymphocytes and monocytes: Virus expression is restricted to immediate-early gene products. *Proc. Natl. Acad. Sci. USA*, **81**, 6134.

31. Einhorn, L. and Ost, A. (1984) Cytomegalovirus infection of human blood cells. *J. Infect. Dis.*, **149**, 207.

32. Lathey, J. L. and Spector, S. A. (1991) Unrestricted replication of human cytomegalovirus in hydrocortisone-treated macrophages. *J. Virol.*, **65**, 6371.

33. Ibanez, C. E., Schrier, R., Ghazal, P., Wiley, C., and Nelson, J. A. (1991) Human cytomegalovirus productivity infects primary differentiated macrophages. *J. Virol.*, **65**, 6581.

34. Drew, W. L., Mintz, L., Hoo, R., and Finley, T. N. (1979) Growth of herpes simplex and cytomegalovirus in cultured human alveolar macrophages. *Am. Rev. Respir. Dis.*, **119**, 287.

35. Young, L. S., Lau, R., Rowe, M., Niedobietek, G., Packham, G., Shanahan, F., *et al.* (1991) Differentiation-associated expression of the Epstein–Barr virus BZLF1 transactivator protein in oral hairy leukoplakia. *J. Virol.*, **65**, 2868.

36. Metcalf, D. (1989) The molecular control of cell division, differentiation commitment and maturation in hemopoietic cells. *Nature*, **339**, 27.

37. Katz, F. E., Tindle, R. W., Sutherland, S. R., and Greaves, M. F. (1985) Identification of a membrane glycoprotein associated with hemopoietic progenitor cells. *Leuk. Res.*, **9**, 191.

38. Adams, D. O. and Hamilton, T. A. (1987) Molecular transduction mechanisms by which Interferon-γ and other signals regulate macrophage development. *Immunol. Rev.*, **97**, 5.

39. Reiser, H., Kuhn, J., Doerr, H. W., Kirchner, H., Munk, K., and Braun, R. (1986) Human cytomegalovirus replicates in primary human bone marrow cells. *J. Gen. Virol.*, **67**, 2595.

40. Sing, G. K. and Ruscetti, F. W. (1990) Preferential suppression of myelopoiesis in normal human bone marrow cells. *Blood*, **75**, 1965.

41. Preiksaitis, J. K. and Janowska, A. (1991) Persistence of human cytomegalovirus in human long-term bone marrow culture: relationship to hemopoiesis. *J. Med. Virol.*, **35**, 76.

42. Minton, E. J., Tysoe, C., Sinclair, J. H., and Sissons, J. G. P. (1994) Human cytomegalovirus infection of the monocyte/macrophage lineage. *J. Virol.*, **68**, 4017.

43. Maciejewski, J. P., Bruening, E. E., Donahue, R. E., Mocarski, E. S., Young, N. S., and St. Jeor, S. C. (1991) Infection of hematopoietic cells by human cytomegalovirus. *Blood*, **80**, 170.

44. Apperley, J. F., Dowding, C., Hibbin, J., Buiter, J., Matutes, E., Sissons, J. G. P., *et al.* (1989) The effect of cytomegalovirus on hemopoiesis: *in vitro* evidence for selective infection of marrow stromal cells. *Exp. Hematol.*, **17**, 38.

45. Wills, M. R., Carmichael, A. J., Mynard, K., Jin, X., Weekes, M., and Sissons, J. G. P. (1996) The human cytotoxic T lymphocyte response to cytomegalovirus is dominated by the structural protein pp65: specificity and T cell receptor usage of pp65-specific CTL. *J. Virol.*, **70**, 7569.

45. Riddell, S. R., Rabin, M., Gaballe, A., Britt, W., and Greenberg, P. D. (1991) Class I MHC-restricted cytotoxic T lymphocyte recognition of cells infected with human cyto-megalovirus does not require endogenous viral gene expression. *J. Immunol.*, **146**, 2795.

46. Revello, M. G., Pecivalle, P., Matteo, A. D., Morino, F., and Gerna, G. (1992) Nuclear expression of the lower matrix protein of human cytomegalovirus in peripheral blood leukocytes of immunocompromised viraemic patients. *J. Gen. Virol.*, **73**, 437.

47. Minton, E. J. (1992) Infection of the monocytic cell lineage by HCMV. Ph. D. Thesis. University of Cambridge.

48. Simmons, P., Kaushansky, K., and Torok-Storb, B. (1990) Mechanisms of cyto-megalovirus-mediated myelosuppression: perturbation of stromal function versus direct infection of stromal cells. *Proc. Natl. Acad. Sci. USA*, **87**, 1386.

49. Cha, T. A., Tom. E., Kemble, G. W., Duke, G. M., Mocarski, E. S., and Spaete, R. R. (1996) HCMV clinical isolates carry at least 19 genes not found in laboratory strains. *J. Virol.*, **70**, 78.

50. Sinclair, J. H. and Sissons, J. G. P. (1996) Latent and persistent infections of monocytes and macrophages. *Intevirology*, **39**, 293.

51. Gonczol E., Andrews, P. W., and Plotkin, S. A. (1984) Cytomegalovirus replicates in differentiated cells but not in undifferentiated human embryonal carcinoma cells. *Science*, **224**, 159.

52. Nelson, J. A. and Groudine, M. (1986) Transcriptional regulation of the human cyto-megalovirus major-immediate early gene is associated with induction of DNase 1 hypersensitive sites. *Mol. Cell. Bol.*, **6**, 452.

53. LaFemina, R. and Hayward, G. S. (1986) Constitutive and retinoic acid inducible expression of cytomegalovirus immediate early genes in human teratocarcinoma cells. *J. Virol.*, **58**, 434.

54. Shelbourn, S. L., Kothari, S. K., Sissons, J. G. P., and Sinclair, J. H. (1989) Repression of human cytomegalovirus gene expression associated with a novel immediate early regulatory region binding factor. *Nucleic Acids Res.*, **17**, 9165.

55. Weinshenker, B. G., Wilton, S., and Rice, G. P. A. (1988) Phorbol ester-induced differentiation permits productive human cytomegalovirus infection in a monocytic cell line. *J. Immunol.*, **140**, 1625.

56. Nelson, J. A., Reynolds-Kohler, C., and Smith, B. (1987) Negative and positive regulation by a short segment in the 5'-flanking region of the human cytomegalovirus major immediate-early gene. *Mol. Cell. Biol.*, **7**, 4125.

57. Lubon, H., Ghazal, P., Henninghausen, L., Reynolds-Kohler, C., Lockshin, C., and Nelson, J. A. (1989) Cell specific activity of the modulator region in the human cytomegalovirus major immediate-early gene. *Mol. Cell. Biol.*, **9**, 1342.

58. Boshart, M. F., Weber, J., Jahn, G., Dorsch-Hasler, K., Fleckenstein, B., and Schaffner, W. (1985) A very strong enhancer is located upstream of an immediate early gene of human cytomegalovirus. *Cell*, **41**, 521.

59. Stamminger, T. H. and Fleckenstein, B. (1990) Immediate early transcription regulation of human cytomegalovirus. *Curr. Top. Microbiol. Immunol.*, **154**, 3.

60. Stinski, M. F., Macias, M. P., Malone, C. L., Thrower, A. R., and Huang, L. (1993) Regulation of transcription from the cytomegalovirus major immediate early promoter by cellular and viral proteins. In *Multidisciplinary approach to understanding cytomegalovirus disease* (ed. S. Michelson and S. A. Plotkin), pp. 3–12. Elsevier Science.

61. Kothari, S., Baillie, J., Sissons, J. G. P., and Sinclair, J. H. (1991) The 21 bp repeat element of the human cytomegalovirus major immediate early enhancer is a negative regulator of gene expression in undifferentiated cells. *Nucleic Acids Res.*, **19**, 1767.

62. Sinclair, J. H., Ballie, J., Bryant, L. A., Taylor-Wiedeman, J. A., and Sissons, J. G. P. (1992) Repression of human cytomegalovirus major immediate early gene expression in a monocytic cell line. *J. Gen. Virol.*, **73**, 433.

63. Huang, T. H., Oka, T., Asai, T., Okada, T., Merrills, B. W., Gertson, P. N., *et al.* (1996) Repression by a differentiation-specific regulator of the HCMV enhancer. *Nucleic Acids Res.*, **24**, 1695.

64. Mendelson, M., Bain, M., Sissons, P., and Sinclair, J. Manuscript in preparation.

65. Liu, R., Baillie, J. E., Sissons, J. G. P., and Sinclair, J. (1994) The transcription factor YY1 binds to negative regulatory elements in the HCMV Major immediate early promoter/enhancer and mediates repression in non-permissive cells. *Nucleic Acids Res.*, **22**, 2435.

66. Pizzorno, M. C., Nastanski, F., and Shenk, T. (1993) The transcription factor YY1 represses the major immediate early protein of human cytomegalovirus in undifferentiated human teratocarcinoma cells. *XVIII International Herpes Virus Workshop*, Pittsburgh, USA.

67. Liu, R., Sissons, J. G. P., and Sinclair, J. Manuscript in preparation.

68. Murphy, J., Sissons, J. G. P., and Sinclair, J. Manuscript in preparation.

69. Bauknecht, T., Angel, P., Royer, D., and zur Hausen, H. (1992) Identification of a negative regulatory domain in the human papillomavirus type 18 promoter: interaction with the transcriptional repressor YY1. *EMBO J.*, **11**, 4607.

70. Montalvo, E. A., Shi, Y., Shenk, T., and Levine, A. J. (1991) Negative regulation of the BZLF1 promoter of Epstein–Barr virus. *J. Virol.*, **65**, 3647.

71. Meier, J. L. and Stinski, M. F. (1997) Effect of modulator deletion on transcription of the HCMV major immediate early genes in infected undifferentiated and differentiated cells. *J. Virol.*, **71**, 1246.

72. Lafemina, R. L. and Hayward, G. S. (1988) Differences in cell type specific blocks to

immediate early gene expression and DNA replication of human simian and murine cytomegalovirus. *J. Gen. Virol.*, **69**, 355.

73. Ho, M. (1991) *Cytomegalovirus biology and infection*, 2nd edn. Plenum, New York.

74. Spector, S. A., Merril, R., Wolf, D., and Dankner, W. M. (1992) Detection of human cytomegalovirus in plasma of AIDS patients during acute visceral disease by DNA amplification. *J. Clin. Microbiol.*, **30**, 2359.

75. Greft, A., van der Giesse, M., van Son W., and The, T. H. (1993) Circulating cytomegalovirus (CMV)-infected endothelial cells in patients with active CMV infection. *J. Infect. Dis.*, **167**, 270.

76. Sinzger, C., Plachter, B., Grefte, A., The, T. H., and Jahn, G. (1996) Tissue macrophages are infected by HCMV *in vivo*. *J. Infect. Dis.*, **173**, 240.

77. Grefte, J. M. M., van der Gun, B. T. F., Schmolke, S., van der Geissen, M., van Son, W. J., Plachter, B., *et al.* (1992) The lower matrix protein pp65 is the principle viral antigen present in peripheral blood leukocytes during active cytomegalovirus infection. *J. Gen. Virol.*, **73**, 2923.

78. Bitsch, A., Kirchner, H., Dupke, R., and Bein, G. (1993) Cytomegalovirus transcripts in peripheral blood leukocytes of actively infected transplant patients detected by reverse transcription-polymerase chain reaction. *J. Infect. Dis.*, **167**, 740.

79. Bonlaer, D., Serr, A., Meyerkonig, U., Kirste, G., Huffert, F. T., and Haller, O. (1995) HCMV immediate early and late transcripts are expressed in all major leukocyte populations *in vivo*. *J. Infect. Dis.*, **172**, 365.

80. Grefte, A., Hamsen, M. C., van der Giessen, M., Knollema, S., van Son W. J., and The, T. H. (1994) Presence of HCMV immediate early mRNA but not ppL83 mRNA in polymorphonuclear and mononuclear leukocytes during active infection. *J. Gen. Virol.*, **75**, 1989.

81. Lathey, J. L., Wiley, C. A., Verity, A. M., and Nelson, J. A. (1990) Cultured human brain capillary endothelial cells are permissive for human cytomegalovirus. *Virology*, **176**, 266.

82. Waldman, W. J., Knight, D. A., Huang, E. H., and Sedmark, D. D. (1995) Bidirectional transmission of infectious cytomegalovirus between monocytes and vascular endothelial cells—an *in vitro* model. *J. Infect. Dis.*, **17**, 263.

83. Speir, E., Modali, R., Huang, E. S., Leon, M. B., Shawl, F., Finkel, T., *et al.* (1994) Potential role of HCMV and p53 interaction in coronary restenosis. *Science*, **265**, 391.

84. Garcia-Blanco, M. A. and Cullen, B. R. (1991) Molecular biology of latency in pathogenic human viruses. *Science*, **254**, 815.

85. Gendelman, H. E., Naryan, O., Kennedy-Stopskof, S., Kennedy, P. G., Ghotbi, Z., Clements, J. E., *et al.* (1986) Tropism of sheep lentiviruses for monocytes: susceptibility to infection and virus gene expression increase during maturation of monocytes to macrophages. *J. Virol.*, **58**, 67.

7 | Control of transcription by adenovirus-E1A proteins

ALT ZANTEMA AND ALEX J. VAN DER EB

1. Introduction

Human adenoviruses induce productive infection in their natural host, but in rodent cells they usually cause only abortive infection. Some abortively infected rodent cells will undergo transformation, in which case adenovirus (Ad) DNA is integrated into the host genome and at least some of the Ad genes are expressed. Early region 1 (E1) of Ad DNA plays an essential role, both in lytic infection and transformation. In either case, it affects the cell by altering cellular gene expression and, in the case of lytic infection, it also stimulates transcription of other Ad genes. The E1 region consists of two transcriptional units, E1A and E1B, which respectively encode two related proteins of about 45 kDa and two unrelated proteins of 19 kDa and 55 kDa. E1A is the gene that induces most of the alterations of transcription and is able to immortalize cells (1). When E1B is also present, a distinct morphological transformation takes place and transformation occurs more efficiently due to the fact that the E1B region protects the cells from apoptosis, one of the (side)-effects of E1A (2): the E1B-55 kDa protein inactivates p53 and thereby blocks the p53-dependent apoptosis (for a review on the role of p53 see ref. 3), while the E1B-19 kDa protein binds to and inactivates the proapoptotic Bax, Bak and Bhk/Bik proteins (4–8).

E1A alters the transcription of genes by binding to proteins of the cellular transcription regulation process, and this review will focus on these regulatory aspects of the E1A proteins. First, the E1A proteins themselves and some general aspects of transcription will be described and then the cellular proteins with which the E1A proteins associate.

2. The adenovirus E1A proteins

The E1A region is transcribed into a number of overlapping mRNAs which vary as a result of differential splicing. In transformed cells, a 12S and 13S mRNA is synthesized, while in lytic infection mRNAs of 9S, 10S, and 11S are also produced (9, 10). The role of the latter mRNAs has not been elucidated. The 12S and 13S mRNAs are translated into two proteins that differ in that the smallest protein lacks an internal

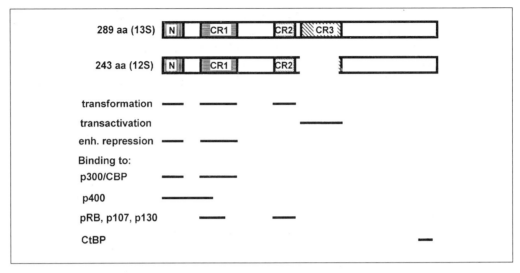

Fig. 1 Two Ad5 E1A proteins are synthesized from the E1A region due to differential splicing. The proteins differ by the lack of an internal stretch of amino acids in the smaller protein. CR1, CR2, and CR3 are the domains that are conserved among various adenovirus serotypes; N is the N terminal domain. The regions required for transformation, *trans*-activation, and enhancer repression and for associations to some proteins are indicated.

stretch of amino acids which is present in the larger protein. Comparison of the protein sequences of E1A proteins of various Ad serotypes has revealed the existence of three conserved domains: conserved region 1 (CR1), CR2, and CR3 (Fig. 1). The conservation within these domains is 50–60% while outside these domains the sequence is hardly conserved (11, 12). The CR3 region is present only in the 13S E1A protein and is involved in transcription activation of other viral genes and of some cellular genes (13, 14). The 12S E1A protein can transform cells, which implies that CR3 is not required for transformation. Deletion studies have shown that CR1 and CR2 together with the non-conserved N terminal domain are essential for the transforming capacity of E1A (15). These same regions are required for the association of E1A to a number of cellular proteins. It has become clear that the mechanism by which E1A executes its functions is by these associations.

Since most of these effects deal with transcriptional regulation, a short overview of the main aspects of eukaryotic transcription will be given first. Subsequently, the regulation of transcription by cellular proteins to which the E1A proteins associate will be discussed, i.e. the so-called pocket proteins, such as the retinoblastoma protein and the transcription co-activators, such as p300. Finally, other associating proteins and their role in transcription regulation will be discussed.

3. Transcriptional regulation

Messenger RNA (mRNA) synthesis requires the binding of RNA polymerase II at the site of transcription initiation, followed by the initiation and elongation of transcription. The process requires a number of specific factors, the main function of which is to bring about proper assembly of the RNA polymerase II complex at the site of transcription initiation (16). Studies on the *in vitro* assembly of a transcription-competent RNA polymerase II complex have identified several general transcription factors (GTFs). The transcription initiation complex has to assemble at the core promoter element that may consist of either a TATA element located approximately 25–30 bp upstream of the transcription initiation site or an initiator element, a pyrimidine-rich sequence around the transcription initiation site.

In vitro studies have shown a stepwise assembly of the GTFs, initiated by TFIID binding to the TATA element. TFIID consists of a TATA-binding protein (TBP) and TBP-associating factors (TAF$_{II}$s). Subsequently, TFIIB and TFIIA binds, then the TFIIF/pol II complex, followed by TFIIE and TFIIH. This pre-initiation complex is able to separate the DNA strands at the transcription initiation site and to synthesize the first phosphodiester bond of the nascent mRNA transcript. This stepwise assembly was demonstrated in *in vitro* studies but more recently, it has been shown that *in vivo* RNA polymerase II is already part of a complex with several GTFs (17). This complex also contains additional factors, like the so-called 'SRBs'. These suppressers of RNA polymerase B were cloned because mutations in these genes could rescue a yeast strain from its cold-sensitive phenotype caused by the removal of 15 of the 26 heptapeptide repeats of the C terminus of RNA polymerase II (18). The SRB factors were found in a multisubunit complex and could be displaced from the complex by antibodies against the C terminal domain of RNA polymerase II (19). In mammalian cells RNA polymerase II is also present in a multiprotein complex which includes SRB proteins (20).

3.1 Regulation of transcription initiation

DNA in eukaryotes is organized in nucleosomes, consisting of a histone octamer with a double set of each of the four histone proteins H2A, H2B, H3, and H4, and 146 bp of DNA wrapped around the octamer in about 1.8 left-handed superhelical turns. The histone H1 protein binds to the DNA at the entry and exit of the nucleosome core and may be involved in the next level of organization of nucleosomal DNA. The packaging of DNA with histones and other proteins forms a barrier for the transcription of the DNA (21). The histones have to be displaced either permanently in the case of constitutively expressed genes, or in an induced manner in the case of termporarily expressed genes (22). The displacement of nucleosomes is influenced by certain *cis*-acting elements, which may lie up to a few hundred bp upstream of the transcription initiation site, and by 'enhancers', sequences found up to several thousands of bp up or downstream of the transcription initiation site. Both types of DNA sequences contain one or more elements that can bind transcription factors. Transcription factors generally have a DNA-binding domain and a transcription-activation domain, while some can also bind a ligand, such as steroid-hormone receptors. Transcription factors in general can stimulate transcription, probably by stabilizing the transcription initiation complex via interaction of the transcription-activation domain with the RNA polymerase complex. This binding may be modulated by post-translational modification of the transcription factor, in particularly phosphorylation, and may be either a direct binding, or an indirect binding via so-called co-activators; it also may involve binding to both GTF's like the $TAF_{II}s$ and to other RNA polymerase II-associated factors, like the SRBs (17, 23).

Replacement of nucleosomes may not always be possible, even when sufficient amounts of transcription factors are present. In such cases, transcription may require catalytic processes. An example of this is the SWI/SNF complex which is able to disrupt the chromatin structure in an ATP-dependent manner and, thereby, enables the binding of transcription factors. This SWI/SNF complex is a 2 MDa complex composed of 9–12 subunits (24). The SWI2/SNF2 subunit, which was originally identified in *S. cerevisiae* catalyses the DNA-dependent ATPase activity. The homologous protein in *Drosophila* is brm (brahma), while in humans two homologous proteins are present: BRG1 and hbrm (25). Other ATPase-dependent nucleosome disruption complexes have been found in *Drosophila*. These complexes, NURF, CHRAC, and ACF, have an ISWI subunit that has homology with SWI2/SNF2 (26, 27).

The chromatin structure clearly represses transcription and additional modifications may occur that resist transcription even further, as has been found for mating switch genes and telomere sequences in yeast and for homeotic genes in *Drosophila* and mammals (26, 28). This will not be discussed here, as there is no information whether viral transforming genes can affect these events.

One of the mechanisms that can affect the nucleosome structure is histone acetylation. Specific acetylation of the N termini of histones also plays a role in histone assembly (29). Acetylation results in a reduction of the positive charge and, thereby, a weaker interaction of the nucleosomes with DNA, and possibly causes easier shifts in nucleosome position. Already a decade ago, the correlation between histone acetylation and transcription activity was described (30), but direct proof was obtained only after cloning of enzymes with histone acetyltransferase (HAT) activity. In yeast, protein GCN5, which is part of a transcription activation complex with the co-activators ADA2 and ADA3, has HAT activity. Also, the human homologue hGCN5 was cloned and, in addition, P/CAF which has a strong homology with hGCN5 in its HAT domain (31). P/CAF can associate with the p300/CBP co-activators (see below). Furthermore, p300/CBP itself has HAT activity (32, 33), just as the transcription co-activators SRC1 (34) and ACTR (35) and the TFIID component $TAF_{II}250$ (36). Not only histones, but also transcription factors might be acetylated, as has been shown for p53 (37) and for TFIIE and TFIIF (38).

Regulation of transcription by HATs will also require deacetylation of histones. Three mammalian histone deacetylases have been cloned, HDAC1, HDAC2, and HDAC3 (29, 39). Tethering of a histone deacetylase to the DNA represses transcription. The binding of HDAC to transcriptional repressors requires multiprotein complexes which have been identified only partially. HDAC has been shown to be involved in the repression of transcription by Mad/Max in which case the homologue of the yeast sin3 is part of the complex, and also in the repression by ligand-free hormone receptors, in which case co-repressors (N-CoR and SMRT) and sin3 are involved. So far, no deacetylation reactions for acetylated transcription factors have been described.

4. Binding of E1A to the retinoblastoma and related proteins

The E1A protein binds to a large set of proteins (40), of which the first one identified was the retinoblastoma-tumour suppressor gene product pRb (41). Retinoblastoma is a childhood tumour of the retina, which arises when both alleles of the retinoblastoma gene located at chromosome 13p14 are inactivated. Studies on the mechanism of action of pRb and the effect of E1A on this protein have provided insight into how E1A can overcome cell cycle arrest. This induction of the cell cycle by E1A is a hallmark of cell transformation and is essential for the infection of quiescent cells by adenoviruses, since virus replication requires gene activities found only in the S-phase of the cell cycle. The elucidation of the role of pRb in transcription resulted from studies on the transcription factor E2F which is involved in the E1A-mediated activation of transcription of the adenovirus E2 region. E2F was found to be complexed to cellular proteins that prevent the participation of E2F in E2 transcription. The E1A protein was shown to disrupt these complexes and to release E2F in a form that could activate E2 transcription (42). Two other proteins that can bind to E1A were identified, p107 and p130 (43, 44), which were found to have homology to pRb and to be able to interact with E2F. The p107 and p130 proteins can also bind to cyclin A-cdk2 and cyclin E-cdk2, which therefore are also present in E1A and E2F complexes. The E1A protein interacts with these cellular proteins via the E1A domains CR2 and CR1. CR1 and CR2 can independently interact with pRb, the CR2 domain having the highest affinity (45). The CR2 domain contains the amino acid sequence LxCxE, which is also present in other viral proteins interacting with pRb: the human papillomavirus (HPV) E7 protein and the SV40 large tumour antigen. Two domains in pRb are required for binding to the viral proteins (Fig. 2A), and homologues of the pRb domains are also present in p107 and p130. These two domains, A and B, were suggested to form a binding pocket, from which the name pocket proteins was derived.

The crystal structure of the pocket domain, without the spacer that is normally present between the two domains, has been resolved (46). The two subdomains of the pocket efficiently interact with each other and the amino acids in the interaction domain are highly conserved between pRb, p107, and p130. A viral peptide containing the LxCxE motif binds to the B domain in the pocket, suggesting that the A domain is required for stabilization of the pocket or for additional interactions. The CR2 domain of E1A is sufficient for binding to pRb, while the CR1 domain is required for the disruption of the E2F-pRb complex (47, 48). Consistently, a CR1 mutant in E1A can form a multimeric complex with pRb and E2F. The pRb-interaction domain in E2F does not contain an LxCxE motif (49, 50), and binding of the E2F peptide to the pRb pocket is not impaired by saturating amounts of a CR2

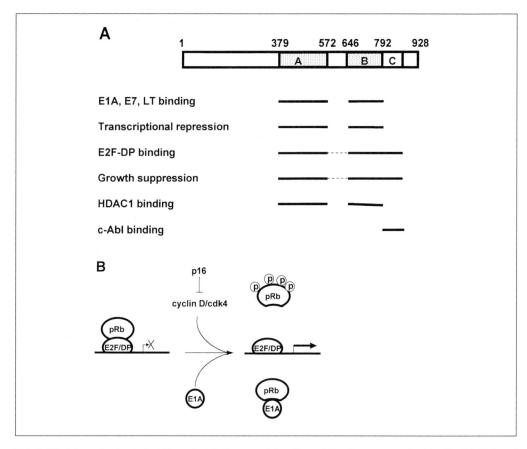

Fig. 2 (A) Schematic drawing of the retinoblastoma protein pRb and the domains required for the binding and effects shown. The A and B domain form the pocket region, together with the C domain it forms the large pocket. A dashed line indicates that the involvement of the spacer region between domain A and B has not been investigated. It should be noted that in some studies the size of the domains studied were slightly different from the ones indicated in the figure. (B) The release of repression of the transcription by E2F/DP transcription factors. In the upper half the normal cell cycle regulated phosphorylation of pRb is illustrated by its most essential kinase, the cyclin D/cdk4 kinase, which can inhibit by the cdk inhibitor p16^{INK4A}. In the lower half the effect of E1A is illustrated, showing the sequestration of pRb. In both cases E2F/DP is freed from its inhibitor and can activate transcription.

peptide (46), indicating that at least two interaction surfaces exist in the pocket domain.

Transfection of the pocket proteins into cells can lead to arrest in the G_0/G_1 phase and some cell types, like the osteosarcoma cell line SAOS2, show a flat phenotype, indicative for cell differentiation. This arrest can be overcome by phosphorylation of the pocket proteins through cyclin/cdk kinases. The pRb protein is increasingly phosphorylated during the G_1 phase and microinjection studies showed that pRb cannot arrest cells during the last part of the G_1 phase when pRb is efficiently phosphorylated (51), indicating a growth-suppressive function for the non- or under-phosphorylated pRb. The activation of cell cycling by viral tumour antigens is also achieved by association of the tumour antigens to particularly the under-phosphorylated pRb. During the cell cycle, the other pocket proteins also become increasingly phosphorylated. Although all three pocket proteins seem to function in a similar manner, in human tumours mutations have been found only in pRb but not in p107 or p130.

In addition, only disruption of the pRb gene in mice leads to a tumour-prone phenotype. pRb+/− mice do not develop retinoblastomas, but pituitary tumours in which the remaining wild-type allele of the retinoblastoma gene is lost (52, 53). In mice, retinoblastomas do develop when pRb is specifically inactivated by expression of an SV40-LT transgene in the retina (54). Knock-out pRb−/− mice are non-viable and die at day 13.5 after gestation with defects in nervous tissue and in haemato-poiesis (52, 55, 56). Both tissues seem to be deficient in terminal differentiation. Knock-outs for p107 or p130 do not have a phenotype, while the double knock-outs die at birth with severe bone malformations (57, 58), suggesting that p107 and p130 may be functionally redundant. Some functions of pRb can also be compensated by p107, as pRb−/−, p107−/− mice die at day 11.5 of gestation instead of at day 13.5, and pRb+/−, p107−/− mice have reduced body weight and viability, in contrast to pRb+/− or p107−/− mice (58).

It is not well established whether transformation by Ad-E1A requires the binding of both pRb and p107/p130. The E1A gene with point mutations in CR2 disrupting binding of the E1A protein to pRb but not to p107/p130 is, when transfected together with E1B, defective in transformation of baby rat kidney cells while they can still activate E2F-dependent transcription (see below) (59). On the other hand, an E1A deletion mutant in the C terminal part of CR2, which still can bind pRb but has a reduced affinity for p107/p130 is defective, wen transfected together with ras, in transformation of baby rat kidney cells (60). Further experiments under identical conditions should confirm whether, indeed, all pocket proteins have to be inactivated by E1A in order to achieve transformation.

4.1 Effect of pocket proteins on E2F

The effect of E1A on the function of the pocket proteins has been focused mainly on the effect of pRb in relation to the E2F transcription factors (61–67). The first E2F transcription factor cloned, E2F1, was found to be capable of activating transcription, and to be repressed by overexpression of pRb (49, 68). Only non- or under-phosphorylated pRb binds to E2F1. During cell cycle progression through the G_1 phase, pRb becomes phosphorylated by cyclin D-cdk4 and cyclin E-cdk2. Essential for phosphorylation is the cyclin D-cdk4 activity. The cdk-inhibitor p16^{INK4A}, which forms an inactive p16^{INK4A}-cdk4 complex, can prevent phosphorylation of pRb (Fig. 2B). Hence, both phosphorylation of pRb and sequestering of pRb by viral oncoproteins will lead to free E2F, which can then stimulate the transcription of certain genes required in S-phase, like dihydrofolate reductase, thymidine kinase, DNA polymerase-α and components of the Origin Recognition Complex, and also a number of genes involved in cell cycle control, like cyclin E, cyclin A, c-Myc, and N-

Myc (67). E2F1 alone can induce S-phase (69), suggesting that the induction of E2F1 target genes is essential for cell cycle progression, although uncontrolled E2F1 overexpression will lead to apoptosis (70).

The actual process is much more complicated, since six members of the E2F family have been cloned. In order to bind to DNA, E2F has to form a heterodimer with a so-called DP factor. Two genes of the DP family have been cloned, DP1 and DP2 (also called DP3). E2F6 (71), also called EMA (72), does not have the transcription-activation domain nor a pRb-binding domain and its role in transcription regulation by E2Fs still has to be determined.

The pocket proteins bind to the C terminal part of the E2F factors and have different affinities for the various E2F factors: pRb binds preferentially to E2F1, -2, and -3, p107 to E2F4, and p130 to both E2F4 and -5. In all cases, binding of the pocket protein leads to inhibition of the *trans*-activation function of E2F and to stabilization of E2F proteins. E2F is also stabilized in the presence of E1A, although E1A releases E2F from binding to the pocket proteins (73). In the S-phase of the cell cycle E2F1, -2 and -3 are inactivated by the interaction of cyclin A/cdk2 with a domain in the N terminal part of these E2Fs. The bound cyclin A/cdk2 is able to phosphorylate DP-1, with ensuing decreased DNA binding (74). When E2F4 and -5 are transiently expressed in cells, they are localized in the cytoplasm which explains the lack of S-phase induction by these E2Fs. When E2F4 is co-expressed with DP-2 or p107 or p130, the location of E2F4 in nuclear. The location of the endogenous E2F4 is found to be regulated during the cell cycle, being nuclear in G_0 and early G_1 and mainly cytoplasmic in S-phase (75, 76).

The binding of pRb to E2F1 not only shields the transcription-activation domain of E2F1 but also represses the activity of the promoter via other transcription factor-binding sites (77). This indicates that pRb is a real repressor of transcription and not only blocks the transcription activation function of E2F. This is consistent with the observation that E2F1 knock-out mice, although viable and showing initially normal development (78, 79), at later stages develop various tumours (79). This suggests that the repressive effect via E2F1 is essential for the normal development which implies that E2F1 could also act as a tumour-suppressor gene. Knock-out E2F5 mice show that E2F5 is not involved in proliferation. Cells grow normally *in vitro* and mice develop normally until birth, but after birth they develop hydrocephali due to increased production of cerebrospinal fluid by the choroid plexus which, in wild-type mice, is one of the tissues that expresses E2F5 (80).

The various pocket proteins might be involved in the regulation of different E2F target genes. Indeed, it has been found that, in knock-out cells, gene expression is differentially affected (81), i.e. in cells lacking either p107 or p130 no changes were found in the expression of E2F target genes, while in cells lacking both p107 and p130 and in cells lacking pRb, some of the E2F-dependent genes are derepressed. However, the genes deregulated in these two cases were completely different, showing that pRb and p107/p130 provide different functions in E2F regulation.

The capacity of pRb to repress promoter activity and not only the activity of the transcription factor to which pRb binds requires additional effects apart from shielding the transcription-activation domain of the transcription factor to which pRb is bound. One possibility is that pRb, once bound to DNA via E2F, can interact with other transcription factors and thereby blocks the transcription-activating capacity of these factors (82). Another possibility might be that it causes alteration of the chromatin structure in such a way that transcription is hampered. Recently, evidence in favour of such a mechanism has been published (83–85). It was shown that pRb is present in a complex with histone deacetylase(s) and, via pRb histone deacetylase activity, can be found in E2F complexes.

Binding to the histone deacetylase HDAC1 was found to require a LxCxE0like sequence in HDAC1 and the A/B pocket in pRb. Viral peptides with the LxCxE sequence can impair the binding of HDAC1 to pRb. The binding of HDAC1 and pRb has never been identified in yeast two-hybrid screens and therefore might be indirect or requires additional factors. One candidate is the pRb-binding protein Rbp48, which also binds to HDAC. Both HDAC1 and pRb were found to repress transcription of E2F-dependent promoters, while combination of the two proteins results in a stronger effect. Furthermore, it was demonstrated that repression correlates with a decreased amount of acetylated histone H3 at the promoter site (85). However, the repression of a promoter by tethering pRb to the DNA via a heterologous DNA-binding domain, is only occasionally relieved by TSA, a specific inhibitor of histone deacetylases (85). TSA-resistant repression is most likely to occur via other mechanisms. Therefore, either deacetylases are recruited, leading to a decreased acetylation of histones and thereby stabilization of the nucleosome structure, or pRb, when tethered to DNA, interacts with other transcription factors and prevent the transcription activity of these factors. The latter mechanism might explain the TSA-independent transcription repression of for example PUκ1 and NFκB-p65. Also, no association of HDAC1 and p107 could be demonstrated and the repression by p107 could not be relieved by TSA, indicating that p107 uses the mechanism of shielding only, to block the transcription activity (85).

The fact that, in contrast to pRb, p107 does not associate the HDAC1 illustrates that the different pocket proteins function in a different fashion when they inhibit the cell cycle. Another well-established difference between p107/p130 and pRb is the spacer region between the two pocket domains in p107 and p130 which is much lager than that of pRb. This spacer domain in p107 and p130 binds cyclin A/cdk2 or cyclin E/cdk2 and thus leads to the presence of cyclin A, cyclin E, and cdk2 in an E1A immunoprecipitate. The binding of cyclin A or cyclin E complexes to the spacer of p107 has been shown to give a growth-suppressor effect independent of the effect via E2F (86). The spacer of p107 inactivates the cyclin A/cdk2 and cyclin E/cdk2 kinase activity in the same way as the cdk inhibitor p21^{CIP1} (87).

4.2   Binding of pRb to other cellular proteins

Apart from the factors mentioned above, the retinoblastoma protein binds and affects other transcription factors and interacts with proteins known to be involved in transcription (65). There is at least one other transcription factor that seems to be affected in a way similar to E2F: Elf-1, a lymphoid-specific member of the Ets family. Only the hypophosphorylated form of pRb binds to Elf-1 and inhibits its transcription activity (88). Therefore, Elf-1 seems to be regulated like the E2F factors, although effects of E1A or other viral oncogenes have not been described. As mentioned above, pRb when tethered to the DNA can inactivate some additional transcription factors like PU-1 and cMyc, presumably by direct interaction, and this effect seems to be specific since, for example, VP16 and Sp1 are not inhibited (82).

Other transcriptional factors, such as ATF2 and cJun, are activated by pRb. A recent study on the function of various pRb mutants has made it clear that pRb exerts these different functions via different (sub)domains of its pocket (89). The pRb mutants that are able to induce a G_1 arrest, are always able to bind to E2F and to repress transcription when these mutants were tethered to the DNA via a heterologous DNA-binding domain. The capacity of the pRb mutants to induce a G_1 arrest does not correlate with the induction of a flat cell phenotype in SAOS2 cells. Induction of the flat cell phenotype did correlate, however, with the pRb-dependent induction of the transcription of a MyoD-dependent promoter in the presence of MyoD. The flat cells show gene expression indicative for bone differentiation. Therefore, pRb seems to have two effects: a transcription-repressive effect via the E2F family of transcription factors (and also via other factors like Elf-1) and a stimulative effect on the transcription of factors like MyoD, glucocorticoid receptor α, NF-IL6, and C/EBP. It is tempting to speculate that this induction of transcription is required for the differentiation of the cells.

As mentioned, cJun is a transcription factor activated by pRb. An interaction of cJun and pRb has been demonstrated in keratinocytes in the G_1 phase of the cell cycle and upon differentiation (90). This interaction takes place between the zipper domain of cJun and the B domain of the pocket or pRb. The cJun-activated collagenase promoter is further activated by pRb, whereas the activity of a cJun- plus pRb-activated minimal c-*jun* promoter could be inhibited by the HPV16 E7 protein. The interaction of pRb with the zipper domain of cJun indicates that pRb might stimulate transcription by enhancing the DNA-binding capacity of cJun; indeed, it was demonstrated that pRb could stimulate *in vitro* binding of cJun to an AP1-binding site. This transcription-stimulatory effect of pRb on cJun-dependent promoters may be restricted to some cell types or to the G_1 stage of the cell cycle, since in other studies it has been demonstrated that the repression of cJun-activated collagenase promoter by E1A does not depend on the binding of E1A to pRb (91). The activation of transcription by pRb is not always dependent on increased DNA binding: the activation of Sp1 by pRb also occurs on a Gal4-fusion protein with the *trans*-activation domain of Sp1 (92).

Interactions of pRb with other factors involved in transcription have also been described, for example with Mdm2, BRG1, hbrm, and TAF$_{II}$ 250. Mdm2 is involved in transcription regulation of p53 and also of E2F (93–95). BRG1 and hbrm are the mammalian counterparts of SWI2/SNF2 and therefore part of the SWI/SNF complex involved in transcription activation via chromatin remodelling (see above). pRb and BRG1 co-operate in the induction of a flat cell phenotype (96) and are also both required to stimulate the transcription activity of the glucocorticoid receptor (97) and to repress the activation functions of E2F1 (98). The interaction is restricted to the hypophosphorylated form of pRb (96) and can also occur with the other pocket proteins (99). Interaction with $TAF_{II}250$ would be more expected for transcription activators then for transcription suppressors. Nevertheless pRb is found to interact with $TAF_{II}250$ via the N terminus and the large pocket of pRb (Fig. 2A), although the domains required for interaction are not identical to those for binding to the viral proteins (100). In a cell line with temperature-sensitive $TAF_{II}250$, the pRb-dependent transcription activation of Sp1 is impaired at the non-permissive temperature, while it can be restored by transfection of wild-type $TAF_{II}250$, indicating that this interaction contributes to transcription activation by pRb (101).

pRb does not only regulate transcription by RNA polymerase II, but also transcription by polymerases I and III (102). The repression of the polymerase I transcription is attributed to the binding of pRb and UBF, an essential factor for transcription by RNA polymerase I. Transcription by polymerase III might be impaired by an interaction of pRb and TFIIIB.

The increasing number of proteins found to be associated with pRb suggests that pRb is part of a large multisubunit complex. Another observation that is consistent with this view is the binding to the nuclear tyrosine kinase oncoprotein c-Abl (103–106). This binding involves the C pocket in pRb and can occur simultaneously with the binding to E2F, and is not disrupted by viral oncoproteins. c-Abl binds only to the hypophosphorylated form of pRb; phosphorylation of pRb releases c-Abl and allows it to bind to DNA.

The LxCxE motif in the viral proteins is essential for interaction with pRb, and it might be expected that there are cellular counterparts with a similar motif. Of the proteins discussed so far, HDAC and BRG1/hbrm have a similar motif. Other cellular proteins that bind to pRb and have an LxCxE motif are the D-type cyclins, indicating that these cyclins not only affect pRb via the stimulation of phosphorylation but also by complex formation (107, 108). Other proteins were cloned that bind to pRb, like HBP1, a HMG box transcriptional repressor (109) and two proteins of approximately 200 kDa, RBP1 and RBP2 (110, 111). In these cases, the homology is restricted to the LxCxE motif, but another gene has been cloned that also has homology to E1A outside the pentapeptide motif. This RIZ gene is a 250 kDa nuclear protein that can bind to pRb but not to p107 or 130 (112, 113). The interaction of RIZ and pRb is disrupted by E1A. Further studies are required to elucidate the role of these pRb-interacting proteins.

In conclusion, it is clear that pRb plays various roles in gene regulation, where its function in various multisubunit complexes might depend on cell type and the phase of the cell cycle. It is not known yet whether viral proteins like E1A hamper the effect of pRb in all these different processes. Furthermore, it is clear that the other pocket proteins do not perform the same roles as pRb. Prevention of association of the pocket proteins to the E2F transcription factors by viral oncogenes is an essential step in transformation and viral infection, but it remains unclear whether the viral proteins affect other functions of the pocket proteins.

5. Binding of E1A to p300 and related proteins

Another domain of E1A that is essential for transformation, S-phase induction, repression of enhancers, and prevention of cell differentiation consists of the N terminal sequence together with the C terminal part of CR1. It is required for the binding of a protein of 300 kDa, called p300. After the p300 cDNA was cloned (114), it became apparent that it was a homologue of CBP, a protein binding specifically to the phosphorylated CREB transcription factor and thus essential for the transcription activation by CREB (115). Subsequently, it was found that CBP also binds to E1A and that E1A represses the transcription co-activator function of both p300 and CBP (116). The amino acid sequences of p300 and CBP contain a number of cysteine/ histidine-rich regions that are depicted as hallmarks in Fig. 3. Furthermore, both proteins contain a bromodomain, an amino acid sequence with unknown function that is also found in other gene products involved in transcription regulation like BRG1 and TAF$_{II}$250. In addition, p300 and CBP have histone acetyltransferase activity (see below).

The initial data on the function of CBP as a co-activator for CREB showed that one domain of CBP binds to phosphorylated CREB while another domain of CBP binds to the general transcription factor TFIIB (117), suggesting an adaptor or bridging-like

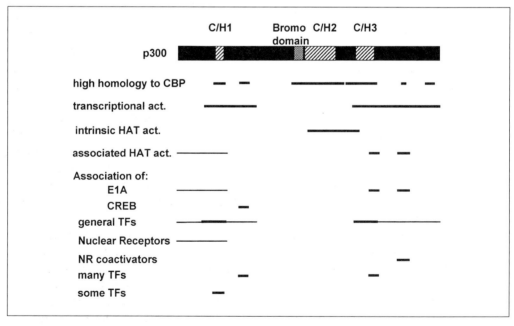

Fig. 3 Schematic drawing of p300 (2414 amino acids), with the three cysteine/histidine-rich regions C/H1, C/H2, and C/H3 and the bromo domain. As an indication for the most conserved regions, the domains in p300 with a contiguous stretch of at least 40 amino acids with over 80% identity with CBP are indicated. The domains involved in the indicated activities and in the association to proteins are summarized. When the characterization has not yet been very detailed a thin line is given, for further explanation see text.

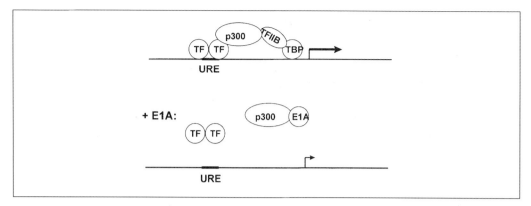

Fig. 4 Schematic view of the function of p300 as an adaptor between upstream transcription factors binding to the upstream regulatory element (URE) and general transcription factors (depicted are TFIIB and TBP, the TATA-binding protein). Disruption of interaction by E1A might lead to an impairment of the formation of a stable complex.

mechanism (Fig. 4). The repressive effect of E1A could be explained by disruption of essential interactions which would prevent the formation of a transcription initiation complex. In addition to the interaction of p300/CBP with TFIIB, interaction of p300/CBP with TBP and RNA helicase A, that is associated with RNA polymerase II, have also been described (118–120). The binding to these factors (referred to as general transcription factors in Fig. 3) correlate well with the transcription activation capacity of various p300/CBP domains when fused to a heterologous DNA-binding domain (118, 119, 121) (Fig. 3). The identification of p300/CBP as a transcription co-activator has led to a search for transcription factors that can bind to p300/CBP and are stimulated in transcriptional activity. Many factors were found to bind to p300/CBP (122). A simplified summary is given in Fig. 3, illustrating the domains of p300/CBP that are involved in these interactions. Most of these protein–protein interactions were established using *in vitro* binding while the effects on transcription were shown by transient transfection assays. Therefore, the relevance of the interactions between p300 and many of these transcription factors at physiological protein concentrations and within the cellular environment remains to be established.

The interaction of p300/CBP with many different transcription factors has led to a model in which activation of transcription requiring p300/CBP sequesters most of the p300/CBP available, and thus impairs the activity of other factors that also de-pend on a co-operation with p300/CBP (123). Further data are required to substanti-ate this model and to determine its contribution to gene regulation. Another role of p300/CBP might be its simultaneous interaction with the transcription factors bind-ing to an enhanceosome. An enhanceosome is a transcription-stimulatory sequence that interacts with a number of transcription factors, the positions of which are important for the effectiveness of the enhanceosome. A well-studied example is the IFNβ enhanceosome that binds the transcription factors NFκB, IRF1, ATF2/cJun and the architectural protein HMG I(Y). The synergistic activation by these transcription factors depends on the precise position of the *trans*-activation domains and on the recruitment of CBP (124). The N terminal 771 amino acids of CBP can form a complex with the transcription factors bound to the enhanceosome. Therefore, in the case of the IFNβ enhanceosome, the N terminal part of CBP seems to be required for complexing with factors binding to the enhanceosome and the C terminal part of CBP for binding to the general transcription factors.

The E1A protein binds to p300 via a domain overlapping the C/H3 domain (114). Recent *in vitro* interaction studies showed that E1A can also bind to two other domains in p300: a domain in the C terminal part, which also interacts with co-activators for nuclear receptors and an N terminal domain (125). The interaction of E1A with the C terminal domain has been well characterized and has been found to compete for the binding to co-activators of nuclear receptors. The binding of E1A to the C terminal domain is weaker than that to the C/H3 domain and still occurs when the third amino acid of E1A is mutated, a mutation that severely impairs the binding to the C/H3 domain. This mutation of the third amino acid of E1A also impairs the transforming capacity of E1A (126), indicating that the binding of E1A to the C/H3 domain of p300 is the essential interaction for transformation.

p300 protein is a stable protein, with a half-life of about ten hours (127, 128), suggesting that its function is not regulated at the level of transcription. p300 is subject to post-translational modifications, it is increasingly phosphorylated when cells progress from the G_1 to the M-phase (129) and can be modified with ubiquitin (128). E1A binds to p300 irrespective of phosphorylation or modification with ubiquitin (128), but SV40 large T binds exclusively to the unphosphorylated but ubiquitinated form of p300. Further evidence that p300 is regulated by phosphorylation came from studies on the differentiation of F9 cells by retinoic acid and E1A which was accompanied by the induction of phosphorylation of p300 (130). Further studies are required to identify and mutate the phosphorylation sites in order to establish the contribution of phosphorylation to the function of p300.

5.1 Role of p300/CBP in transcription regulation

To establish whether a transcription factor is dependent on p300/CBP for its activity, the activation of this factor by p300/CBP and the repression by E1A mutants is often studied. In addition, the repression of transcription by E1A should be released when p300 is overexpressed, as has been described for the SV40 enhancer/promoter (114), MyoD-dependent transcription 9118), α- and β-myosin heavy chain promoters (131, p21^{CIP1} promoter (132), cJun-dependent collagenase promoter (133), and p53-dependent transcription (134). The activation of a transcription factor by p300/CBP does not require the complete p300/CBP molecule: when the C/H3 domain is deleted, p300/CBP is in general equally effective as a co-activator. This C/H3 domain binds not only E1A but also many transcription factors, histone acetyltransferase P/CAF (31), and kinase p90^{rsk1} (135). Only in a few cases, deletion of he C/H3 domain was established to lead to an impaired transcriptional co-activator function of p300/CBP, as for example for p53 (134). The minimal domain that was shown to release the repression of MyoD-dependent transcription by E1A consists of a p300 mutant lacking the E1A-binding domain and all C terminal sequences (118). The effectiveness of this p300 mutant probably can be explained from the observation that MyoD not only interacts with p300 via the E1A-binding domain but also via the N terminal domain of p300 (136). The interaction of transcription factors with various domains in p300 has been found more often, but so far little is known about the relative binding affinities and the significance of these interactions.

It is clear that p300 is a complex regulator of transcription and the model that p300 can activate transcription factors and that this activation can be impaired by E1A may be an oversimplification. For example, the repression by E1A of the cJun-activated collagenase promoter occurs via the DNA-binding domain of cJun (137). This repressive effect on the DNA-binding domain can be released by over-expression of p300. This is surprising since a transcriptional co-activator is expected to function via the *trans*-activation domain of a transcription factor. It has been also suggested that stimulation of the c-*jun* promoter during differentiation of F9 cells by E1A might involve p300 (130, 138). In this case, the stimulation of the c-*jun* promoter by E1A (139) is enhanced by p300, indicating that E1A and p300 co-operate instead of counteract each other's effect (M. C. A. Duynda, *et al.*, unpublished).

5.2 The intrinsic and associated acetyltransferase activity of p300/CBP

The original idea that the p300/CBP co-activator functions as an adaptor between upstream transcription factors and the basal transcription machinery, as depicted in Fig. 4, was adapted when it was found that p300/CBP binds to P/CAF (31). P/CAF was cloned in a search for proteins that bind to p300, in analogy to the interaction between GCN5 and ADA2 in yeast. GCN5 was already known to have histone acetyltransferase activity and thus to be involved in chromatin remodelling. P/CAF is similar to GCN5 in its histone acetyltransferase activity domain, while it has an N terminal extension that interacts with p300/CBP (140). P/CAF binds to the C/H3 domain in p300/CBP like E1A, and the two proteins compete for binding with p300/CBP (31, 141). In addition, P/CAF also interacts with the N terminal part of CBP (140).

P/CAF is not the only factor with HAT activity that associates with p300/CBP. The nuclear receptor co-activators SRC1 (34) and ACTR (35) which bind to the C terminal part of p300/CBP also have HAT activity. Furthermore, p300 and CBP also have an intrinsic histone-acetyltransferase activity (32, 33). Remarkably, there is hardly any sequence similarity to other histone acetyltransferase enzymes (142). This may be due to the fact that the acetyl donor, acetyl CoA, has diverse binding motifs (143) and p300/CBP may form its own particular type.

The HAT activity of p300/CBP is assumed to destabilize the nucleosome structure and thereby stimulate transcription. As is shown in Fig. 3, no stimulation of transcription was found for that domain of p300/CBP. However, nucleosome structure may depend on the promoter used and, indeed, the HAT activity of CBP can stimulate some promoters and has no effect on others (142). Studies so far have shown that MyoD-dependent transcription requires the HAT activity of P/CAF and not that of p300 (141), while CREB-dependent transcription requires the HAT activity of CBP (140). In addition, acetylation of non-histone proteins can occur. P300 acetylates p53 at a lysine residue in the C terminal domain, leading to a stabilization of the tetrameric structure of p53 and, thereby, to increased DNA binding (144). Furthermore, p300/CBP and P/CAF can acetylate the general transcription factors TFIIE and TFIIF (38). Future studies are therefore needed to clarify more precisely the requirement of the various HAT activities for the diverse promoters and, more importantly, to show the actual target for acetylation.

5.3 Role of p300/CBP in cell growth and differentiation

The differentiation of a number of cell types was found to be repressed by E1A, depending on its capacity to interact with p300. This suggests that p300 stimulates expression of genes required for differentiation. Indeed, p300 is involved in expression of $p21^{CIP1}$ (132), which is required for arresting cells from further cell divisions, and also in the expression of other genes involved in differentiation, e.g. MyoD in the case of muscle-specific differentiation (145–148). The induction of these genes by p300 can be repressed by E1A. Therefore, it is clear that, at least in cultured cells, p300/CBP is a co-activator for various transcription factors among which factors involved in cell differentiation.

Information on the role of p300/CBP in whole organisms comes from both patients and knock-out studies. In humans, Rubinstein–Taybi syndrome, characterized by skeletal abnormalities and mental retardation, is related to a haploid mutation of CBP (149). Mice lacking a single CBP gene show some abnormalities in the skeletal patterning and, therefore, are somewhat similar to the Rubinstein–Taybi syndrome (150). A homozygous deletion of CBP is lethal, as is a homozygous deletion of p300 (151). In both cases, lethality is around day 10 of gestation, with defects in neurulation and heart development. The double heterozygous p300+/−, CBP+/− mice are not viable either, indicating that sufficient total levels of p300 and CBP are required for development. The disruption of CBP in *Drosophila* shoes its importance in pattern formation. CBP is required as a cofactor for the activity of *cubitus interruptus (ci)*, a transcription factor required for the *hedgehog* signalling pathway that is an important determinant of pattern formation (152). Furthermore, decreased levels in *twist* cause a deficient differentiation of the mesoderm, due to the requirement of CBP for activating *dorsal* (the *Drosophila* homologue of NFκB) for the transcription of *twist* (153). Also in *Caenorhabditis elegans*, the disruption of CBP (via RNA) leads to the lack of all differentiated cell types apart from neuronal cells. Histone deacetylase genes also repress all somatic differentiation, suggesting that CBP activates transcription and differentiation in *C. elegans* by antagonizing a repressive effect of histone de-acetylase activity (154). The involvement of p300 in differentiation would indicate that p300 has a general repressive effect on cell growth. However, mouse embryo fibroblasts (MEFs) from p300−/− embryo's are impaired in their growth capacity, indicating that p300 may also contribute to cell proliferation (151).

5.4 Role of the individual p300/CBP members

Until recently, no clear differences had been discovered between the interactions and functions of p300 and CBP and in this review data concerning either protein are considered as a common property. It may be expected, however, that p300 and CBP have different roles and the identical results so far may be due to the fact that most studies were performed with overexpression and *in vitro* association. Indeed, it was shown recently that inactivation of p300 or CBP with ribozymes has different effects on the RA-induced differentiation of F9 cells and that p300, in particular, is important for the induction of transcription of the RARβ and p21^{CIP1} promoters, while CBP is needed for the activation of the p27^{KIP1} promoter (155). Studies with mouse embryo fibroblasts from p300 knock-out mice have also shown that the transcriptional activity of the retinoic acid receptor is impaired while the CREB-dependent transcription is not (151). Therefore, the p300 protein seems to be a critical cofactor for the retinoic acid receptor, while it is dispensable for activation by CREB, and it is tempting to speculate that CBP is critical for CREB-dependent transcription. Further studies with cells in which p300 and/or CBP have been deleted should reveal the relative contributions of these proteins to the activity of various transcription factors.

There are some indications that the family of p300/CBP co-activators has additional members. Monoclonal antibodies raised against p300 recognize specifically p300 or p300 together with CBP (156). Furthermore, the monoclonal antibodies fall into two categories, one group does not co-precipitate cellular proteins, but does co-precipitate E1A when it is present, and the other group shows co-precipitation of cellular proteins and cross-reacts with a 270 kDa protein. This p270 does not associate to E1A and seems to have hardly any homology with p300/CBP (157). It has been suggested to be a SWI1 homologue. Indeed, p270 has been found in a complex with BRG1, BAF170, BAF155, and hSNF5, which are all components of the SWI/SNF complex. This complex has ATPase activity (as expected in the presence of BRG1) and no HAT activity, indicating that p270 itself has no HAT activity. p300 associates with proteins normally present in a TBP complex and it was found that both p300 and CBP were present in a TBP complex apparently as alternates to TAF$_{II}$ 250 (156, 158). Such a complex does have HAT activity but no ATPase activity (157). Therefore, p270 does not seem to be a homologue of p300/CBP although some antibodies do cross-react.

Monoclonal antibodies have also been raised against p300 that cross-react with p400 (127). The p400 protein does bind to the E1A protein, via the N terminus from amino acid 1 to 50 (159). Although p400 may be a possible homologue of p300/CBP, binding of E1A to p400 is not required for transformation of cells: an E1A mutant with a deletion of amino acids 26–35 does not bind to p400 but is not impaired in its transformation capacity (60).

6. Other proteins associating with E1A

The binding of the E1A proteins to pRb and p300 seems to be most crucial for the capacity of E1A to transform cells and therefore this review has focused on these two proteins. These targets are involved in transcription regulation, at least in part via effects on chromatin structure and, therefore, binding of E1A will affect chromatin structure. Evidence that E1A affects chromatin structure comes also from studies in yeast, in which the N terminal 82 amino acids of E1A were shown to impair cell growth. E1A blocks SWI/SNF-dependent transcription activation, probably by inhibition of the ability of SWI/SNF to remodel chromatin (160).

E1A binds in its N terminal domain to several additional proteins and some of these interactions will be shortly mentioned in this section. E1A is able to activate the hsp70 promoter via the TATA element. The N terminal domain of the E1A protein can bind to Dr1, a factor that upon binding to the TATA-binding protein (TBP) represses transcription. E1A disrupts the complex of Dr1 and TBP and thereby releases the repression by Dr1 (161). On the other hand, the N terminal domain of E1A can repress transcription by interaction of the N terminal domain with TBP. E1A can dissociate TBP from the TATA element and TFIIB can prevent this dissociation (162).

The CR3 region of E1A is required for the activation of other viral promoters and can also activate a number of cellular promoters. This region is able to bind to several factors among which TBP, cJun, ATF2, and Sp1 (163, 164). The precise contribution of these interactions and of the interaction with TAF_{II} 110 (165) for the stimulation of transcription has to be established further.

The C terminal domain of E1A contains the nuclear localization domain, but it also important for immortalization of primary epithelial cells, of which the mechanism is not yet established (166). Transformation of cells with E1A together with ras is impaired by sequences in the C terminal part of E1A. These sequences in the 12S E1A protein, amino acid 225–238, were found to bind to a 48 kDa protein, called CtBP, C terminal binding protein (167) and this protein might have effects in transcription regulation (168).

In conclusion, it has become clear that the E1A proteins can cause drastic alterations in cells. E1A mediates these changes by binding to several cellular proteins which, in most cases, results in alterations in transcriptional control of a large variety of cellular genes.

Acknowledgements

The authors thank Erik Kalkhoven, Petra Keblusek, and Hans van Ormondt for critically reading this manuscript. The work of the authors is supported by grants from the Dutch Cancer Society and The Netherlands Organisation for Scientific Research.

References

1. Houweling, A., van den Elsen, P. J., and van der Eb, A. J. (1980) Partial transformation of primary rat cells by the leftmost 4.5% fragment of adenovirus 5 DNA. *Virology*, **105**, 537.
2. White, E., Sabbatini, P., Debbas, M., Wold, W. S. M., Kusher, D. I., and Gooding, L. R. (1992) The 19-kilodalton adenovirus E1B transforming protein inhibits programmed cell death and prevents cytolysis by tumor necrosis factor-alpha. *Mol. Cell. Biol.*, **12**, 2570.
3. Gottlieb, T. M. and Oren, M. (1996) p53 in growth control and neoplasia. *Biochem. Biophys. Acta*, **1287**, 77.
4. Debbas, M. and White, E. (1993) Wild-type p53 mediates apoptosis by E1A, which is inhibited by E1B. *Genes Dev.*, **7**, 546.
5. Han, J. H., Sabbatini, P., Perez, D., Rao, L., Modha, D., and White, E. (1996) The E1B 19K protein blocks apoptosis by interacting with and inhibiting the p53-inducible and death-promoting Bax protein. *Genes Dev.*, **10**, 461.
6. Farrow, S. N., White, J. H. M., Martinou, I., Raven, T., Pun, K. T., Crinham, C. J., et al. (1995) Cloning of a bcl-2 homologue by interaction with adenovirus E1B 19K. *Nature*, **374**, 731.
7. Boyd, J. M., Gallo, G. J., Elangovan, B., Houghton, A. B., Malstrom, S., Avery, B. J., et al. (1995) Bik, a novel death-inducing protein shares a distinct sequence motif with Bcl-2 family proteins and interacts with viral and cellular survival-promoting proteins. *Oncogene*, **11**, 1921.
8. Han, J., Sabbatini, P., and White, E. (1996) Induction of apoptosis by human Nbk/Bik, a BH3-containing protein that interacts with E1B 19K. *Mol. Cell. Biol.*, **16**, 5857.
9. Svensson, C., Pettersson, U., and Akusjarvi, G. (1983) Splicing of adenovirus 2 early region 1A mRNAs is non-sequential. *J. Mol. Biol.*, **165**, 475.
10. Stephens, C. and Harlow, E. (1987) Differential splicing yields novel adenovirus 5 E1A mRNAs that encode 30 kd and 35 kd proteins. *EMBO J.*, **6**, 2027.
11. Van Ormondt, H., Maat, J., and Dijkema, R. (1980) Comparison of nucleotide sequences of the early E1a regions for subgroups A, B and C of human adenoviruses. *Gene*, **12**, 63.
12. Kimelman, D., Miller, J. S., Porter, D., and Roberts, B. E. (1985) E1a regions of the human adenoviruses and of the highly oncogenic simian adenovirus 7 are closely related. *J. Virol.*, **53**, 399.
13. Flint, J. and Shenk, T. (1989) Adenovirus E1A protein paradigm viral transactivator. *Annu. Rev. Genet.*, **23**, 141.
14. Berk, A. J. (1986) Adenovirus promoters and E1A transactivation. *Annu. Rev. Genet.*, **2-**, 45.
15. Peeper, D. S. and Zantema, A. (1993) Adenovius-E1A proteins transform cells by sequestering regulatory proteins. *Mol. Biol. Rep.*, **17**, 197.
16. Orphanides, G., Lagrange, R., and Reinberg, D. (1996) The general transcription factors of RNA polymerase II. *Genes Dev.*, **10**, 2657.
17. Koleske, A. J. and Young, R. A. (1995) The RNA polymerase II holoenzyme and its implications for gene regulation. *Trends Biochem., Sci.*, **20**, 113.
18. Thompson, C. M., Koleske, A. J., Chao, D. M., and Young, R. A. (1993) A multisubunit complex associated with the RNA polymerase II CTD and TATA-binding protein in yeast. *Cell*, **73**, 1361.
19. Kim, Y. J., Bjorklund, S., Li, Y., Sayre, M. H., and Kornberg, R. D. (1994) A multiprotein mediator of transcriptional activation and its interaction with the C-terminal repeat domain of RNA polymerase II. *Cell*, **77**, 599.

20. Maldonado, E., Shiekhattar, R., Sheldon, M., Cho, H., Drapkin, R., Rickert, P., *et al.* (1996) A human RNA polymerase II complex associated with STB and DNA-repair proteins. *nature*, **381**, 86.

21. Felsenfeld, G. (1992) Chromatin as an essential part of the transcriptional mechanism. *Nature*, **355**, 219.

22. Workman, J. L. and Buchman, A. R. (1993) Multiple functions of nucleosomes and regulatory factors in transcription. *Trends Biochem. Sci.*, **18**, 90.

23. Struhl, K. (1996) Chromatin structure and RNA polymerase II connection: implications for transcription. *Cell*, **84**, 179.

24. Wang, W., Xue, Y., Zhou, S., Kuo, A., Cairns, B. R., and Crabtree, G. R. (1996) Diversity and specialization of mammalian SWI/SNF complexes. *Genes Dev.*, **10**, 2117.

25. Peterson, C. L. and Tamkuin, J. W. (1995) The SWI-SNF complex: a chromatin remodeling machine? *Trends Biochem. Sci.*, **20**, 143.

26. Felsenfeld, G. (1996) Chromatin unfolds. *Cell*, **86**, 13.

27. Kadonaga, J. T. (1998) Eukaryotic transcription: an interlaced network of transcription factors and chromatin-modifying machines. *Cell*, **92**, 307.

28. Pirrotta, V. (1998) Polycombing the genome: PcG, trxG, and chromatin silencing. *Cell*, **93**, 333.

29. Grunstein, M. (1997) Histone acetylation in chromatin structure and transcription. *nature*, **389**, 349.

30. Hebbes, T. R., Throne, A. W., and Crane-Robinson, C. (1988) A direct link between core histone acetylation and transcriptionally active chromatin. *EMBO J.*, **7**, 1395.

31. Yang, X. J., Ogryzko, V. V., Nishikawa, J., Howard, B. H., and Nakatani, Y. (1996) A p300/CBP-associated factor that competes with the adenoviral oncoprotein E1A. *Nature*, **382**, 319.

32. Ogryzko, V. V., Schiltz, R. L., Russanova, V., Howard, B. H., and Nakatani, Y. (1996) The transcriptional coactivators p300 and CBP are histone acetyltransferases. *Cell*, **87**, 953.

33. Bannister, A. J. and Kouzarides, T. (1996) The CBP co-activator is a histone acetyltransferase. *Nature*, **384**, 641.

34. Spencer, T. E., Jenster, G., Burcin, M. M., Allis, C. D., Zhou, J., Mizzen, C. A., *et al.* (1997) Steroid receptor coactivator-1 is a histone acetyltransferase. *nature*, **389**, 194.

35. Chen, H., Lin, R. J., Schiltz, R. L., Chakravarti, D., Nash, A., Nagy, L., *et al.* (1997) Nuclear receptor coactivator ACTR is a novel histone acetyltransferase and forms a multimeric activation complex with P/CAF and CBP/p300. *Cell*, **90**, 569.

36. Mizzen, C. A., Yang, X. J., Kokubo, T., Brownell, J. E., Bannister, A. J., Owen-Hughes, T., *et al.* (1996) The TAF(II)250 subunit of TFIID has histone acetyltransferase activity. *Cell*, **87**, 1261.

37. Avantaggiati, M. L., Ogryzko, V., Gardner, K., Giordano, A., Levine, A. S., and Kelly, K. (1997) Recruitment of p300/CBP in p53-dependent signal pathways. *Cell*, **89**, 1175.

38. Imhof, A., Yang, X. J., Ogryzko, V. V., Nakatani, Y., Wolffe, A. P., and Ge, H. (1997) Acetylation of general transcription factors by histone acetyltransferases. *Curr. Biol.*, **7**, 689.

39. Emiliani, S., Fischle, W., Van Lint, C., Al-Abed, Y., and Verdin, E. (1998) Characterization of a human RPD3 ortholog, HDAC3. *Proc. Natl. Acad. Sci. USA*, **95**, 2795.

40. Harlow, E., Whyte, P., Franza, B. R., Jr., and Schley, C. (1986) Association of adenovirus early-region 1A proteins with cellular polypeptides. *Mol. Cell. Biol.*, **6**, 1579.

41. Whyte, P., Buchkovich, K. J., Horowitz, J. M., Friend, S. H., Raybuck, M., Weinberg, R. A., *et al.* (1988) Association between an oncogene and an anti-oncogene: the adenovirus E1A proteins bind to the retinoblastoma gene product. *nature*, **334**, 124.

42. Nevins, J. R. (1992) E2f—a link between the rb tumor suppressor protein and viral oncoproteins. *Science*, **258**, 424.

43. Ewen, M. E., Zing, Y. G., Lawrence, J. B., and Livingston, D. M. (1991) Molecular cloning, chromosomal mapping, and expression of the cDNA for p107, a retinoblastoma gene product-related protein. *Cell*, **66**, 1155.

44. Li, Y., Graham, C., Lacy, S., Duncan, A. M. V., and Whyte, P. (1993) The adenovirus-E1A-associated 130-kD protein is encoded by a member of the retinoblastoma gene family and physically interacts with cyclin-A and cyclin-E. *Genes Dev.*, **7**, 23766.

45. Dyson, N., Guida, P., McCall, C., and Harlow, E. (1992) Acenovirus-E1A makes 2 distinct contacts with the retinoblastoma protein. *J. Virol.*, **66**, 4606.

46. Lee, J. O., Russo, A. A., and Pavletich, N. P. (1998) Structure of the retinoblastoma tumour-suppressor pocket domain bound to a peptide from HPV E7. *Nature*, **391**, 859.

47. Ikeda, M. A. and Nevins, J. R. (1993) Identification of distinct roles for separate E1A domains in disruption of E2F complexes. *Mol. Cell. Biol.*, **13**, 7029.

48. Gould, K. L. and Nurse, P. (1989) Tyrosine phosphorylation of the fission yeast cdc2+ protein kinase regulates entry into mitosis. *Nature*, **342**, 39.

49. Helin, K., Lees, J. A., Vidal, M., Dyson, N., Harlow, E., and Fattaey, A. (1992) A cDNA encoding a pRB-binding protein with properties of the transcription factor E2F. *Cell*, **70**, 337.

50. Shan, B., Durfee, T., and Lee, W. H. (1996) Disruption of RB/E2F-1 interaction by single point mutations in E2F-1 enhances S-phase entry and apoptosis. *Proc. Natl. Acad. Sci. USA*, **93**, 679.

51. Goodrich, D. W., Wang, N. P., Qian, Y. W., Lee, E. Y. H. P., and Lee, W. H. 91991) The retinoblastoma gene product regulates progression through the G1 phase of the cell cycle. *Cell*, **67**, 293.

52. Jacks, T., Fazeli, A., Schmit, E. M., Bronson, R. T., Goodell, M. A., and Weinberg, R. A. (1992) Effects of an Rb mutation in the mouse [see comments]. *Nature*, **359**, 295.

53. Harrison, D. J., Hooper, M. L., Armstrong, J. F., and Clarke, A. R. (1995) Effects of heterozygosity for the Rb-1t19neo allele in the mouse. *Oncogene*, **10**, 1615.

54. Windle, J. ., Albert, D. M., O'Brien, J. M., Marcus, D. M., Disteche, C. M., Bernards, R., et al. (1990) Retinoblastoma in transgenic mice. *Nature*, **343**, 665.

55. Lee, E. Y., Chang, C. Y., Hu, N., Wang, Y. C., Lai, C. C., Herrup, K., et al. (1992) Mice deficient for Rb are nonviable and show defects in neurogenesis and haematopoiesis. *Nature*, **259**, 288.

56. Clarke, A. R., Maandag, E. R., van Roon, M., van der Luft, N. M., van der Valk, M., Hooper, M. L., et al. (1992) Requirement for a functional Rb-1 gene in murine development. *Nature*, **359**, 328.

57. Cobrinik, D., Lee, M. H., Hannon, G., Mulligan, G., Bronson, R. T., Dyson, N., et al. (1996) Shared role of the pRB-related p130 and p107 proteins in limb development. *Genes Dev.*, **10**, 1633.

58. Lee, M. H., Williams, B. O., Mulligan, G., Mukai, S., Bronson, R. T., Dyson, N., et al. (1996) Targeted disruption of p107: functional overlap between p107 and Rb. *Genes Dev.*, **10**, 1621.

59. Corbeil, H. B. and Branton, P. E. (1994) Functional importance of complex formation between the retinoblastoma tumor suppressor family and adenovirus E1A proteins as determined by mutational analysis of E1A conserved region 2. *J. Virol.*, **68**, 6697.

60. Egan, C., Bayleu, S. T., and Branton, P. E. (1989) Binding of the Rb1 protein to E1A products is required for adenovirus transformation. *Oncogene*, **4**, 383.

61. Weinberg, R. A. (1995) The retinoblastoma protein and cell cycle control. *Cell*, **81**, 323.

62. Sellers, W. R. and Kaelin, W. G., Jr. (1997) Role of the retinoblastoma protein in the pathogenesis of human cancer. *J. Clin. Oncol.*, **15**, 3301.

63. Herwig, S. and Strauss, M. (1997) The retinoblastoma protein: A master regulator of cell cycle, differentiation and apoptosis. *Eur. J. Biochem.*, **246**, 581.

64. Beijersbergen, R. L. and Bernards, R. (1996) Cell cycle regulation by the retinoblastoma family of growth inhibitory proteins. *Biochem. Biophys. Acta*, **1287**, 103.

65. Sellers, W. R. and Kaelin, W. G. (1996) RB as a modulator of transcription. *Biochim. Biophys. Acta*, **1288**, M1.

66. Bernards, R. (1997) E2F: a nodal point in cell cycle regulation. *Biochim. Biophys. Acta*, **1333**, M33.

67. Helin, K. (1998) Regulation of cell proliferation by the E2F transcription factors. *Curr. Opin. Genet. Dev.*, **8**, 28.

68. Kaelin, W. G., Jr., Krek, W., Sellers, W. R., DeCaprio, J. A., Ajchenbaum, F., Fuchs, C. S., *et al.* (1992) Expression cloning of a cDNA encoding a retinoblastoma-binding protein with E2F-like properties. *Cell*, **70**, 351.

69. Johnson, D. G., Schwarz, J. K., Cress, W. D., and Nevins, J. R. (1993) Expression of transcription factor E2F1 induces quiescent cells to enter S-phase. *Nature*, **365**, 349.

70. Kowalik, T. F., Degregori, J., Schwarz, J. K., and Nevins, J. R. (1995) E2F1 overexpression in quiescent fibroblasts leads to induction of cellular DNA synthesis and apoptosis. *J. Virol.*, **69**, 2491.

71. Trimarchi, J. M., Fairchild, B., Verona, R., Moberg, K., Andon, N., and Lees, J. A. (1998) E2F-6, a member of the E2F family that can behave as a transcriptional repressor. *Proc. Natl. Acad. Sci. USA*, **95**, 2850.

72. Morkel, M., Wenkel, J., Bannister, A. J., Kouzarides, T., and Hagemeier, C. (1997) An E2F-like repressor of transcription. *Nature*, **390**, 567.

73. Hateboer, G., Kerkhoven, R. M., Shvarts, A., Bernards, R., and Beijersbergen, R. L. (1996) Degradation of E2F by the ubiquitin-proteasome pathway: Regulation by retinoblastoma family proteins and adenovirus transforming proteins. *Genes Dev.*, **10**, 2960.

74. Krek, W., Ewen, M. E., Shirodkar, S., Arany, Z., Kaelin, W. G., Jr., and Livingston, D. M. (1994) Negative regulation of the growth-promoting transcription factor E2F-1 by a stably bound cyclin A-dependent protein kinase. *Cell*, **78**, 161.

75. Muller, H., Moroni, M. C., Vigo, E., Petersen, B. O., Bartek, J., and Helin, K. (1997) Induction of S-phase entry by E2F transcription factors depends on their nuclear localization. *Mol. Cell. Biol.*, **17**, 5508.

76. Lindeman, G. J., Gaubatz, S., Livingston, D. M., and Ginsberg, D. (1997) The subcellular localization of E2F-4 is cell-cycle dependent. *Proc. Natl. Acad. Sci. USA*, **94**, 5095.

77. Weintraub, S. J., Prater, C. A., and Dean, D. C. (1992) Retinoblastoma protein switches the E2F site from positive to negative element. *Nature*, **358**, 259.

78. Field, S. J., Tsai, F. Y., Kuo, F., Zubiaga, A. M., Kaelin, W. G., Jr., Livingston, D. M., *et al.* (1996) E2F-1 functions in mice to promote apoptosis and suppress proliferation. *Cell*, **85**, 549.

79. Yamasaki, L., Jacks, T., Bronson, R., Goillot, E., Harlow, E., and Dyson, N. J. (1996) Tumor induction and tissue atrophy in mice lacking E2F-1. *Cell*, **85**, 537.

80. Lindeman, G. J., Dagnino, L., Gaubatz, S., Xu, Y., Bronson, R. T., Warren, H. B., *et al.* (1998) A specific, nonproliferative role for E2F-5 in choroid plexus function revealed by gene targeting. *Genes Dev.*, **12**, 1092.

81. Hurford, R. K., Jr., Cobrinik, D., Lee, M. H., and Dyson, N. (1997) pRB and p107/p130 are

required for the regulated expression of different sets of E2F responsive genes. *Genes Dev.*, **11**, 1447.

82. Weintraub, S. J., Chow, K. N., Luo, R. X., Zhang, S. H., He, S., and Dean, D. C. (1995) Mechanism of active transcriptional repression by the retinoblastoma protein. *Nature*, **375**, 812.

83. Magnaghi-Jaulin, L., Groisman, R., Naguibneva, I., Robin, P., Lorain, S., Le Villain, J. P., *et al.* (1998) Retinoblastoma protein represses transcription by recruiting a histone daecetylase. *nature*, **391**, 601.

84. Brehm, A., Miska, E. A., McCance, D. J., Reid, J. L., Bannister, A. J., and Kouzarides, T. (1998) Retinoblastoma protein recruits histone deacetylase to repress transcription. *Nature*, **391**, 597.

85. Luo, R. X., Postigo, A. A., and Dean, D. C. (1998) Rb interacts with histone deacetylast to repress transcription. *Cell*, **92**, 463.

86. Zhu, L., Enders, G., Lees, J. A., Beijersbergen, R. L., Bernards, R., and Harlow, E. (1995) The pRB-related protein p107 contains two growth suppression domains: independent interactions with E2F and cyclin/cdk complexes. *EMBO J.*, **14**, 1904.

87. Zhu, L., Harlow, E., and Dynlacht, B. D. (1995) p107 uses a p21CIP1-related domain to bind cyclin/cdk2 and regulate interactions with E2F/ *Genes Dev.*, **9**, 1740.

88. Wang, C. Y., Petryniak, B., Thompson, C. B., Kaelin, W. G., and Leiden, J. M. (1993) Regulation of the Ets-related transcription factor Elf-1 by binding to the retinoblastoma protein. *Science*, **260**, 1330.

89. Sellers, W. R., Novitch, B. G., Miyake, S., Heith, A., Otterson, G. A., Kaye, F. J., *et al.* (1998) Stable binding to E2F is not required for the retinoblastoma protein to activate transcription, promote differentiation, and suppress tumor cell growth. *Genes Dev.*, **12**, 95.

90. Nead, M. A., Baglia, L. A., Antinore, M. J., Ludlow, J. W., and McCance, D. J. (1998) Rb binds c-Jun and activates transcription. *EMBO J.*, **17**, 2342.

91. Offringa, R., Gebel, S., van Dam, H., Timmers, M., Smits, A., Zwart, R., *et al.* (1990) A novel function of the transforming domain of E1a: repression of Ap-1 activity. *Cell*, **62**, 527.

92. Kim, S. J., Onwuta, U. S., Lee, Y. I., Li, R., Botchan, M. R., and Robbins, P. D. (1992) The retinoblastoma gene product regulates Sp1-mediated transcription. *Mol. Cell. Biol.*, **12**, 2455.

93. Piette, J., Neel, H., and Marechal, V. (1997) Mdm2: keeping p53 under control. *Oncogene*, **15**, 1001.

94. Xiao, Z. X., Chen, J., Levine, A. J., Modjtahedi, N., Xing, J., Sellers, W. R., *et al.* (1995) Interaction between the retinoblastoma protein and the oncoprotein MDM2. *Nature*, **375**, 694.

95. Martin, K., Trouche, D., Hagemeier, C., Sorensen, T. S., La Thangue, N. B., and Kouzarides, T. (1995) Stimulation of E2F1/DP1 transcriptional activity by MDM2 oncoprotein. *Nature*, **375**, 691.

96. Dunaief, J. L., Strober, B. E., Guha, S., Khavari, P. A., Alin, K., Luban, J., *et al.* (1994) The retinoblastoma protein and BRG1 form a complex and cooperate to induce cell cycle arrest. *Cell*, **79**, 119.

97. Singh, P., Coe, J., and Hong, W. J. (1995) A role for retinoblastoma protein in potentiating transcriptional activation by the glucocorticoid receptor. *Nature*, **374**, 562.

98. Trouche, D., Le Chalony, C., Muchardt, C., Yaniv, M., and Kouzarides, T. (1997) RB and hbrm cooperate to repress the activation functions of E2F1. *Proc. Natl. Acad. Sci. USA*, **94**, 11268.

99. Strober, B. E., Dunaief, J. L., Guha, S., and Goff, S. P. (1996) Functional interactions

between the hBRM/hBRG1 transcriptional activators and the pRB family of proteins. *Mol. Cell. Biol.*, **16**, 1576.

100. Shao, Z., Siegert, J. L., Ruppert, S., and Robbins, P. D. (1997) Rb interacts with TAF(II)250/TFIID through multiple domains. Oncogene, **15**, 385.

101. Shao, Z., Ruppert, S., and Robbins, P. D. (1995) The retinoblastoma-susceptibility gene product binds directly to the human TATA-binding protein-associated factor TAFII250. *proc. Natl. Acad. Sci. USA*, **92**, 3115.

102. White, R. J. (1997) Regulation of RNA polymerases I and III by the retinoblastoma protein: a mechanism for growth control? *Trends Biochem. Sci.*, **22**, 77.

103. Wen, S. T., Jackson, P. K., and van Etten, R. A. (1996) The cytostatic function of c-Ab1 is controlled by multiple nuclear localization signals and requires the p53 and Rb tumor suppressor gene products. *EMBO J.*, **15**, 1583.

104. Welch, P. J. and Wang, J. Y. (1995) Abrogation of retinoblastoma protein function by c-Ab1 through tyrosine kinase-dependent and -independent mechanisms. *Mol. Cell. Biol.*, **15**, 5542.

105. Welch, P. J. and Wang, J. Y. (1995) Disruption of retinoblastoma protein function by coexpression of its C pocket fragment. *Genes Dev.*, **9**, 31.

106. Welch, P. J. and Wang, J. Y. (1993) A C-terminal protein-binding domain in the retino-blastoma protein regulates nuclear c-Ab1 tyrosine kinase in the cell cycle. *Cell*, **75**, 779.

107. Ewen, M. E., Sluss, H. K., Sherr, C. J., Matsushiome, H., Kato, J. Y., and Livingston, D. M. (1993) Functional interactions of the retinoblastoma protein with mammalian D-type cyclins. *Cell*, **73**, 487.

108. Dowd, S. F., Hinds, P. W., Louie, K., Reed, S. I., Arnold, A., and Weinberg, R. A. (1993) Physical interaction of the retinoblastoma protein with human D-cyclins. *Cell*, **73**, 499.

109. Tevosian, S. G., Shih, H. H., Mendelson, K. G., Sheppard, K. A., Paulson, K. E., and Yee, A. S. (1997) HBp1: a HMG box transcriptional repressor that is targeted by the retinoblastoma family. *Genes Dev.*, **11**, 383.

110. Defeo Jones, D., Huang, P. S., Jones, R. E., Haskell, K. M., Vuocolo, G. A., Hanobik, M. G., *et al.* (1991) Cloning of cDNAs for cellular proteins that bind to the retinoblastoma gene product. *Nature*, **352**, 251.

111. Fattaey, A. R., Helin, K., Dembski, M. S., Dyson, N., Harlow, E., Vuocolo, G. A., *et al.* (1993) Characterization of the retinoblastoma binding proteins RBP1 and RBP2. *Oncogene*, **8**, 3149.

112. Buyse, I. M., Shao, G., and Huang, S. (1995) The retinoblastoma protein binds to RIZ, a zinc-finger protein that shares an epitope with the adenovirus E1A protein. *Proc. Natl. Acad. Sci. USA*, **92**, 4467.

113. Buyse, I. M. and Huang, S. (1997) *In vitro* analysis of the E1A-homologous sequences of RIZ. *J. Virol.*, **71**, 6200.

114. Eckner, R., Ewen, M. E., Newsome, D., Gerdes, M., DeCaprio, J. A., Lawrence, J. B., *et al.* (1994) Molecular cloning and functional analysis of the adenovirus E1A-associated 300-kD protein (P300) reveals a protein with properties of a transcriptional adaptor. *Grenes Dev.*, **8**, 869.

115. Chrivia, J. C., Kwok, R. P., Lamb, N., Hagiwara, M., Montminy, M. R., and Goodman, R. H. 91993) Phosphorylated CREB binds specifically to the nuclear protein CBP. *Nature*, **365**, 855.

116. Lundblad, J. R., Kwok, R. P. S., Laurance, M. E., Harter, M. L., and Goodman, R. H. (1995) Adenoviral E1A-associated protein p300 as a functional homologue of the transcriptional co-activator CBP. *Nature*, **374**, 85.

117. Kwok, R. P., Lundblad, J. R., Chrivia, J. C., Richards, J. P., Bachinger, H. P., Brennan, R. G., *et al.* (1994) Nuclear protein CBP is a coactivator for the transcription factor CREB. *Nature*, **370**, 223.

118. Yuan, W., Condorelli, G., Caruso, M., Felsani, A., and Giordano, A. (1996) Human p300 protein is a coactivator for the transcription factor MyoD. *J. Biol. Chem.*, **271**, 9009.

119. Swope, D. L., Mueller, C. L., and Chrivia, J. C. (1996) CREB-binding protein activates transcription through multiple domains. *J. Biol. Chem.*, **271**, 28138.

120. Nakajima, T., Uchida, C., Anderson, S. F., Lee, C. G., Hurwitz, J., Parvin, J. D., *et al.* (19970 RNA helicase A mediates association of CBP with RNA polymerase II. *Cell*, **90**, 1107.

121. Janknecht, R. and Nordheim, A. (1996) Regulation of the c-fos promoter by the ternary complex factor Sap-1a and its coactivator CBP. *Oncogene*, **12**, 1961.

122. Shikama, N., Lyon, J., and Lathangue, N. B. (1997) The p300/CBP family: Integrating signals with transcription factors and chromatin. *Trends Cell Biol.*, **7**, 230.

123. Janknecht, R. and Hunter, T. (1996) Transcription. A growing coactivator network. *Nature*, **383**, 22.

124. Merika, M., Williams, A. J., Chen, G., Collins, T., and Thanos, D. (1998) Recruitment of CBP/p300 by the IFNβ enhanceosome is required for synergistic activation of transcription. *Mol. Cell*, **1**, 277.

125. Kurokawa, R., Kalafus, D., Ogliastro, M. H., Kioussi, C., Xu, L., Torchia, J., *et al.* (1998) Differential use of CREB binding protein-coactivator complexes. *Science*, **279**, 700.

126. Wang, H. G. H., Rikitake, Y., Carter, M. C., Yaciuk, P., Abraham, S. E., Zerler, B., *et al.* (1993) Identification of specific adenovirus-E1A N-terminal residues critical to the binding of cellular proteins and to the control of cell growth. *J. Virol.*, **67**, 476.

127. Lill, N. L., Tevethia, M. J., Eckner, R., Livingston, D. M., and Modjtahedi, N. (1997) p300 family members associate with the carboxyl terminus of simian virus 40 large tumor antigen. *J. Virol.*, **71**, 129.

128. Avantaggiati, M. L., Carbone, M., Graessmann, A., Nakatani, Y., Howard, B., and Levine, A. S. (1996) The SV40 large T antigen and adenovirus E1a oncoproteins interact with distinct isoforms of the transcriptional co-activator, p300. *EMBO J.*, **15**, 2236.

129. Banerjee, A. C., Recupero, A. J., Mal, A., Piotrkowski, A. M., Wang, D. M., and Harter, M. L. (1994) The adenovirus E1A 289R and 243R proteins inhibit the phosphorylation of p300. *Oncogene*, **9**, 1733.

130. Kitabayashi, I., Eckner, R., Aray, Z., Chiu, R., Gachelin, G., Livingston, D. M., *et al.* (1995) Phosphorylation of the adenovirus E1A-associated 300 kDa protein in response to retinoic acid and E1A during the differentiation of F9 cells. *EMBO J.*, **14**, 3496.

131. Hasegawa, K., Meyers, M. B., and Kitsis, R. N. (1997) Transcriptional coactivator p300 stimulates cell type-specific gene expression in cardiac myocytes. *J. Biol. Chem.* **272**, 20049.

132. Missero, C., Calautti, E., Eckner, R., Chin, J., Tsai, L. H., Livingston, D. M., *et al.* (19950 Involvement of the cell-cycle inhibitor Cip1/WAF1 and the E1A-associated p300 protein in terminal differentiation. *Proc. Natl. Acad. Sci. USA*, **92**, 5451.

133. Smits, P. H. M., Dewit, L., Vandereb, A. J., and Zantema, A. (1996) The adenovirus E1A-associated 300 kDa adaptor protein counteracts the inhibition of the collagenase promoter by E1A and represses transformation. *Oncogene*, e**12**, 1529.

134. Lill, N. L., Grossman, S. R., Ginsberg, D., Decaprio, J., and Livingston, D. M. 91997) Binding and modulation of p53 by p300/CBP coactivators. *Nature*, **387**, 823.

135. Nakajima, T., Fukamizu, A., Takahashi, J., Gage, F. H., Fisher, T., Blenis, J., *et al.* (1996) The signal-dependent coactivator CBP is a nuclear target for pp90RSK. *Cell*, **86**, 465.

136. Sartorelli, V., Huang, J., Hamamori, Y., and Kedes, L. (1997) Molecular mechanisms of myogenic coactivation by p300: Direct interaction with the activation domain of MyoD and with the MADS box of MEF2C. *Mol. Cell. Biol.*, **17**, 1010.

137. Hagmeyer, B. M., Konig, H., Herr, I., Offringa, R., Zantema, A., Vendereb, A. J., *et al.* (1993) Adenovirus E1A negatively and positively modulates transcription of AP-1 dependent genes by dimer-specific regulation of the DNA binding and transactivation activities of Jun. *EMBO J.*, **12**, 3559.

138. Kawasaki, H., Song, J., Eckner, R>, Ugai, H., Chiu, R., Taira, K., *et al.* (1998) p300 and ATF-2 are components of the DRF complex, which regulates retinoic acid- and E1A-mediated transcription of the c-jun gene in F9 cells. *Genes Dev.*, **12**, 233.

139. Duyndam, M. C. A., Vandam, H., Vandereb, A. J., and Zantema, A. (1996) The CR1 and CR3 domains of the adenovirus type 5 E1A proteins can independently mediate activation of ATF-2. *J. Virol.*, **70**, 5852.

140. Korzus, E., Torchia, J., Rose, D. W., Xu, L., Kurokawa, R., McInerney, E. M., *et al.* (1998) Transcription factor-specific requirements for coactivators and their acetyltransferase functions. *Science*, **279**, 703.

141. Puri, P. L., Sartorelli, V., Yang, X. J., Hamamori, Y., Ogryzko, V. V., Howard, B. H., *et al.* (1998) Different roles of p300 and PCAF acetyltransferases in muscle differentiation. *Mol. Cell*, **1**, 35.

142. Martinez-Balbas, M. A., Bannister, A. J., Martin, K., Haus-Seuffert, P., Meisterernst, M., and Kouzarides, T. (1998) The acetyltransferase activity of CBP stimulates transcription. *EMBO J.*, **17**, 2886.

143. Engel, C. and Wierenga, R. (1996) The diverse world of coenzyme A binding proteins. *Curr. Opin. Struct. Biol.*, **6**, 790.

144. Gu, W. and Roeder, R. G. (1997) Activation of p53 sequence-specific DNA binding by acetylation of the p53 C-terminal domain. *Cell*, **90**, 595.

145. Mymryk, J. S., Lee, R. W. H., and Bayley, S. T. (1992) Ability of adenovirus 5 E1A proteins to suppress differentiation of BC3h1 myoblasts correlates with their binding to a 300-kDa cellular protein. *Mol. Biol. Cell*, **3**, 1107.

146. Heasley, L. E., Benedict, S., Gleavy, J., and Johnson, G. L. (1991) Requirement of the adenovirus E1A transformation domain 1 for inhibition of PC12 cell neuronal differentiation. *Cell Regul.*, **2**, 479.

147. Caruso, M., Martelli, F., Giordano, A., and Felsani, A. (1993) Regulation of MyoD gene transcription and protein function by the transforming domains of the adenovirus E1A oncoprotein. *Oncogene*, **8**, 267.

148. Shrivastava, A., Yu, J., Artandi, S., and Calame, K. (1996) YY1 and c-Myc associate *in vivo* in a manner that depends on c-Myc levels. *Proc. Natl. Acad. Sci. USA*, **93**, 10638.

149. Petrij, F., Giles, R. H., Dauwerse, H. G., Saris, J. J., Hennekam, R. C., Masuno, M., *et al.* (1995) Rubinstein–Taybi syndrome caused by mutations in the transcriptional co-activator CBP. *Nature*, **376**, 348.

150. Tanaka, Y., Naruse, I., Maekawa, T., Masuya, H., Shiroishi, T., and Ishii, S. (1997) Abnormal skeletal patterning in embryos lacking a single Cbp allele: a partial similarity with Rubinstein–Taybi syndrome. *Proc. Natl. Acad. Sci. USA*, **94**, 10215.

151. Yao, T. P., Oh, S. P., Fuchs, M., Zhou, N. D., Ch'ng, L. E., Newsome, D., *et al.* (1998) Gene dosage-dependent embryonic development and proliferation defects in mice lacking the transcriptional integrator p300. *Cell*, **93**, 361.

152. Spector, D. J., McGrogan, M., and Raskas, H. J. (1978) Regulation of the appearance of cytoplasmic RNAs from region 1 of the adenovirus 2 genome. *J. Mol. Biol.*, **126**, 395.

153. Akimaru, H., Hou, D. X., and Ishii, S. (1997) Drosophila CBP is required for dorsal-dependent twist gene expression. *Nature Genet.*, **17**, 211.

154. Shi, Y. and Mello, C. (1998) A CBP/p300 homolog specifies multiple differentiation pathways in Caenorhabditis elegans. *Genes Dev.*, **12**, 943.

155. Kawasaki, H., Eckner, R., Yao, T. P., Taira, K., Chiu, R., Livingston, D. M., *et al.* (1998) Distinct roles of the co-activators p300 and CBP in retinoic-acid-induced F9-cell differentiation. *Nature*, **393**, 284.

156. Dallas, P. B., Yaciuk, P., and Moran, E. (1997) Characterization of monoclonal antibodies raised against p300: Both p300 and CBP are present in intracellular TBP complexes. *J. Virol.*, **71**, 1726.

157. Dallas, P. B., Wayne Cheney, I., Liao, D., Bowrin, V., Byam, W., Pacchione, S., *et al.* (1998) p300/CREB binding protein-related protein p270 is a component of mammalian SWI/SNF complexes. *Mol. Cell. Biol.*, **18**, 3596.

158. Abraham, S. E., Lobo, S., Yaciuk, P., Wang, H. G. H., and Moran, E. (1993) p300, and p300-associated proteins, are components of TATA-binding protein (TBP) complexes. *Oncogene*, **8**, 1639.

159. Barbeau, D., Charbonneau, R., Whalen, S. G., Bayley, S. T., and Branton, P. E. (1994) Functional interactions within adenovirus E1A protein complexes. *Oncogene*, **9**, 359.

160. Miller, M. E., Cairns, B. R., Levinson, R. S., Yamamoto, K. R., Engel, D. A., and Smith, M. M. (1996) Adenovirus E1A specifically blocks SWI/SNF-dependent transcriptional activation. *Mol. Cell. Biol.*, **16**, 5737.

161. Kraus, V. B., Inostroza, J. A., Yeung, K., Reinberg, D., and Nevins, J. R. (1994) Interaction of the Dr1 inhibitory factor with the TATA binding protein is disrupted by adenovirus E1A. *Proc. Natl. Acad. Sci. USA*, **91**, 6279.

162. Song, C. Z., Loewenstein, P. M., Toth, K., Tang, Q. Q., Nishikawa, A., and Green, M. (1997) The adenovirus E1A repression domain disrupts the interaction between the TATA binding protein and the TATA box in a manner reversible by TFIIB. *Mol. Cell. Biol.*, **17**, 2186.

163. Lee, W. S., Kao, C. C., Bryant, G. O., Liu, X., and Berk, A. J. 91991) Adenovirus E1A activation domain binds the basic repeat in the TATA box transcription factor. *Cel*, **67**, 365.

164. Liu, F. and Green, M. R. (1994) Promoter targeting by adenovirus-E1a though interaction with different cellular DNA-binding domains. *Nature*, **368**, 520.

165. Mazzarelli, J. M., Atkins, G. B., Geisberg, J. V., and Ricciardi, R. P. (1995) The viral oncoproteins Ad5 E1A, HPV16E7 and SV40 TAg bind a common region of the TBP-associated factor-110. *Oncogene*, **11**, 1859.

166. Gopalakrishnan, S., Douglas, J. L., and Quinlan, M. P. (1997) Immortalization of primary epithelial cells by E1A 12S requires late, second exon-encoded functions in addition to complex formation with pRB and p300. *Cell Growth Differentiation*, **8**, 541.

167. Schaeper, U., Boyd, J. M., Verma, S., Uhlmann, E., Subramanian, T., and Chinadurai, G. (1995) Molecular cloning and characterization of a cellular phosphoprotein that interacts with a conserved C-terminal domain of adenovirus E1A involved in negative modulation of oncogenic transformation. *Proc. Natl. Acad. Sci. USA*, **92**, 10467.

168. Sollerbrant, K., Chinnadurai, G., and Svensson, C. (1996) The CtBP binding domain in the adenovirus E1A protein controls CR1-dependent transactivation. *Nucleic Acids Res*, **24**, 2578.

8 | Adenovirus proteins that regulate apoptosis

WILLIAM S. M. WOLD AND G. CHINNADURAI

1. Introduction

Viruses and the hosts they infect are involved in an evolutionary battle. The host has physical barriers that protect itself from viruses. Infected cells can commit suicide to preclude virus replication. Virus-infected cells are destroyed by the innate and adaptive arms of the immune system, and virions are eliminated by specific antibodies. Viruses, in turn, have mechanisms that allow the virus to infect cells and survive attack by the host. Virus proteins inhibit intrinsic cellular apoptosis, and block killing of infected cells by cytokines and cells of the immune system. These viral proteins are interesting tools to understand virus pathogenesis, cellular apoptosis, and the immune response. They also provide novel insights into many key regulatory aspects of molecular and cellular biology. In this article we describe human adenovirus (Ad) proteins that counteract host antiviral responses. Aspects of this topic have been reviewed elsewhere (1–5). There are 51 serotypes of human Ads, which fall into six subgroups, A–F. We will limit our discussion to the commonly studied serotypes 5 (Ad5) and Ad2, members of subgroup C.

2. Course of adenovirus infection

The Ad genome is a linear duplex DNA of 36 000 base pairs organized into six major transcription units (Fig. 1). All viral genes are transcribed by host proteins, but certain Ad proteins, e.g. from the E1A region, regulate the transcription machinery. When Ad infects cultured cells, the virion binds via the Ad-coded fibre protein to a specific cellular receptor named CAR, the penton base interacts with integrins such as $\alpha_V\beta_5$, then the virion enters the cell via endocytosis (6–9). The viral DNA–protein core exits the endosomes and is transported to the nucleus (10, 11). The first Ad proteins to be expressed are the 'immediate early' E1A proteins, which drive quiescent cells from G_0 into S-phase, and which induce transcription of the 'delayed early' genes in the E1B, E2, E3, and E4 transcription units (12).

There are about 23 early genes which play a role in synthesis of viral mRNAs, proteins, and DNA, in usurping the host cell, and in counteracting host defenses. These genes are grouped in transcription units roughly according to their functions

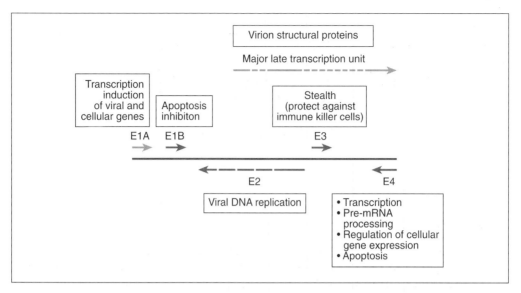

Fig. 1 Schematic of the Ad5 genome (simplified version). The long black bar represents the duplex DNA genome of 36 kb. The arrows indicate transcription units controlled by a specific promoter for each unit. With all transcription units, multiple mRNAs are generated by alternative splicing. The genes tend to be grouped according to general function, as indicated within the boxed areas.

(Fig. 1). For example, E1A proteins regulate transcription of viral and cellular genes, E1B proteins inhibit intrinsic cellular apoptosis, E2 proteins are involved in Ad DNA replication, and E3 proteins confer stealth properties that hide the infected cell from killer cells of the immune system. Viral DNA replication begins at 7–8 hours post-infection, then late genes are expressed, nearly all of which are located in the major late transcription unit. Most late genes (approximately 13) code for structural proteins of the virion. After about one day virions begin to assemble in the nucleus, then beginning at about two to three days the cell dies and releases virus particles.

Ad5 causes mild acute infections of the upper respiratory tract in young children, often asymptomatic, and may form persistent infections in lymphoid cells (13). Specific data for the course of Ad respiratory infections in humans are sparse. Most humans have been infected as infants as indicated by the presence of antibodies, and appear to develop life-long immunity to acute infections. Antibodies to virion structural proteins are found in most adults, as are peripheral blood lymphocytes that proliferate in response to Ad (14, 15). In a phase I study of human cancer (mesothelioma) patients infected intrapleurally with a replication-defective Ad vector expressing the herpes virus thymidine kinase as a therapeutic agent, strong antibody and cellular responses were observed to Ad proteins (15). In a phase I cystic fibrosis gene therapy trial, there was localized production of the cytokine interleukin-6 (IL-6) and the chemokine IL-8 (16). How these specific immune responses develop has not been studied, but we can surmise what probably happens from data generated in animal models, primarily the mouse (17–19). Many studies in the mouse used the type of replication-defective Ad vectors that may be used in human gene therapy. Replication-competent Ads were used in a cotton rat-lung model, where Ad replication occurs and is reported to resemble Ad-induced phenumonia in humans (20, 21). The full spectrum of immune response occurs in these animal models, including an early inflammatory response mediated by macrophages, natural killer (NK) cells, neutrophils, and cytokines, a T cell response including CD4$^+$ T helper cells and CD4$^+$ cytotoxic T lymphocytes (CTL), and a neutralizing antibody response (see ref. 15).

We can further surmise the response to Ad in humans as judged by a general understanding of the host response to viral infections (22). Initially, the cell responds to Ad infection by producing interferon (IFN)-α and IFN-β, these are known to limit Ad infections in cultured cells (23). These interferons, as well as IFN-γ which is produced later by immune cells, induce cellular genes that interfere with Ad replication (see refs 24, 25). Within a day or so following acute infection, an immune response develops and proceeds in two phases, the early innate inflammatory phase (first few days), followed by the late immune-specific phase. In the innate phase, macrophages become activated and secrete cytokines such as tumour necrosis factor-α (TNF), interleukin 1-β (IL-1β), and IL-6 and, no doubt chemokines such as IL-8, that leads to activation and infiltration of other macrophages, neutrophils, and NK cells into the infected tissue. These cells attack and kill the infected cells, both by phagocytosis and by induction of apoptosis through specific apoptotic pathways in the target cells. Dendritic cells are attracted, become antigen-presenting cells, migrate to lymphoid

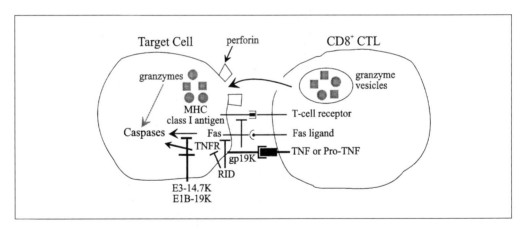

Fig. 2 Schematic illustrating killing of an Ad-infected target cell by an Ad-specific CTL. Also shown are the Ad proteins that function in various ways to block killing by CTL. The MHC class I antigen–virus peptide complex is transported from the ER to the cell surface where it interacts with a specific T cell receptor, resulting in activation of the CTL. E3-gp19K, an ER membrane glycoprotein, binds to class I antigens in the ER, inhibits their transport to the cell surface, and prevents recognition by CTL. When activated, killer mechanisms on the CTL are induced, including up-regulation of Fas ligand and TNF. These ligands interact with their cognate receptors on target cells and induce apoptosis through the caspases. The Ad RID protein clears these receptors from the target cell surface, thereby preventing apoptosis. the E3-14.7K and E1B-19K proteins act downstream of TNFR1 and Fas to block steps in the apoptotic pathways. The RID, E3-14.7K, and E1B-19K proteins may also prevent Fas-mediated killing by NK cells, which recognize target cells lacking MHC class I antigens.

organs, and educate lymphocytes to become CD4$^+$ helper cells and CD8$^+$ CTL which migrate into the infected area. The T cell receptor on the CTL recognizes viral peptides complexed with major histocompatibility complex (MHC) class I antigens on the surface of infected cells. This activates the killer activity of the CTL, including up-regulation of CD95/Fas/APO-1 (Fas) ligand and TNF. TNF may also be secreted in an active form. TNF and Fas ligand are also up-regulated on NK cells and macrophages.

CTL kill targets in large part through the perforin–granzyme pathway (26) (Fig. 2). Perforin expressed by the CTL forms holes in the target and the granzymes are introduced. One of the granzymes, granzyme B, activates the pro-apoptotic caspases (Asp-directed proteases) which induce apoptosis. Another major pathway of CTL killing is mediated by Fas and Fas ligand (26, 27). Fas ligand on the CTL interacts with its receptor, Fas, on the target, initiating a series of protein–protein interactions that results in activation of the caspase cascade and induction of apoptosis (Fig. 3). First, Fas trimerizes, then binds to the protein named Fas associated death domain (FADD). Binding is through a specific domain found in both proteins, called the death domain (DD). FADD has a second domain called the death effector domain (DED). The FADD DED binds a DED in Procaspase 8, an inactive form of Caspase 8.

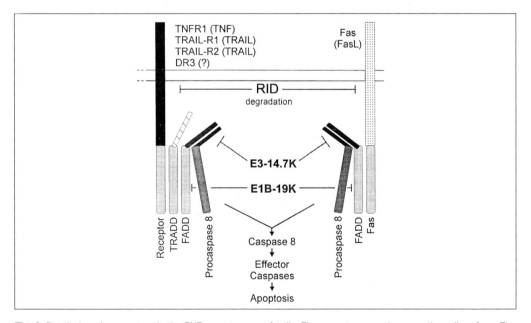

Fig. 3 Death domain receptors in the TNF receptor superfamily. The receptors are shown on the cell surface. The death domains, which are related regions involved in protein–protein interaction, are shown in light stipple. The darkest stipple indicates the death effector domain (another protein–protein interaction domain) for FADD and Procaspase 8. Upon binding of their ligand, the receptors cluster into protein complexes as shown, resulting in *trans*-cleavage and activation of Caspase 8, which in turn cleaves and activates downstream executioner caspases such as Caspase 3. Sites where the Ad RID, E3-14.7K, and E1B-19K proteins block apoptosis are shown, as discussed in the text. TRAIL-R1, TRAIL-R2, and DR3 are members of the TNF receptor superfamily, it is not known whether apoptosis through these receptors is inhibited by the Ad proteins, but this is a likely possibility.

Upon binding, Caspase 8 is activated by autocleavage, and it cleaves and stimulates downstream caspases to put the apoptotic program into effect. A third, probably more minor pathway of CTL-induced cytolysis involves TNF and one of its receptors, TNFR1. Induction of apoptosis through the TNF-TNFR1 system is similar to that with Fas except that the DD of TNFR1 first binds the protein named TRADD before binding FADD (Fig. 3). Further, another DD-containing protein, named RIP, interacts with the DD of the TNFR-TRADD complex and plays a role in inducing apoptosis (28, 29).

The acute Ad infection likely represents only part of the life-style of Ads, the other part is persistent low level asymptomatic infection. The state of the viral genome in persistently infected cells is not known, but it is probably not latent infection as we understand this term for the herpes simplex viruses. Infected children may shed subgroup C Ads for weeks to even years following infection (13). Lymphoid cells are thought to be the reservoir for the virus. Ads do not infect these cell types very well, probably because they lack the Ad receptor (CAR), but Ad does replicate. Ads have been found in peripheral blood lymphocytes, tonsils, and lungs of asymptomatic adults, and, interestingly in lung eithelial cells (13, 30). Thus, the Ad anti-immune and anti-apoptotic proteins to be discussed below should be considered not only with regard to their ability to prolong acute infections, but also to prevent the elimination of persistently infected cells.

3. E3-coded proteins that inhibit immune-mediated apoptosis

Ad encodes several proteins that modulate immune-mediated apoptotic mechanisms (4, 5, 31). Some of the themes are common to how other DNA viruses evade the immune system (32), but the specific mechanisms and the proteins involved are unique to Ads. Most of these Ad anti-immune proteins are derived from genes in the E3 transcription unit (Fig. 4). There are seven E3 proteins, and functions are known for five of them. None of these proteins is required for Ad replication in cultured cells, as would be expected from their anti-immune functions. The protein named E3-gp19K (a 19 kDa glycoprotein coded by the E3 transcription unit) is the first line of defense against CTL. It is a type I membrane N-linked glycoprotein localized in the membrane of the endoplasmic reticulum (ER). The lumenal domain extends into the lumen of the ER, there is a transmembrane domain near the C terminus, and a short domain that protrudes into the cytoplasm. This lumenal domain binds the lumenal domain of the MHC class I antigen heavy chain (33, 34). The cytoplasmic domain has a signal, KKXX (lysine–lysine–any amino acid–any amino acid), that serves to retain the E3-gp19K–class I antigen complex in the ER (35). Since class I antigens are not transported to the cell surface, the cells are not recognized and therefore not killed by CTL (17, 36, 37). E3-gp19K binds to all haplotypes of human class I antigens, although with significantly differing affinities (38). It binds to some but not all mouse class I antigens. E3-gp19K is conserved in all respiratory Ads (31). Ads that infect the

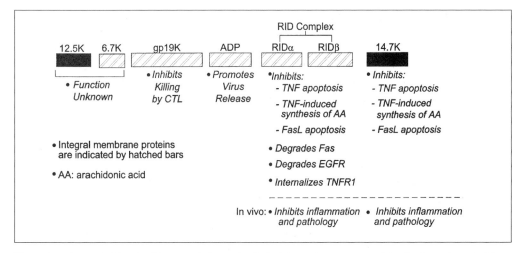

Fig. 4 Proteins coded by the E3 transcription unit. Each Ad protein, represented by the bars, is coded by a specific mRNA generated by alternative splicing of a pre-mRNA derived from the E3 promoter. FasL, Fas ligand.

gastrointestinal tract lack the gene; but they down-regulate class I molecules by repressing at the transcriptional level, repression is mediated by the Ad-coded E1A proteins (1).

The E3-coded proteins named RID and E3-14.7K inhibit some of the killing pathways induced in infected cells by CTL (39–41). As far as is known, the perforin–granzyme pathway is not affected, but both the Fas-Fas ligand and TNF-TNFR1 systems are blocked. RID is a membrane protein complex composed of the RIDα and RIDβ polypeptides (42). Note: RIDα and RIDβ were previously named E3-10.4K and E3.14.5K, and the complex was named E3-10.4K/14.5K (41). RIDα has two forms, one with the N terminal signal sequence cleaved and the other with the signal uncleaved, joined by a disulfide bond (43). RIDβ is a type I membrane polypeptide with a cleaved signal (44). Both RIDα and RIDβ are oriented with their C terminal domains in the cytoplasm. The cytoplasmic domain of RIDβ is phosphorylated on at least one serine residue (45) and the lumenal domain is *O*-glycosylated (46). The significance of these modifications is not known. RID is localized in the plasma membrane, and in cells transiently transfected with RID it can be seen by immunofluorescence in vesicles (41, 47, 48).

RID inhibits apoptosis induced by a Fas agonist, a monoclonal antibody to Fas that triggers apoptosis through the Fas pathway (41, 49, 50). RID causes Fas to be removed from the cell surface into endosomes which are sorted to lysosomes where Fas is degraded (41). This effect on Fas can be observed in both Ad-infected cells and cells transiently transfected with RID, indicating that RID is both necessary and sufficient for this to occur (41). RID exerts a similar effect on the receptor for epidermal growth factor (EGFR) (42, 51), and probably the receptors of insulin and insulin-like growth factor (52). The effect of RID is not a general one, because cell surface transferrin and MHC class I antigens are not affected (41, 53). Also not affected are CD40, a member of the TNF receptor superfamily (49, 50), and HER2, a member of the EGFR family (52).

RID also inhibits apoptosis induced by TNF (54). TNFR1 is cleared from the cell surface in two human cell lines tested, but considerably less efficiently than Fas or EGFR (unpublished results). It is not known if TNFR1 enters endosomes, or if it is degraded. Down-regulation is likely to be one mechanism by which RID blocks the effect of TNF, but may not be the only one. Clearance of TNFR1 was not observed in another study, possibly because mouse cells were examined earlier in infection (49).

As would be expected, RID can inhibit killing of Ad-infected cells by CTL; this was shown using CTL obtained from perforin knock-out mice (41). The Fas pathway was affected, as suggested by the clearance of Fas from the infected cell surface. Preliminary results indicate that RID may also play a role in inhibiting cytolysis of infected cells by NK cells (unpublished results). Although NK cells kill mostly through the perforin–granzyme system, the Fas pathway is also involved.

The mechanism of action of RID is under investigation; a model is shown in Fig. 5, and it applies to both Fas and EGFR. RID may stimulate the internalization of Fas by a mechanism whereby RID and Fas both enter the same endosomes. These endosomes go to lysosomes, Fas is discarded in the lysosomes and destroyed, and RID recycles back to the cell surface to pick up another Fas molecule and repeat the process. There are several lines of evidence that support this model. First, RID and

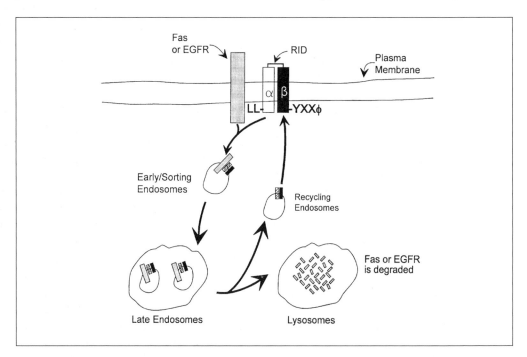

Fig. 5 Proposed model for the mechanism by which RID causes Fas and EGFR to be removed from the cell surface and degraded by lysosomes. (See the text for details.)

Fas are both residents of the plasma membrane (41, 47, 48). Secondly RID and Fas can sometimes be seen in the same vesicles (41). Thirdly, RID is stable as indicated by pulse-chase experiments (48), while Fas is unstable when RID is present. Fourthly, RID has motifs that are known to function in the sorting of receptors (55). RIDα has one or two dileucine motifs, Ll or IL, at its extreme C terminus, and RIDβ has a YXXϕ sequence, YFNL, near its C terminus (56, 57), with ϕ being a large aliphatic or a bulky aromatic amino acid. Deletion of the dileucine motif or mutation of the Y to A abrogates the function of RID (unpublished results).

E3-14.7K is a non-membrane protein of 128 amino acids (58) which is found in the cytoplasm and nucleus (59, unpublished results). E3-14.7K inhibits TNF-induced apoptosis of Ad-infected cells as well as cells transiently and stably transfected with E3-14.7K (40, 60–62). It was reported to also inhibit apoptosis through the Fas pathway (63), although this was not observed in another study (49). Yeast two-hybrid experiments have identified three proteins, FIP-1, FIP-2, and FIP-3 that bind to E3-14.7K (59, 64). FIP-1 (14.7K interacting protein-1) is a member the Ras family of small GTPases; it is not known whether FIP-1 affects the ability of 14.7K to inhibit TNF-induced apoptosis (64). FIP-2 appears to be a member of a novel family of proteins that is involved in TNF signalling (59). It has two leucine zipper domains that are often found in transcription factors. Immunofluorescence indicates that it co-localizes with E3-14.7K in the cytoplasm, especially in the perinuclear region. FIP-2 does not cause apoptosis upon transient transfection into cells, but it does overcome the ability of E3-14.7K to inhibit apoptosis induced by overexpression of TNFR1 or RIP. FIP-3 is homologous to FIP-2 (cited in ref. 59). Unlike FIP-2, FIP-3 binds directly to RIP. Thus, FIP-2 and FIP-3 may be part of a TNF-induced apoptotic pathway that operates through the TNFR1-TRADD-RIP complex. E3-14.7K may inhibit TNF-induced apoptotis, at least in part, by targeting the FIP components of this pathway (59).

In addition to inhibiting TNF-induced apoptosis, RID and E3-14.7K also inhibit TNF-induced synthesis of arachidonic acid (AA) (40, 65, 66). The AA is synthesized by the 85 kDa cytosolic phospholipase A_2 ($cPLA_2$) (67, 68). There is considerable evidence that activation of $cPLA_2$ is necessary for TNF-induced apoptosis, although the explanation for this observation is not understood (40, 67–69). $cPLA_2$ normally exists in the cytosol in an inactive state. In response to TNF (and probably Fas ligand), $cPLA_2$ translocates to membranes where it cleaves AA from membrane phospholipids. RID inhibits TNF-induced translocation of $cPLA_2$ to membranes (which could help explain how RID inhibits TNF-induced apoptosis) (70). The RID-mediated inhibition of $cPLA_2$ translocation can be observed at eight hours post-infection, which is well before RID has cleared TNFR1 from the cell surface (70, unpublished); therefore, the ability of RID to block $cPLA_2$ translocation is probably independent of its ability to down-regulate TNFR1. It is unclear how E3.14.7K prevents TNF-induced synthesis of AA. E3-14.7K presumably inhibits TNF-induced activation of caspases (Fig. 3), and caspases are reported to cleave and activate $cPLA_2$ (71). However, another study did not conclude that cleavage of $cPLA_2$ is necessary for TNF-induced apoptosis (72).

AA is the precursor to the pro-inflammatory eicosinoids. RID and E3-14.7K, by preventing TNF-induced synthesis of cPLA$_2$, may inhibit a TNF-induced eicosinoid-mediated inflammatory response (66, 70). Indeed, there is considerable evidence from animal model studies that RID and E3-14.7K do inhibit inflammation *in vivo*.

The functions of the E3 proteins *in vivo* were first addressed in a cotton rat lung model, where a late phase inflammatory response was observed in animals infected with an Ad mutant lacking E3-gp19K (73). This result was attributed to the ability of Ad to prevent attack by CTL. However, in a mouse lung model, mutants lacking E3-gp19K had a wild-type phenotype, i.e. modest inflammation, infiltration of leucocytes, and pathology (18). In another study where E3-gp19K was expressed together with the β-glucuronidase reporter from a replication-defective Ad vector in the mouse lung and liver, there was reduced inflammation and prolonged reporter expression, provided that E3-gp19K had good affinity for the products from both MHC class I alleles (74). This latter result makes sense, and it argues that E3-gp19K does, in fact, play a role in combating the immune response *in vivo*.

RID and E3-14.7K would be expected to exert effects in both the early and late phases of the immune response. In cotton rat lungs infected with an Ad mutant deleting both RID and E3-14.7K, there was an increase in neutrophils (73). In mouse lungs infected with mutants that lack RID, E3-14.7K, or both proteins, the presence of either RID or E3-14.7K independently inhibited inflammation and pathology (18). In a transgenic mouse that expresses E3-14.7K specifically in the lung from the surfactant protein C promoter, there was strong inhibition as compared to wild-type mice in inflammation and pathology in response to infection with an Ad vector expressing luciferase as a reporter, and luciferase persisted for an extended period (75). In another model where E3-14.7K was expressed together with TNF from a vacinia virus vector, E3-14.7K overcame the antiviral effects of TNF (76, 77). Thus, there is considerable evidence that the E3 proteins have anti-immune effects *in vivo*, as would be expected from their properties in cell culture.

The idea that Ad forms persistent infections in lymphoid cells was mentioned earlier. In this regard, it is interesting that the E3 promoter is the only Ad promoter that contains NFκB sites (78), and that transcription can be induced through these sites by TNF (79). This induction occurred independently of the E1A proteins. NFκB is a key transcription factor induced in the inflammatory and immune responses. Thus, a scenario might obtain where the Ad genome is more or less dormant in resting lymphoid cells; when these cells become activated and synthesize NFκB, the E3 promoter is induced and the cells are protected from immune-mediated destruction (78, 79).

4. Induction of apoptosis by E1A

The E1A region codes for two major proteins, of 289 (289R) and 243 amino acids (243R) (Fig. 6), that play a central role in the replication of Ads. Since the natural targets for Ad infection are terminally differentiated quiescent epithelial cells, these viruses induce transient proliferation of infected cells in order to facilitate efficient synthesis of viral DNA. Viral induction of cellular DNA synthesis appears to contribute to a cytotoxic effect in infected cells resulting in a cell death program that resembles apoptosis. The E1A proteins are essential for induction of both cell proliferation and apoptosis. Recognition that Ads induce an apoptosis-like process comes from studies of viral mutants defective in the E1B-19K protein. Infection of permissive human cells by Ad E1B-19K mutants induces enhanced cytopathic effect in infected cells (80–82). Cellular DNA extracted from cells infected with these mutants exhibits a characteristic pattern of fragmentation (83–85) similar to that observed during apoptosis (86). Expression of either the 289R or the 243R protein in the absence of E1B-19K induces DNA fragmentation while expression of a minor 28 kDa E1A protein in the absence of E1B-19K does not result in such a phenotype (87). These results suggest that expression of the major E1A proteins induces DNA fragmentation during Ad infection. Expression of E1A proteins alone in the absence of other viral proteins (mediated by DNA transfection) in quiescent primary rat kidney cells has also been shown to induce a process that resembles apoptosis (88).

The mechanism(s) by which the E1A proteins induce apoptosis is not fully understood. Since E1A promotes the accumulation of the p53 tumour suppressor protein by extension of protein stability, it has been believed that apoptosis induced during Ad infection may be the result of elevated levels of p53 (89). However, infection of p53-lacking human and mouse cells by E1B-19K mutants revealed that

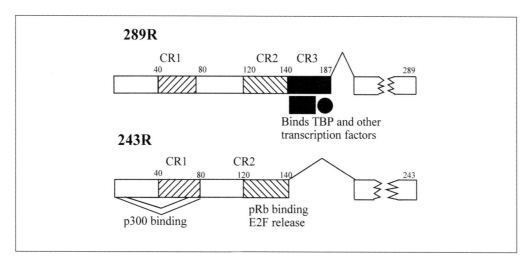

Fig. 6 Schematic of the E1A 289R and 243R proteins. The proteins are synthesized from alternatively spliced mRNAs, and are identical except that 289R contains a domain, termed conserved region 3 (CR3), that is absent from 243R. Conserved region 1 (CR1) and conserved region 2 (CR2) are present in both proteins. CR1, CR2, and CR3 are conserved among the different subgroups of Ads. CR3 binds the TATA box binding protein (TBP) and many other transcription factors, and stimulates transcription by causing the assembly of transcription factors onto promoters. The 289R protein is thought to be responsible for inducing transcription from Ad promoters. The 243R protein is thought to play a role in driving quiescent cells from G_0 into S-phase. As discussed in the text, 243R functions by interacting with p300 at sites near the N terminus and in CR1, and with pRb and other pocket proteins at CR2.

apoptosis induced during viral infection is not fully dependent on p53 (90). Infection of a mouse embryo fibroblast cell line containing chromosomal deletions of both p53 alleles by Ad E1B mutants expressing the E1A-243R protein alone did not induce significant DNA fragmentation while a mutant expressing the 289R protein did (91). Introduction of wild-type p53 into cells that express the 243R protein through infection with a second Ad vector resulted in apoptosis. These results suggest that he 243R protein mediates p53-dependent apoptosis while the 289R protein mediates p53-independent apoptosis (91). However, another study, which employed an E1B-19K mutant that also expresses only the E1A-243R protein, reported apoptosis in the p53-deficient human Saos2 cell line (92).Thus, the direct involvement of p53 in apoptosis during Ad infection remains uncertain. It may play a role in a cell type-dependent manner or the specific genetic background of the cell lines used might influence the outcome. However, there is strong evidence that E1A proteins sensitize cells for apoptosis by various DNA damaging agents such as anti-cancer agents by increasing accumulation of p53 (89).

Among the two major E1A proteins, it appears that 289R induces apoptotis by activating expression of the E4 region during viral infection (Section 7). The mechanism by which expression of the E1A proteins in the absence of other viral proteins induces apoptosis remains unresolved. Mutational analysis of E1A suggests that the ability to induce apoptosis during viral infection is linked to the ability of E1A to induce cellular DNA synthesis (93). Two N terminal domains of the E1A proteins which bind with the p300/CBP (p300) transcriptional co-activator and pRb (as well as the related 'pocket' proteins p107 and p130) are sufficient in mediating these activities. p300 binds to two general areas of the E1A 243R protein, a site very near the N terminus and a site at amino acids 40–80 known as conserved region 1 (CR1) (Fig. 6) (94). pRb and its family members bind to a second site, CR2 (Fig. 6). Although binding of E1A to both p300 and pRb is essential for induction of apoptosis in rodent cells (93), binding to pRb-related proteins may be sufficient for induction of apoptosis in human cells (92).

In addition to inducing apoptosis directly, the 243R protein also sensitizes cells to apoptosis in response to the death receptor ligands TNF and Fas ligand (95, 96). Studies with virus mutants in the E1A region indicate that binding of 243R to either p300 or pRb sensitizes cells to TNF-induced apoptosis. Mutants that bind neither p300 nor pRb do not sensitize the cells. The 243R protein also sensitizes cells to killing by NK cells, a property that maps to the p300 binding region of 243R (96). This susceptibility to killing by NK cells could reflect the ability of 243R to sensitize cells to apoptosis through the Fas (and possibly TNF) pathways (96). Altogether, these various studies raise the possibility that there may be a common underlying theme that connects the 243R-mediated induction of DNA synthesis, induction of apoptosis, stabilization of p53, sensitization to TNF- and Fas-induced apoptosis, and sensitization to NK cell killing.

pRb normally exists in cells in the G_1 state of the cell cycle as a complex with the transcriptional factor E2F-1 (and E2F family members), tethered via E2F-1 to E2F-binding sites in promoters of genes expressed in the S-phase (94, 97–100). Here, pRb acts as a transcriptional co-repressor. As part of the repression mechanism, pRb recruits a histone deacetylase to promoters containing E2F sites (101, 102). Presumably, removal of the highly charged acetyl groups allows the positively charged histones to interact more tightly with DNA, keeping the chromatin in a closed state and repressing transcription. Phosphorylation of pRb by cyclin-dependent kinases inactivates pRb, releasing E2F-1. Free E2F-1 can bind and activate S-phase promoters with E2F sites. E1A binding to pRb is thought to mimic the effects of pRb phosphorylation, causing release of E2F. This may also affect recruitment of histone deacetylase. E2F-1 possesses a potent apoptotic activity when overexpressed (103, 104). Delivery of E2F-1 through an Ad vector into mouse and human cells induced efficient apoptosis (105).

The other cellular protein bound by the N terminal region of E1A, p300, is a transcriptional co-activator that binds many different transcription factors and is accordingly targeted to promoters, some of which, like pRb, control entry of the cell into S-phase (94, 98). p300 and the highly related CBP have intrinsic histone deacetylase activity. CBP (and also probably p300) is activated by phosphorylation by cyclin-dependent kinases, presumably resulting in activation (i.e. derepression) of transcription by acetylating histones and opening chromatin to the transcription apparatus. Interaction of E1A appears to inhibit the HAT activity of p300. Thus, in part, E1A modulates the expression of S-phase genes by chromatin remodelling through pRb and p300/CBP.

Two recent observations may provide a link between E1A binding to pRb and p300, up-regulation of p53, and apoptosis (107, 108) (Fig. 7). p53 levels are regulated by MDM2, a novel nuclear ubiquitin ligase that binds and destabilizes p53. It is now known that p300 binds both MDM2 and p53 and participates in the degradation of p53, possibly by forming a scaffold for assembly of factors required for p53 ubiquitination and degradation (108). E1A interaction with p300 could interfere with

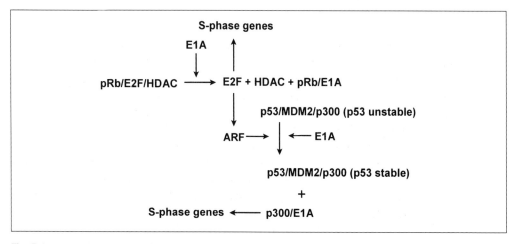

Fig. 7 Interactions that may occur linking E1A induced regulation of expression of cellular genes, stabilization of p53, and apoptosis.

the scaffold, preventing degradation of p53. Another player in this story is a protein named ARF (107). ARF binds to MDM2 and p53, stabilizing p53. ARF is induced in cells by E2F-1 which, as discussed, is regulated by pRb (109). E1A induces ARF expression, probably by interacting with pRb/E2F-1 complexes and liberating E2F-1 (110). Thus, E1A stabilizes p53 through its interactions with both pRb and p300 (Fig. 7).

In addition to p53 stabilization, there are likely to be other mechanisms by which E1A induces apoptosis. Recent biochemical studies have identified a novel apoptosis-inducing factor designated Oncogene Generated Activity (OGA) in cells that express E1A constitutively (111). OGA is a complex (112) that consists of apoptosis-inducing factors Apaf-1, Caspase 9, and cytochrome *c* (113, 114). Expression of either the 243R or the 289R proteins also activates the expression of a pro-apoptotic form of a poly-tropic integral ER membrane protein, Bap31 (115). In cells undergoing apoptosis, Bap31 associates with Procaspase 8 and is processed into a 20 kDa (p20) apoptotic form. In cells infected with Ad E1B mutants that express either the 243R or the 289R protein, Bap31 is processed into the p20 apoptotic form.

E1A is an extensively studied viral oncogene and serves as a model for the trans-forming genes of several small DNA tumour viruses. E1A can immortalize primary cultured cells and can co-operate with other viral and cellular genes in oncogenic transformation. These activities of E1A are controlled either by the 243R or the 289R protein. The transforming activities of E1A have been extensively studied using primary epithelial cells prepared from baby rat kidneys (BRK) as the model system. Introduction of E1A into these cells induces transient proliferation and terminal proliferation arrest. The majority of the proliferated cells appear to die by apoptosis (88). Co-expression of the E1B (116) or the E4 (117) regions results in oncogenic transformation. At least part of the transforming activity of E1A may require the inactivation of p53, inasmuch as the E1B-55K and E4orf6 proteins that inactivate p53 can co-operate with the E1A proteins in transformation of BRK cells. However, it appears that E1A-expressing BRK cells may also undergo p53-independent apoptosis since E1B-19K and E1B-55K can additively co-operate with E1A in transformation (118, 119).

5. Inhibition of apoptosis by the E1B-19K protein

Among the various Ad proteins, the E1B-19K protein plays an important role in suppression of apoptosis during viral infection. Viral mutants defective in the 19K protein produce extensive cytopathic effect in infected cells. These infected cells exhibit cellular changes generally associated with apoptosis. This apoptotic pheno-type is observed in spite of the expression of other apoptosis inhibitory proteins such as E1B-55K and E4orf6, highlighting the dominant role that 19K plays in suppression of virally-induced apoptosis. In addition, the 19K protein can also efficiently block apoptosis induced by a number of heterologous stimuli in virus-infected cells when expressed from the viral chromosome or alone. These stimuli include TNF (120, 121), anti-Fas antibodies (41, 122), and DNA damaging agents such as UV (123) and the anti-cancer agent cisplatin (124). The anti-apoptosis activity of E1B-19K appears to be related to its ability to co-operate with E1A in oncogenic transformation of primary cells (124, 125).

It is now established that the 19K protein is a distantly related member of the BCL-2 family and anti-apoptosis proteins. The functional similarity between these two proteins was established from mutant complementation studies. During infection by Ad 19K mutants in Chinese hamster ovary cells and in human cells that ectopically overexpress BCL-2, the cytopathic phenotype of the viral mutants is suppressed (123, 126). Similarly, Ad recombinants that express BCL-2 instead of both E1B proteins (19K and 55K) do not exhibit the enhanced cytopathic effect and do not induce fragmentation of cellular and viral DNA, suggesting that BCL-2 can functionally substitute for the E1B-19K protein during viral replication (90). The E1B-19K and BCL-2 proteins share patchy amino acid homology (126, 127) and at least some of the homologous sequences can be functionally exchanged between these two proteins (128). The most compelling evidence that BCL-2 and E1B-19K proteins are structural and functional homologues comes from the observation that these two proteins interact with a set of common cellular proteins (127, 129–132).

The mechanism(s) by which 19K and BCL-2 suppress apoptosis is not known. However, it has become apparent that the BCL-2 family proteins control apoptosis pathways at a point upstream of activation of Caspase 3 (and related caspases) involved in execution of cell death (133–135). Solution and X-ray crystallographic structural studies have revealed that the structure of BCL-x_L resembles the structure of certain bacterial toxins that form membrane pores (136). However, it is unclear how this property may be related to the anti-apoptotic activity. Recent biochemical studies with CED-9, the *Caenorhabditis elegans* homologue of BCL-2, have provided a biochemical framework for understanding at least some aspects of apoptosis suppression by mammalian BCL-2 family proteins (137). CED-9 forms complexes with a membrane associated adapter protein CED-4. This interaction appears to prevent proteolytic activation of CED-3 (a caspase) by CED-4 (Fig. 8). In a manner analogous to CED-9, BCL-x_L has also been shown to complex with Apaf-I, a

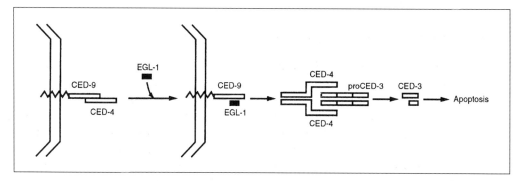

Fig. 8 Initiation and execution of apoptosis in *C. elegans*. Apoptosis is initiated by the BH3 protein EGL-1. EGL-1 is postulated to release CED-4 from a complex with CED-9. CED-4 (as dimers) then associates with Pro-CED-3 and activates CED-3 protease leading to apoptosis.

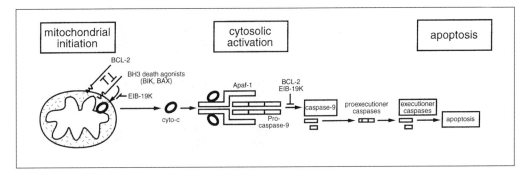

Fig. 9 BH3 domain-dependent apoptosis in mammalian cells. BH3 proteins such as BIK and BAX may induce mitochondrial damage resulting in release of cytochrome *c* into the cytosol. Binding of cytochrome *c* and ATP with the cytosolic adaptor protein Apaf-I results in aggregation of Apaf-I. Apaf-I then recruits and activates initiator Procaspase 9. Caspase 9 then sequentially activates executioner caspases (e.g. Caspase 3 and Caspase 7) resulting in apoptosis. BCL-2 family anti-apoptosis proteins may inhibit this apoptosis paradigm at the initiation level by complexing with the BH3 death agonists as well as by complexing with Apaf-I and preventing Caspase 9 activation.

mammalian homologue of CED-4 (138, 139). Apaf-I (113, 114) binding with Procaspase 9 results in activation of Caspase 9 (one of the upstream caspases involved in initiation of apoptosis) (Fig. 9). BCL-x_L binds to a region of Apaf-I where Procaspase 9 also binds.

Thus, BCL-x_L/Apaf-I interaction appears to prevent Caspase 9 association with Apaf-I, thereby preventing recruitment of Procaspase 9 by Apaf-I and auto-proteolytic activation of Caspase 9 (138, 139). Although it is believed that other mammalian BCL-2 family proteins may also function in a similar fashion, this remains to be investigated experimentally. This may be particularly important for E1B-19K since it does not appear to contain sequences similar to the BCL-x_L domain (BH4) implicated in interaction with Apaf-I (140). E1B-19K and other BCL-2 family anti-apoptosis proteins interact with a number of BCL-2 family pro-apoptotic proteins such as BAX, BAK, and BIK. While BAX and BAK share extensive homology with BCL-2, BIK shares only a single domain (BH3) with other BCL-2 proteins (Figs 10, 11). Since the discovery of BIK, a large number of BH3-alone pro-apoptotic proteins have been identified (137). The recognition that the BH3 domain pro-apoptotic proteins may be key regulators of apoptosis comes from the identification of the *C. elegans* BH3-containing proapoptotic protein named EGL-1 (141). EGL-1 is essential for apoptosis in *C. elegans* and is genetically located upstream of CED-9 in the apoptotic pathway. The activity of EGL-1 requires the activities of CED-4 and CED-3. It has been proposed that EGL-1 binding to CED-9 may displace CED-4 from CED-9 allowing CED-4 to recruit Pro-CED-3 and initiate apoptosis by activation of CED-3 (Fig. 8). Other investigators have also proposed a similar model since mammalian BH3-containing anti-apoptotis proteins BAX, BAK, and BIK precluded interaction of CED-4 with BCL-x_L (142). Based on these models, it has been recently suggested that interaction of BH3 proteins such as BIK may release Apaf-I to form a complex with BCL-2 (137). These models suggest that apoptosis is initiated by

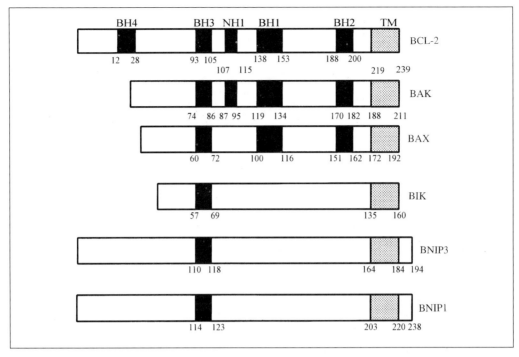

Fig. 10 BCL-2 family pro-apoptotic proteins. Domain structures of BCL-2 and selected proapoptotic proteins that interact with E1B-19K. BH, BCL-2 homology domain; NH, E1B-19K homology domain; TM, transmembrane domain.

interaction of the BH3-domain death agonists with BCL-2 and related endogenous anti-apoptosis proteins (Fig. 9).

However, it has been observed that certain mutants of BIK defective in induction of cell death efficiently form complexes with BCL-2 and BCL-x_L (143). This would be consistent with a model in which, at least in the case of BIK, binding with BCL-2/BCL-x_L is insufficient for induction of cell death. The possibility that the BH3-containing death agonists induce apoptosis independently of interaction with BCL-2 and related proteins remains to be further investigated. Recently, it has been observed that a member of the BH3 death agonists, BID, is proteolytically processed during Fas-induced apoptosis and the BH3-containing C terminal fragment is translocated to mitochondria (144, 145). The activated form of BID has been shown to induce release of cytochrome *c* from isolated mitochondria in a BH3-dependent fashion (144). BID-mediated release of cytochrome *c* from mitochondria is suppressed by BCL-2 (144). It is possible that other BH3 domain proteins such as BIK may induce similar damage to organelles or subcellular architectures. E1B-19K and other anti-apoptotic proteins may protect the organelles and subcellular structures by neutralizing the BH3-proteins by protein complex formation (Fig. 9). Anti-apoptotic proteins, in addition to targeting BH3-containing death agonists, may target other components of the apoptosis paradigm. Recently, it has been shown that E1B-19K prevents Fas-

```
      53    D A L A L R L A C I G D E M D V      68   BIK
      65    Q L T A A R L K A L G D E L D S      90   HRK
      72    G Q V G R Q L A I I G D D I N R      87   BAK
      84    R N I A R H L A Q V G D S M D R      99   BID
      57    K K L S E C L K R I G D E L D S      72   BAX
     108    N L E K A E L L Q G G D L L R Q     123   BNIP1
     104    K E V E S I L K K N S D W L W D     119   BNIP3
     145    Q R Y G R E L R R M S D E F E G     160   BAD
      91    P V V H L A L R Q A G D D F S R     106   BCL-2
      84    A A V K Q A L R E A G D E F E L      99   BCL-x_L
      52    Y E I G S K L A A M C D D F D A      67   EGL-1

                            BH3
```

Fig. 11 The BH3 domains of various mammalian BH3 proteins and EGL-1 of *C. elegans*,

mediated apoptosis by preventing oligomerization of FADD (146). Since E1B-19K and other BCL-2 family proteins suppress apoptosis induced by a multitude of different stimuli, continued investigation should unearth other potential targets for these proteins.

An overall view of how E1B-19K, RID, and E3-14.7K are thought to work in conjunction to prevent apoptosis through the death receptors Fas and TNFR1 is depicted in Fig. 12. E1B-19K is proposed to function at two levels, one preventing the oligomerization of FADD, and the second to sequester the BH3-containing proteins BAX, BIK, BAK, BNIP1, and BNIP3. These BH3-containing proteins are proposed not to function in the direct pathway where Caspase 8 activates downstream executioner caspases, rather in the release of cytochrome *c* from mitochondria and the activation of Caspase 9 (see above). RID clears Fas and TNFR1 from the cell surface. RID also prevents TNF-induced activation of cPLA$_2$ (not shown), which may contribute to its ability to inhibit TNF-induced apoptotis. E3-14.7K functions at two levels, by interacting with Procaspase 8 and by interacting with FIP-2 and FIP-3.

6. Inhibition of apoptosis by the E1B-55K protein

The E1B-55K protein is essential for viral replication in normal cells. It also efficiently co-operates with E1A in oncogenic transformation. E1B-55K mediates its effects on viral replication and in cell transformation by forming a specific complex with the p53 tumour suppressor protein (147). The 55K protein binds to the N terminal *trans*-activation domain of p53 (148) resulting in inhibition of p53's transcriptional activation function (149). The association between E1B-55K and p53 appears to be important for overcoming the growth-suppressive function of p53 during viral replication. Inactivation of p53 is critical for the replication of Ad which requires quiescent epithelial target cells to enter the S-phase, otherwise p53 would induce

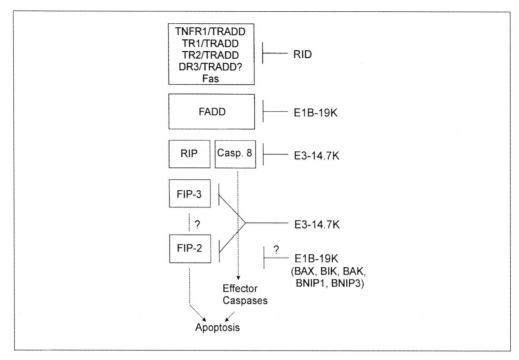

Fig. 12 Schematic showing where the Ad RID, E3-14.7K, and E1B-19K proteins are proposed to function in inhibiting apoptosis induced through the death receptors TNFR1 and Fas. Data have been published for Fas and TNFR1, but not TRAIL receptor 1 (TR1), TRAIL receptor 2 (TR2), or death receptor 3 (DR3).

growth arrest and apoptosis. This phenomenon has been exploited for developing Ad vectors that are defective in the 55K protein, and are unable to replicate in normal human cells (which contain functional p53) but are able to replicate in tumour cells deficient in p53 (150).

The 55K protein also possesses an intrinsic transcriptional repressor activity that suppresses the basal transcription machinery when localized to a promoter as a fusion protein with the heterologous (GAL4) DNA binding domain (151). *In vitro* transcription studies with purified 55K protein have also indicated that 55K functions as specific repressor for templates containing p53-binding sites (152). Thus, complex formation with p53 is believed to target the repressor activity of 55K to promoters containing p53-binding sites. Mutants of 55K defective in phosphorylation of three C terminal Ser/Thr residues (Ser$_{490}$, Ser$_{491}$, and Thr$_{495}$) are defective in the repressor activity. Interestingly, these mutants also are defective in suppression of p53-induced apoptosis. These results suggest that the transcriptional repressor activity conferred by the p53/55K complex may be important for suppression of apoptosis (153). The phosphorylation-defective 55K mutants were also defective in transformation of primary cells in co-operation with E1A, suggesting that the transforming and anti-apoptosis activities of 55K may be linked (153).

7. Apoptosis regulation by E4 proteins

The E4 gene region is required for productive viral multiplication in permissive cells. The E4 transcription unit is complex and encodes a number of differentially spliced mRNA species which code for at least seven different proteins (154, 155). The E4 proteins provide functions essential for efficient viral DNA replication, shut-off of host protein synthesis, and viral cytopathicity (156, 157). These functions may be related to the known activities of E4 proteins. For example, a 36 kDa protein coded by open reading frame 6 (E4orf6) mediates selective nuclear export of viral mRNAs (158). The proteins coded by E4orf3, E4orf4, and E4orf6 also control alternative splicing of late viral mRNAs (159, 160). In addition to their role during productive viral replication, E4 proteins also exhibit oncogenic properties. The E4orf6 protein which binds to the p53 tumour suppressor protein co-operates with E1A in onco- genic transformation of primary cells and converts E1A-E1B-transformed human 293 cells into an oncogenic state (117, 161). The E4orf1 gene also encodes a transforming protein that is distantly related to dUTP pyrophosphatase (162). The E4orf1 gene of Ad9 is responsible for induction of mammary tumours in rats (163).

The E4 gene products also modulate apoptosis in a way that may contribute to their oncogenic activities and for the cytopathic effects required for efficient release of virus particles during productive viral infection. In common with the E1B-55K protein, the E4orf6 protein can also efficiently bind to the p53 tumour suppressor protein (164). The E4orf6 protein binds near the C terminus of p53 close to the oligomerization domain. The association of E4orf6 with p53 blocks the interaction of a component of the basal transcription machinery, $TAF_{II}31$ (a subunit of transcription factor TFIID) with the N terminal transcriptional activation domain of p53, thereby blocking p53-mediated transcriptional activation. The E4orf6 association with p53, in concert with EB-55K, has also been shown to target p53 for active degradation in Ad- infected cells (165). The biological consequence of the interaction of E4orf6 with p53 appears to be inhibition of p53-induced apoptosis (117). The anti-apoptosis activity of E4orf6 is specific for apoptosis induced by p53 since E4orf6 is unable to block apoptosis induced by TNF. This anti-apoptosis activity of E4orf6 may be linked to its oncogene co-operation activity observed in primary rodent epithelial cells. In primary rat kidney cells, E4orf6 co-operates with E1A in oncogenic transformation, presumably by inhibiting E1A-induced p53-dependent apoptosis (117).

The E4 gene region has also been shown to modulate apoptosis induced by E1A through p53-independent mechanisms. Infection of p53-null cells by Ad mutants lacking the E1B region induces apoptosis (90, 91), indicating that Ad infection induces p53-independent apoptosis. Mutational analysis has ascribed this activity of E1A to the 289R protein (91). Infection of p53-null mouse cells that constitutively express E1A with Ad mutants lacking both the E1A and E1B regions but containing other early gene regions (E2, E3, and E4) induced rapid apoptosis (91). These results suggested that E1A might *trans*-activate one or more viral early genes that induce apoptosis (i.e. p53-independent). Infection of human p53-deficient Saos2 cells with mutant viruses lacking both E1B and E4 remained viable longer than cells infected with wild-type Ad5. In addition, an Ad vector lacking both the E1 and E4 gene regions was unable to induce apoptosis, suggesting that the E4 region may be responsible for induction of apoptosis. Further mapping studies have revealed that the E4orf4 protein contributes to the apoptotic activity of E4 (166–168). Over-expression of E4orf4 induces cell death that is characterized by various apoptotic features such as cytoplasmic vacuolation, condensation of chromatin, internucleosomal DNA fragmentation, and externalization of phosphoytidylserine. The cytotoxic effect of E4orf4 is significantly suppressed by BCL-2 or BCL-x_L (166). Interestingly, E4orf4-induced apoptosis is independent of known caspases since it is not suppressible by the broad spectrum caspase inhibitor zVAD-fmk (166). Although the mechanism by which E4orf4 induces apoptosis remains to be elucidated, it may be related to the ability of E4orf4 to interact with protein phosphatase 2A (169). A mutant of E4orf4 defective in interaction with PP2A is defective in induction of apoptosis (167). It is possible that E4orf4 induces apoptosis by interfering with the MAP kinase signalling pathway (170). The association between E4orf4 and PP2A may lead to apoptosis by alteration of the phosphorylation status of cellular apoptosis-regulating proteins such as the pro- and anti-apoptotic BCL-2 family members.

8. Inhibition of the antiviral effects of interferons

Another aspect of the host response that is targeted by Ad are the interferons (IFNs). We will not discuss this topic extensively (for access to this literature, see refs 24, 25, 171). IFNs stimulate transcription of genes that limit virus replication. Both IFN-α/β and IFN-γ induce signal transduction from the cell surface to promoters via the JAK-STAT pathways. For IFN-α, a transcription factor named interferon-stimulated gene factor 3 (ISGF3) is formed which consists of the STAT1α and STAT2 proteins, that binds to the interferon-stimulated response element (ISRE). For IFN-γ, the gamma-stimulated factor (GAF) is formed, consisting of the protein tyrosine kinases JAK1 and JAK2 and the DNA binding protein STAT1α, which binds and stimulates transcription through the gamma-activated sequence (GAS) (25). The Ad E1A proteins interfere with the formation of the ISGF3 and GAF transcription factors. Both the 243R and 289R proteins appear to be involved, although the 243R protein can be sufficient; in the latter case, 243R exerts its effect by binding to p300 (171).

A second level of protection against IFNs is afforded by a small Ad-coded RNA (159 nucleotides) named VA-RNA$_I$ (24). (Ad also codes another small RNA, VA-RNA$_{II}$, whose function is unknown.) VA-RNA$_I$ is synthesized in very large amounts at late stages of infection. The target of VA-RNA$_I$ is PKR, an IFN-inducible kinase that, when activated by double-stranded RNA, phosphorylates eukaryotic initiation factor 2 (eIF2), thereby precluding translation. Since Ad proteins cannot be made, Ad replication is blocked. VA-RNA$_I$ binds to PKR, prevents its activation by double-stranded RNA, and allows protein synthesis to continue (24). Thus, Ad overcomes IFN-mediated antiviral responses at two levels: E1A prevents transcription of IFN-inducible genes, and VA-RNA$_I$ prevents the IFN/PKR-mediated shut-off of protein synthesis.

9. Induction of cell lysis and virus release by the adenovirus death protein (ADP)

Ad devotes considerable attention to keeping the infected cell alive so that virus replication can occur. However, once replication is complete, the virus needs to find its way out of the cell so that it can infect other cells. The E4orf4 protein may mediate one such mechanism. Another mechanism is mediated by ADP (172, 173). ADP was previously named E3-11.6K according to its predicted molecular mass (174). ADP is an integral membrane protein of 101 amino acids, with an internal (about amino acids 40–60) uncleaved signal-anchor sequence (175). When examined at different times post-infection, the Ad2 version of ADP can be found in the ER and Golgi, but eventually it accumulates predominantly in the nuclear membrane. The N terminal lumenal portion of the protein is glycosylated with complex Asn-linked oligo-saccharides (175) as well as O-linked oligosaccharides (unpublished). The C terminal cytoplasmic (or possibly nucleoplasmic) region is palmitoylated (176).

The gene for ADP is located in the E3 transcription unit (Fig. 4), and ADP is expressed in small amounts from the E3 promoter at early stages of infection (177). However, the gene for ADP is primarily a late gene, part of the major late tran-scription unit where ADP is expressed in large amounts via splicing from the major late promoter (173). In fact, the peak period of ADP synthesis occurs at very late stages (about 24–30 hours post-infection) when much of the virus assembly is complete. This observation is in accord with the idea that ADP mediates virus release from cells.

Evidence that ADP mediates cell lysis and virus release came from studies with Ad mutants that lack ADP. Cells infected with ADP-positive Ads begin to lyse and release virus after about two days, and lysis is complete by about three days (172, 173). With mutants that lack ADP, cell lysis does not occur until about five days. The cells show typical Ad cytopathic effect (swollen nuclei cells rounded up and de-tached into clusters), but they remain intact with their nuclei packed full with virus. The mechanism of action of ADP is unknown. The ADP-mediated cell death does not display typical features of apoptosis, but it seems unlikely that it is non-specific necrosis because that would not be expected to be mediated by a specific protein (172, 173). Given that ADP is abundant and hydrophobic, it is possible that it acts as a detergent to form holes in membranes that Ad can penetrate. Alternatively, ADP could be a specific channel that mediates transport of virions and/or ions across membranes. However, there are no data on these ideas, and many other mechanisms are possible.

10. Concluding remarks

Ads and the cells and host they infect are in a continuous struggle over cell death (Fig. 13). Shortly following infection, the E1A proteins take control of the cell, deregulating the cell cycle and forcing the cell to make the tools for DNA synthesis.

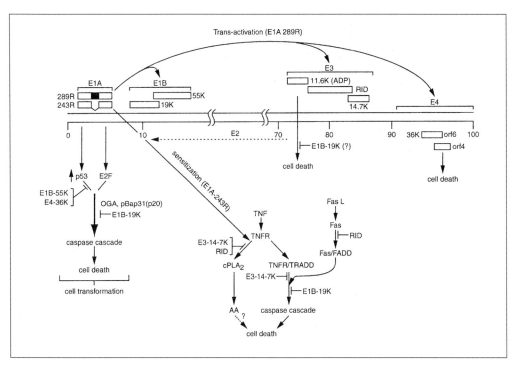

Fig. 13 Control of apoptosis by Ad genes. The Ad5 genome map that shows early gene regions and the major proteins involved in apoptosis regulation are indicated. OGA, oncogene generated activity; cPLA$_2$, cytosolic phospholipase A$_2$; AA, arachidonic acid.

The cell responds to this insult by attempting to commit suicide. Ad responds by synthesizing the E1B-19K and E1B-55K proteins which prevent apoptosis. The cell synthesizes IFNs and attempts to make a variety of proteins that prevent virus replication. Ad responds by having E1A block the activity of transcription factors required for IFN-inducible genes, and by producing VA-RNA$_I$ which inhibits the activity of one of the key IFN-inducible proteins, PKR, whose job is to shut down protein synthesis. The host immune system attempts to eliminate the infected cell at both the early innate and late immune-specific stages of the immune response. Again the virus fights back, producing RID, E3-14.7K, and E1B-19K which prevent apoptosis induced by death ligands from the TNF receptor family, which are expressed on macrophages, NK cells, and CTL. In the immune-specific phase, E3-gp19K hides the infected cell from CTL by preventing the transport of MHC class I antigen–viral peptide complexes to the cell surface. Finally, when replication is all over and Ad has assembled in the cell nucleus, the ADP protein of Ad, and perhaps also the E4orf4 protein, turns the table on the cell and causes it to lyse and release Ad particles. Although there is a great deal of violence at the level of the cell, the outcome at the level of the host is relatively benign. Rarely do subgroup C Ad infections get out of control. Thus, we can consider the battle between the virus and the host to be a draw.

If Ad were to predominate in this battle, death would ensue, and that would not be good for anybody.

Acknowledgements

W. S. M. W. was supported by grants CA21470, CA58538, and CA71704 from the National Institutes of Health, G. C. was supported by grants CA33616 and CA73803 from the National Institutes of Health, and grant VM-174 from the American Cancer Society, and a grant from the Charlotte Geyer Foundation. W. S. M. W. thanks Ann Tollefson for her outstanding contributions to the research described and for her comments on this article.

References

1. Blair, G. E. and Hall, K. T. (1998) Human adenoviruses: evading detection by cytotoxic T lymphocytes. *Semin. Virol.*, **8**, 387.
2. Chinnadurai, G. (1998) Control of apoptosis by human adenovirus genes. *Semin. Virol.*, **8**, 399.
3. Teodoro, J. G. and Branton, P. E. (1997) Regulation of apoptosis by viral gene products. *J. Virol.*, **71**, 1739.
4. Krajsci, P. and Wold, W. S. M. (1998) Inhibition of tumor necrosis factor and interferon triggered responses by DNA viruses. *Cell Dev. Biol.*, **9**, 351.
5. Wold, W. S. M. and Tollefson, A. E. (1998) Adenovirus E3 proteins: 14.7K, RID, and gp19K inhibit immune-induced cell death, adenovirus death protein promotes cell death. *Semin. Virol.*, **8**, 515.
6. Wickham, T. J., Mathias, P., Cheresh, D. A., and Nemerow, G. R. (1993) Integrins $\alpha_v\beta_3$ and $\alpha_v\beta_5$ promote adenovirus internalization but not virus attachment. *Cell.* **73**, 309.
7. Bergelson, J., Cunningham, J. A., Droguett, G., Kurt-Jones, E., Krithivas, A., Hong, J. S., *et al.* (1997) Isolation of a common receptor for coxsackie B viruses and adenoviruses 2 and 5. *Science*, **275**, 1320.
8. Li, E., Stupack, D., Klemke, R., Cheresh, D. A., and Nemerow, G. R. (1998) Adenovirus endocytosis via α_v integrins requires phosphoinositide-3-OH kinase. *J. Virol.*, **72**, 2055.
9. Tomko, R. P., Xu, R., and Philipson, L. (1997) HCAR and MCAR: The human and mouse cellular receptors for subgroup C adenoviruses and group B coxsackieviruses. *Proc. Natl. Acad. Sci. USA*, **94**, 3352.
10. Greber, U. F., Willetts, M., Webster, P., and Helenius, A. (1993) Stepwise dismantling of adenovirus 2 during entry into cells. *Cell*, **75**, 477.
11. Greber, U. F., Webster, P., Weber, J., and Helenius, A. (1996) The role of the adenovirus protease in virus entry into cells. *EMBO J.*, **15**, 1766.
12. Shenk, T. (1996) Adenoviridae: the viruses and their replication. In *Fields virology* (ed. B. N. Fields, D. M. Knipe, and P. M. Howley), pp. 2111–48. Philadelphia: Lippincott-Raven.
13. Horwitz, M. S. (1996) Adenoviruses. In *Fields virology* (ed. B. N. Fields, D. M. Knipe, and P. M. Howley), pp. 2149–71. Philadelphia: Lippincott-Raven Publishers.
14. Flomenberg, P., Piaskowski, V., Truitt, R. L., and Caspter, J. T. (1995) Characterization of human proliferative T cell responses to adenovirus. *J. Infect. Dis.*, **171**, 1090.

15. Sterman, D. H., Treat, J., Liszky, L. A., Amin, K. M., Coonrod, L., Molnar-Kimber, J., *et al.* (1998) Adenovirus-mediated herpes simplex virus thymidine kinase/ganciclovir gene therapy in patients with localized malignancy: results of a phase I clinical trial in malignant mesothelioma. *Hum. Gene Ther.*, **9**, 1083.

16. Knowles, M. R., Hohneker, K. W., Zhou, Z., Olsen, J. C., Noan, T. L., Hu, P. C., *et al.* (1995) A controlled study of adenoviral-vector-mediated gene transfer in the nasal epithelium of patients with cystic fibrosis. *N. Engl. J. Med.*, **333**, 871.

17. Rawle, F. C., Tollefson, A. E., Wold, W. S. M., and Gooding, L. R. (1989) Mouse anti-adenovirus cytotoxic T lymphocytes. Inhibition of lysis by E3 gp19K but not E3 14.7K. *J. Immunol.*, **143**, 2031.

18. Sparer, T., Tripp, R. A., Dillehay, D. L., Hermiston, T. W., Wold, W. S. M., and Gooding, L. R. (1996) The role of adenovirus early region 3 proteins (gp19K, 10.4K, 14.5K, and 14.7K) in a murine pneumonia model. *J. Virol.*, **70**, 2431.

19. Wolff, G., Worgall, S., vanRooijen, N., Song, W.-R., Harvey, B.-G., and Crystal, R. G. (1997) Enhancement of *in vivo* adenovirus-mediated gene transfer and expression by prior depletion of tissue macrophages in the target organ. *J. Virol.*, **71**, 624.

20. Ginsberg, H. S., Moldawer, L. L., Sehgal, P. B., Redington, M., Kilian, P. L., Chanock, R. M., *et al.* (1991) A mouse model for investigating the molecular pathogenesis of adenovirus pneumonia. *Proc. Natl. Acad. Sci. USA*, **88**, 1651.

21. Prince, G. A., Porter, D. A., Jenson, A. B., Horswold, R. L., Chanock, R. M., and Ginsberg, H. S. (1993) Pathogenesis of adenovirus type 5 pneumonia in cotton rats (Sigmodon hispidus). *J. Virol.*, **67**, 101.

22. O'Garra, A. (1998) Cytokines induce the development of functionally heterogeneous T helper cell subsets. *Immunity*, **8**, 275.

23. Anderson, K. P. and Fennie, E. H. (1987) Adenovirus early region 1A modulation of interferon antiviral activity. *J. Virol.*, **61**, 787.

24. Mathews, M. B. and Shenk, T. (1991) Adenovirus virus-associated RNA and translation control. *J. Virol.*, **65**, 5657.

25. Leonard, G. T. and Sen, G. C. (1996) Effects of adenovirus E1A protein on interferon-signaling. *Virology*, **224**, 25.

26. Nagata, S. (1997) Apoptosis by death factor. *Cell*, **88**, 355.

27. Ashkenazi, A. and Dixit, V. M. (1998) Death receptors: signaling and modulation. *Science*, **281**, 1305.

28. Stanger, B. Z., Leder, P., Lee, T. H., Kim, E., and Seed, B. (1995) RIP: a novel protein containing a death domain that interacts with Fas/APO-1 (CD95) in yeast and causes cell death. *Cell*, **81**, 513.

29. Hsu, H., Huang, J., Shu, H.-B., Baichwal, V., and Goeddel, D. V. (1996) TNF-dependent recruitment of the protein kinase RIP to the TNF receptor-1 signaling complex. *Immunity*, **4**, 387.

30. Elliott, W. M., Hayashi, S., and Hogg, J. C. (1995) Immunodetection of adenoviral E1A proteins in human lung tissue. *Am. J. Respir. Cell Mol. Biol.*, **12**, 642.

31. Wold, W. S. M., Tollefson, A. E., and Hermiston, T. W. (1995) Strategies of immune modulation by adenoviruses. In *Viroreceptors, virokines, and related mechanisms of immune modulation by DNA viruses* (ed. G. McFadden), pp. 145–83. Texas: R.G. Landes Co.

32. Ploegh, H. L. (1998) Viral strategies of immune evasion. *Science*, **280**, 248.

33. Andersson, M., Paabo, S., Nilsson, T., and Peterson, P. A. (1985) Impaired intracellular transport of class I MHC antigens as a possible means for adenoviruses to evade immune surveillance. *Cell*, **43**, 215.

34. Burgert, H. G. and Kvist, S. (1985) An adenovirus type 2 glycoprotein blocks cell surface expression of human histocompatibility class I antigens. *Cell*, **41**, 987.

35. Jackson, M. R., Nilsson, T. and Peterson, P. A. (1993) Retrieval of transmembrane proteins to the endoplasmic reticulum. *J. Cell. Biol.*, **121**, 317.

36. Andersson, M., McMichael, A., and Peterson, P. A. (1987) Reduced allorecognition of adenovirus-2 infected cells. *J. Immunol.*, **138**, 3960.

37. Burgert, H. G. and Kvist, S. (1987) The E3/19K protein of adenovirus type 2 binds to the domains of histocompatibility antigens required for CTL recognition. *EMBO J.*, **6**, 2019.

38. Beier, D. C., Cox, J. H., Vining, D. R., Cresswell, P., and Engelhard, V. H. (1994) Association of human class I MHC alleles with the adenovirus E3/19K protein. *J. Immunol.*, **152**, 3862.

39. Gooding, L. R., Elmore, L. W., Tollefson, A. E., Brady, H. A., and Wold, W. S. M. (1988) A 14,700 MW protein from the E3 region of adenovirus inhibits cytolysis by tumor necrosis factor. *Cell*, **53**, 341.

40. Krajcsi, P., Dimitrov, T., Hermiston, T. W., Tollefson, A. E., Ranheim, T. S., Vande Pol, S. B., *et al.* (1996) The adenovirus E3-14.7K protein and the E3-10.4K/14.5K complex of proteins, which independently inhibit tumor necrosis factor (TNF)-induced apoptosis, also independently inhibit TNF-induced release of arachidonic acid. *J. Virol.*, **70**, 4904.

41. Tollefson, A. E., Hermiston, T. W., Lichtenstein, D. L., Colle, C. F., Tripp, R. A., Dimitrov, T., *et al.* (1998) Forced degradation of Fas inhibits apoptosis in adenovirus-infected cells. *Nature*, **392**, 726.

42. Tollefson, A. E., Stewart, A. R., Yei, S. P., Saha, S. K., and Wold, W. S. M. (1991) The 10,400- and 14.500-dalton proteins encoded by region E3 of adenovirus form a complex and function together to down-regulate the epidermal growth factor receptor. *J. Virol.*, **65**, 3095.

43. Krajcsi, P., Tollefson, A. E., Anderson, C. W., Stewart, A. R., Carlin, C. R., and Wold, W. S. M. (1992) The E3-10.4K protein of adenovirus is an integral membrane protein that is partially cleaved between Ala$_{22}$ and Ala$_{23}$ and has a C$_{cyt}$ orientation. *Virology*, **187**, 131.

44. Krajcsi, P., Tollefson, A. E., Anderson, C. W., and Wold, W. S. M. (1992) The adenovirus E3 14.5-kilodalton protein, which is required for down-regulation of the epidermal growth factor receptor and prevention of tumor necrosis factor cytolysis, is an integral membrane protein oriented with its C terminus in the cytoplasm. *J. Virol.*, **66**, 1665.

45. Krajcsi, P. and Wold, W. S. M. (1992) The adenovirus E3-14.5K protein which is required for prevention of TNF cytolysis and for down-regulation of the EGF receptor contains phosphoserine. *Virology*, **187**, 492.

46. Krajcsi, P., Tollefson, A. E., and Wold, W. S. M. (1992) The E4-14.5K integral membrane protein of adenovirus that is required for down-regulation of the EGF receptor and for prevention of TNF cytolysisis O-glycosylated but not N-glycosylated. *Virology*, **188**, 570.

47. Hoffman, P., Yaffe, M. B., Hoffman, B. L., Yei, S., Wold, W. S. M., and Carlin, C. (1992) Characterization of the adenovirus E3 protein that down-regulates the epidermal growth factor receptor. Evidence for intermolecular disulfide bonding and plasma membrane localization. *J Biol. Chem.*, **267**, 13480.

48. Stewart, A. R., Tollefson, A. E., Krajcsi, P., Yei, S. P., and Wold, W. S. M. (1995) The adenovirus E3 10.4K and 14.5K proteins, which function to prevent cytolysis by tumor necrosis factor and to down-regulate the epidermal growth factor receptor, are localized in the plasma membrane. *J. Virol.*, **69**, 172.

49. Shisler, J., Yang, C., Walter, B., Ware, C. F., and Gooding, L. R. (1997) The adenovirus E3-10.4K/14.5K complex mediates loss of cell surface Fas (CD95) and resistance to Fas-induced apoptosis. *J. Virol.*, **71**, 8299.

50. Elsing, A. and Burgert, H.-G. (1998) The adenovirus E3/10.4K-14.5K proteins down-modulate the apoptosis receptor Fas/Apo-1 by inducing its internalization. *Immunology*, **95**, 10072.

51. Carlin, C. R., Tollefson, A. E., Brady, H. A., Hoffman, B. L., and Wold, W. S. M. (1989) Epidermal growth factor receptor is down-regulated by a 10,400 MW protein encoded by the E3 region of adenovirus. *Cell*, **57**, 135.

52. Kuivinen, E., Hoffman, B. L., Hoffman, P. A., and Carlin, C. R. (1993) Structurally related class I and class II receptor protein tyrosine kinases are down-regulated by the same E3 protein coded for by human group C adenoviruses. *J. Cell Biol.*, **120**, 1271.

53. Hermiston, T. W., Tripp, R. A., Sparer, T., Gooding, L. R., and Wold, W. S. M. (1993) Deletion mutation analysis of the adenovirus type 2 E3-gp19K protein: identification of sequences within the endoplasmic reticulum lumenal domain that are required for class I antigen binding and protection from adenovirus-specific cytotoxic T lymphocytes. *J. Virol.*, **67**, 5289.

54. Gooding, L. R., Ranheim, T. S., Tollefson, A. E., Aquino, L., Duerksen-Hughes, P., Horton, T. M., *et al.* (1991) The 10,400- and 14,500-dalton proteins encoded by region E3 of adenovirus function together to protect many but not all mouse cell lines against lysis by tumor necrosis factor. *J. Virol.*, **65**, 4114.

55. Marks, M. S., Ohno, H., Kirchhausen, T., and Bonifacino, J. S. (1997) Protein sorting by tyrosine-based signals: adapting to the Ys and wherefores. *Trends Cell Biol.*, **7**, 124.

56. Tollefson, A. E., Krajcsi, P., Yei, S. P., Carlin, C. R., and Wold, W. S. M. (1990) A 10,400-molecular-weight membrane protein is coded by region E3 of adenovirus. *J. Virol.*, **64**, 794.

57. Tollefson, A. E., Krajcsi, P., Pursley, M. H., Gooding, L. R., and Wold, W. S. M. (1990) A 14,500 MW protein is coded by region E3 of group C human adenoviruses. *Virology*, **175**, 19.

58. Tollefson, A. E. and Wold, W. S. M. (1988) Identification and gene mapping of a 14,700-molecular-weight protein encoded by region E3 of group C adenoviruses. *J. Virol.*, **62**, 33.

59. Li, Y., Kang, J., and Horwitz, M. S. (1998) Interaction of an adenovirus E3 14.7-kilodalton protein with a novel tumor necrosis factor alpha-inducible cellular protein containing leucine zipper domains. *Mol. Cell. Biol.*, **18**, 1601.

60. Gooding, L. R., Sofola, I. O., Tollefson, A. E., Duerksen-Hughes, P., and Wold, W. S. M. (1990) The adenovirus E3-14.7K protein is a general inhibitor of tumor necrosis factor-mediated cytolysis. *J. Immunol.*, **145**, 3080.

61. Horton, T. M., Ranheim, T. S., Aquino, L., Kusher, D. I., Saha, S. K., Ware, C. F., *et al.* (1991) Adenovirus E3 14.7K protein functions in the absence of other adenovirus proteins to protect transfected cells from tumor necrosis factor cytolysis. *J. Virol.*, **65**, 2629.

62. Gooding, L. R., Elmore, L. W., Tollefson, A. E., Brady, H. A., and Wold, W. S. M. (1988) A 14,700 MW protein from the E3 region of adenovirus inhibits cytolysis by tumor necrosis factor. *Cell*, **53**, 341.

63. Chen, P., Tian, J., Kovesdi, I., and Bruder, J. B. (1998) Interaction of the adenovirus 14.7K protein with FLICE inhibits Fas ligand-induced apoptosis. *J. Biol. Chem.*, **273**, 5815.

64. Li, Y., Kang, J., and Horwitz, M. S. (1997) Interaction of an adenovirus 14.7-kilodalton protein inhibitor of tumor necrosis factor alpha cytolysis with a new member of the GTPase superfamily of signal transducers. *J. Virol.*, **71**, 1576.

65. Zilli, DF., Voelkel-Johnson, C., Skinner, T., and Laster, S. M. (1992) The adenovirus E3 region 14.7 kDa protein, heat and sodium arsenite inhibit the TNF-induced release of arachidonic acid. *Biochem. Biophys. Res. Commun.*, **188**, 177.

66. Laster, S. M., Wold, W. S. M., and Gooding, L. R. (1994) Adenovirus proteins that regulate susceptibility to TNF also regulate the activity of PLA$_2$. *Semin. Virol.*, **5**, 431.

67. Voelkel-Johnson, C., Thorne, T. E., and Laster, S. M. (1996) Susceptibility to TNF in the presence of inhibitors of transcription or translation is dependent on the activity of cPLA$_2$ in human melanoma tumor cells. *J. Immunol.*, **156**, 201.

68. Hayakawa, M., Ishida, N., Takeuchi, K., Shibamoto, S., Hori, T., Oku, N., *et al.* (1993) Arachidonic acid-selective cytosolic phospholipase A$_2$ is crucial in the cytotoxic action of tumor necrosis factor. *J. Biol. Chem.*, **268**, 11290.

69. Thorne, T. E., Voelkel-Johnson, C., Case, W. M., Parks, L. W., and Laster, S. M. (1996) The activity of cytosolic phospholipase A$_2$ is required for the lysis of adenovirus-infected cells by tumor necrosis factor. *J. Virol.*, **70**, 8502.

70. Dimitrov, T., Krajcsi, P., Hermiston, T. W., Tollefson, A. E., Hannink, M., and Wold, W. S. M. (1997) Adenovirus E3-10.4K/14.5K protein complex inhibits tumor necrosis factor-induced translocation of cytosolic phospholipase A$_2$ to membranes. *J. Virol.*, **71**, 2830.

71. Wissing, D., Mouritzen, H., Egeblad, M., Poirier, G. G., and Jaattela, M. (1997) Involvement of caspase-dependent activation of cytosolic phospholipase A$_2$ in tumour necrosis factor-induced apoptosis. *Proc. Natl. Acad. Sci. USA*, **94**, 5073.

72. Voelkel-Johnson, C., Entingh, A. J., Wold, W. S. M., Gooding, L. R., and Laster, S. M. (1995) Activation of intracellular proteases is an early event in TNF-induced apoptosis. *J. Immunol.*, **154**, 1707.

73. Ginsberg, H. S., Lundholm-Beauchamp, U., Horswood, R. L., Pernis, B., Wold, W. S. M., Chanock, R. M., *et al.* (1989) Role of early region 3 (E3) in pathogenesis of adenovirus disease. *Proc. Natl. Acad. Sci. USA*, **86**, 3823.

74. Bruder, J. T., Jie, T., McVey, D. L., and Kovesdi, I. (1997) Expression of gp19K increases the persistence of transgene expression from an adenovirus vector in the mouse lung and liver. *J. Virol.*, **71**, 7623.

75. Harrod, K. S., Hermiston, T. W., Trapnell, B. C., Wold, W. S. M., and Whitsett, J. A. (1998) Lung-specific expression of E3-14.7K in transgenic mice attenuates adenoviral vector-mediated lung inflammation and prolongs transgene expression. *Hum. Gene Ther.*, **9**, 1885.

76. Tufariello, J., Cho, S., and Horwitz, M. S. (1994) The adenovirus E3 14.7-kilodalton protein which inhibits cytolysis by tumor necrosis factor increases the virulence of vaccinia virus in a murine pneumonia model. *J. Virol.*, **68**, 453.

77. Tufariello, J. M., Cho, S., and Horwitz, M. S. (1994) Adenovirus E3 14.7-kilodalton protein, an antagonist of tumor necrosis factor cytolysis, increases the virulence of vaccinia virus in severe combined immunodeficient mice. *Proc. Natl. Acad. Sci. USA*, **91**, 10987.

78. Williams, J. L., Garcia, J., Harrich, D., Pearson, L., Wu, F., and Gaynor, R. (1990) Lymphoid specific gene expression of the adenovirus early region 3 promoter is mediated by NF-kappa B binding motifs. *EMBO J.*, **9**, 4435.

79. Deryckere, F. and Burgert, H.-G. (1996) Tumor necrosis factor alpha induces the adenovirus early 3 promoter by activation of NF-κB. *J. Biol. Chem.*, **271**, 30249.

80. Takemori, N., Cladaras, C., Bhat, B., Conley, A. J., and Wold, W. S. M. (1984) cyt gene of adenoviruses 2 and 5 is an oncogene for transforming function in early region E1B and encodes the E1B 19,000-molecular-weight polypeptide. *J. Virol.*, **52**, 793.

81. Chinnadurai, G. (1983) Adenovirus 2 Ip+ locus codes for a 19 kd tumor antigen that plays an essential role in cell transformation. *Cell*, **33**, 759.

82. Subramanian, T., Kuppuswamy, M., Mak, S., and Chinnadurai, G. (1984) Adenovirus cyt+ locus, which controls cell transformation and tumorigenicity, is an allele of lp+ locus, which codes for a 19-kilodalton tumor antigen. *J. Virol.*, **52**, 336.

83. Subramanian, T., Kuppuswamy, M., Gysbers, J., Mak, S., and Chinnadurai, G. (1984) 19-kDa tumor antigen coded by early region E1b of adenovirus 2 is required for efficient synthesis and for protection of viral DNA. *J. Biol. Chem.*, **259**, 11777.

84. Pilder, S., Logan, J., and Shenk, T. (1984) Deletion of the gene encoding the adenovirus 5 early region 1b 21,000-molecular-weight polypeptide leads to degradation of viral and host cell DNA. *J. Virol.*, **52**, 664.

85. White, E., Grodzicker, T., and Stillman, B. W. (1984) Mutations in the gene encoding the adenovirus early region 19,000-molecular-weight tumor antigen cause the degradation of chromosomal DNA. *J. Virol.*, **52**, 410.

86. Wyllie, A. H. (1980) Glucocorticoid-induced thymocyte apoptosis is associated with endogenous endonuclease activation. *Nature*, **284**, 555.

87. White, E. and Stillman, B. (1987) Expression of adenovirus E1B mutant phenotypes is dependent on the host cell and on synthesis of E1A proteins. *J. Virol.*, **61**, 426.

88. Rao, L., Debbas, M., Sabbatini, P., Hockenbery, D., Korsmeyer, S., and White, E. (1992) The adenovirus E1A proteins induce apoptosis, which is inhibited by the E1B 19-kDa and Bcl-2 proteins. *Proc. Natl. Acad. Sci. USA*, **89**, 7742.

89. Lowe, S. W. and Ruley, H. E. (1993) Stabilization of the p53 tumor suppressor is induced by adenovirus 5 E1A and accompanies apoptosis. *Genes Dev.*, **7**, 535.

90. Subramanian, T., Tarodi, B., and Chinnadurai, G. (1995) p53-independent apoptotic and necrotic cell deaths induced by adenovirus infection: suppression by E1B 19K and Bcl-2 proteins. *Cell Growth Differentiation*, **6**, 131.

91. Teodoro, J. G., Shore, G. C., and Branton, P. E. (1995) Adenovirus E1A proteins induce apoptosis by both p53-dependent and p53-independent mechanisms. *Oncogene*, **11**, 467.

92. Chiou, S. K. and White, E. (1997) p300 binding by E1A cosegregates with p53 induction but is dispensable for apoptosis. *J. Virol.*, **71**, 3515.

93. Mymryk, J. S., Shire, K., and Bayley, S. T. (1994) Induction of apoptosis by adenovirus type 5 E1A in rat cells requires a proliferation block. *Oncogene*, **9**, 1187.

94. Moran, E. (1994) Mammalian cell growth controls reflected through protein interactions with the adenovirus E1A gene products. *Semin. Virol.*, **5**, 327.

95. Shisler, J., Duerksen-Hughes, P., Hermiston, T. W., Wold, W. S. M., and Gooding, L. R. (1996) Induction of susceptibility to TNF by adenovirus E1A is dependent upon binding to either p300 or p105-Rb and induction of DNA synthesis. *J. Virol.*, **70**, 68.

96. Cook, J. L., Krantz, C. K., and Routes, B. A. (1996) Role of p300-family proteins in E1A oncogene induction of cytolytic susceptibility and tumor cell rejection. *Proc. Natl. Acad. Sci. USA*, **93**, 13985.

97. Weintraub, S. J., Prater, C. A., and Dean, D. C. (1992) Retinoblastoma protein switches the E2F site from positive to negative element. *Nature*, **358**, 259.

98. Nevins, J. R. (1995) Adenovirus E1A: transcription regulation and alteration of cell growth control. *Curr. Top. Microbiol. Immunol.*, **199**, 25.

99. DePinho, R. A. (1998) Transcriptional repression. The cancer-chromatin connection. *Nature*, **391**, 533.

100. Kouzarides, T. (1998) Transcriptional control by the retinoblastoma protein. *Semin. Cancer Biol.*, **6**, 91.

101. Brehm, A., Miska, E. A., McCance, D. J., Reid, J. L., Bannister, A. J., and Kouzarides, T. (1998) Retinoblastoma protein recruits histone deacetylase to repress transcription. *Nature*, **391**, 597.

102. Luo, R. X., Postigo, A. A., and Dean, D. C. (1998) RB interacts with histone deacetylase to repress transcription. *Cell*, **92**, 463.

103. Kowalik, T. F., DeGregori, J., Schwarz, J. K., and Nevins, J. R. (1995) E2F1 overexpression in quiescent fibroblasts leads to induction of cellular DNA synthesis and apoptosis. *J. Virol.*, **69**, 2491.

104. Shan, B. and Lee, W. H. (1994) Deregulated expression of E2F-1 induces S-phase entry and leads to apoptosis. *Mol. Cell. Biol.*, **14**, 8166.

105. Agah, R., Kirshenbaum, L. A., Abdellatif, M., Truong, L. D., Chakraborty, S., and Schneider, M. L. H. (1997) Adenoviral delivery of E2F-1 directs cell cycle reentry and p53-independent apoptosis in postmitotic adult myocardium *in vivo*. *J. Clin. Invest.*, **100**, 2722.

106. Chakravarti, D., Ogryzko, V., Kao, H.-Y., Nash, A., Chen, H., Nakatani, Y., and Evans, R. M. (1999) A viral mechanism for inhibition of p300 and PCAF acetyltransferase activity. *Cell*, **96**, 393.

107. Prives, C. (1998) Signaling to p53: Breaking the MDM2-p53 circuit. *Cell*, **95**, 5.

108. Grossman, S. R., Perez, M., Kung, A. L., Joseph, M., Mansur, C., Xiao, Z., *et al.* (1998) p300/MDM2 complexes participate in MDM2-mediated p53 degradation. *Mol. Cell*, **2**, 405.

109. Bates, S., Phillips, A. C., Clark, P. A., Stott, F., Peters, G., Ludwig, R. L., *et al.* (1998) p14^ARF links the tumor suppressors RB and p53. *Nature*, **395**, 124.

110. de Stanchina, E., McCurrach, M. E., Zindy, F., Shieh, S. Y., Ferbeyre, G., Samuelson, A. V., *et al.* (1998) E1A signaling to p53 involves the p19(ARF) tumor suppressor. *Genes Dev.*, **12**(15), 2434.

111. Fearnhead, H. O., McCurrach, M. E., O'Neill, J., Zhang, K., Lowe, S. W., and Lazebnik, Y. A. (1997) Oncogene-dependent apoptosis in extracts from drug-resistant cells. *Genes Dev.*, **11**, 1266.

112. Fearnhead, H. O., Rodriguez, J., Govek, E. E., Guo, W. J., Kobayashi, R., Hannon, G., *et al.* (1999) Oncogene-dependent apoptosis is mediated by caspase-9. *Proc. Natl. Acad. Sci. USA*, **95**, 13664.

113. Zou, H., Henzel, W. J., Liu, X., Lutschg, A., and Wang, X. (1997) Apaf-1, a human protein homologous to C. elegans CED-4, participates in cytochrome c-dependent activation of caspase-3. *Cell*, **90**, 405.

114. Li, P., Nijhawan, D., Budihardjo, I., Srinivasula, S. M., Amad, M., Alnemri, E. S., *et al.* (1997) Cytochrome *c* and dATP-dependent formation of Apaf-1/Caspase-9 complex initiates an apoptotic protease cascade. *Cell*, **91**, 479.

115. Ng, F. W. H., Nguyen, M., Kwan, T., Branton, P. E., Nicholson, D. W., Cromlish, J. A., *et al.* (1997) p28 BAP31, a BCL-2/BCL-XL- and Procaspase-8-associated protein in the endoplasmic reticulum. *J. Cell Biol.*, **139**, 327.

116. Bernards, R., Schrier, P. I., Bos, J. L., and Van der Eb, A. J. (1983) Role of adenovirus types 5 and 12 early region 1b tumor antigens in oncogenic transformation. *Virology*, **127**, 45.

117. Moore, M., Horikoshi, N., and Shenk, T. (1996) Oncogenic potential of the adenovirus E4orf6 protein. *Proc. Natl. Acad. Sci. USA*, **93**, 11295.

118. White, E. and Cipriani, R. (1990) Role of adenovirus E1B proteins in transformation: altered organization of intermediate filaments in transformed cells that express the 19-kilodalton protein. *Mol. Cell. Biol.*, **10**, 120.

119. McLorie, W., McGlade, C. J., Takayesu, D., and Branton, P. E. (1991) Individual adenovirus E1B proteins induce transformation independently but by additive pathways. *J. Gen. Virol.*, **72**, 1467.

120. Gooding, L. R., Aquino, L., Duerksen-Hughes, P. J., Day, D., Horton, T. M., Yei, S. P., *et al.* (1991) The E1B 19,000-molecular-weight protein of group C adenoviruses prevents tumor necrosis factor cytolysis of human cells but not of mouse cells. *J. Virol.*, **65**, 3083.

121. White, E., Sabbatini, P., Debbas, M., Wold, W. S. M., Kusher, D. I., and Gooding, L. R. (1992) The 19-kilodalton adenovirus E1B transforming protein inhibits programmed cell death and prevents cytolysis by tumor necrosis factor α. *Mol. Cell. Biol.*, **12**, 2570.

122. Hashimoto, S., Ishii, A., and Yonehara, S. (1991) The E1b oncogene of adenovirus confers cellular resistance to cytotoxicity of tumor necrosis factor and monoclonal anti-Fas antibody. *Int. Immunol.*, **3**, 343.

123. Tarodi, B., Subramanian, T., and Chinnadurai, G. (1993) Functional similarity between adenovirus E1B 19K gene and Bcl2 oncogene: mutant complementation and suppression of cell death induced by DNA damaging agents. *Int. J. Oncol.*, **3**, 467.

124. Subramanian, T., Tarodi, B., Govindarajan, R., Boyd, J. M., Yoshida, K., and Chinnadurai, G. (1993) Mutational analysis of the transforming and apoptosis suppression activities of the adenovirus E1B 175R protein. *Gene*, **124**, 173.

125. White, E., Cipriani, R., Sabbatini, P., and Denton, A. (1991) Adenovirus E1B 19-kilodalton protein overcomes the cytotoxicity of E1A proteins. *J. Virol.*, **65**, 2968.

126. Chiou, S. K., Tseng, C. C., Rao, L., and White, E. (1994) Functional complementation of the adenovirus E1B 19-kilodalton protein with Bcl-2 in the inhibition of apoptosis in infected cells. *J. Virol.*, **68**, 6553.

127. Boyd, J. M., Malstrom, S., Subramanian, T., Venkatesh, L. K., Schaeper, U., Elangovan, B., *et al.* (1994) Adenovirus E1B 19 kDa and Bcl-2 proteins interact with a common set of cellular proteins. *Cell*, **79**, 341.

128. Subramanian, T., Boyd, J. M., and Chinnadurai, G. (1995) Functional substitution identifies a cell survival promoting domain common to adenovirus E1B 19 kDa and Bcl-2 proteins. *Oncogene*, **11**, 2403.

129. Boyd, J. M., Gallo, G. J., Elangovan, B., Houghton, A. B., Malstrom, S., Avery, B. J., *et al.* (1995) Bik, a novel death-inducing protein shares a distinct sequence motif with Bcl-2 family proteins and interacts with viral and cellular survival-promoting proteins. *Oncogene*, **11**, 1921.

130. Farrow, S. N., White, J. H., Martinou, I., Raven, T., Pun, K. T., Grinham, C. J., *et al.* (1995) Cloning of a bcl-2 homologue by interaction with adenovirus E1B 19K. *Nature*, **374**, 731.

131. Han, J., Sabbatini, P., Perez, D., Rao, L., Modha, D., and White, E. (1996) The E1B 19K protein blocks apoptosis by interacting with and inhibiting the p53-inducible and death-promoting Bax protein. *Genes Dev.*, **10**, 461.

132. Han, J., Sabbatini, P., and White, E. (1996) Induction of apoptosis by human Nbk/Bik, a BH3-containing protein that interacts with E1B 19K. *Mol. Cell. Biol.*, **16**, 5857.

133. Boulakia, C. A., Chen, G., Ng, F. W., Teodoro, J. G., Branton, P. E., Nicholson, D. W., *et al.* (1996) Bcl-2 and adenovirus E1B 19 kDa protein prevent E1A-induced processing of CPP32 and cleavage of poly (ADP-ribose) polymerase. *Oncogene*, **12**, 529.

134. Armstrong, R. C., Aja, T., Xiang, J., Guar, S., Krebs, J. F., Hoang, K., *et al.* (1996) Fas-induced activation of the cell death-related protease CPP32 is inhibited by Bcl-2 and by ICE family protease inhibitors. *J. Biol. Chem.*, **271**, 16850.

135. Chinnaiyan, A. M., Orth, K., O'Rourke, K., Duan, H., Poirier, G. G., and Dixit, V. M. (1996) Molecular ordering of the cell death pathway: Bcl-2 and Bcl-XL function upstream of the CED-3-like apoptotic proteases. *J. Biol. Chem.*, **271**, 4573.

136. Muchmore, S. W., Sattler, M., Liang, H., Meadows, R. P., Harlan, J. E., Yoon, H. S., *et al.* (1996) X-ray and NMR structure of human Bcl-XL, an inhibitor of programmed cell death. *Nature*, **381**, 335.

137. Adams, J. M. and Cory, S. (1998) The Bcl-2 protein family: arbiters of cell survival. *Science*, **281**, 1322.

138. Pan, G., O'Rourke, K., and Dixit, V. M. (1998) Caspase-9, Bcl-XL, and Apaf-1 form a ternary complex. *J. Biol. Chem.*, **273**, 5841.

139. Hu, Y., Benedict, M. A., Wu, D., Inohara, N., and Nunez, G. (1998) Bcl-XL interacts with Apaf-1 and inhibits Apaf-1 dependent caspase-9 activation. *Proc. Natl. Acad. Sci. USA*, **95**, 4386.

140. Pan, G., O'Rourke, K., Chinnaiyan, A. M., Gentz, R., Ebner, R., Ni, J., et al. (1997) The receptor for the cytotoxic ligand TRAIL. *Science*, **276**, 111.

141. Conradt, B. and Horvitz, H. R. (1998) The *C. elegans* protein EGL-1 is required for programmed cell death and interacts with the Bcl-2-like protein CED-9. *Cell*, **93**, 519.

142. Chinnaiyan, A. M., O'Rourke, K., Lane, B. R., and Dixit, V. M. (1997) Interaction of CED-4 with CED-3 and CED-9: a molecular framework for cell death. *Science*, **275**, 1122.

143. Elangovan, B. and Chinnadurai, G. (1997) Functional dissection of the pro-apoptotic protein Bik. Heterodimerization with anti-apoptosis proteins is insufficient for induction of cell death. *J. Biol. Chem.*, **272**, 24494.

144. Luo, X., Budihardjo, I., Zou, H., Slaughter, C., and Wang, X. (1998) Bid, a Bcl2 interacting protein, mediates cytochrome *c* release from mitochondria in response to activation of cell surface death receptors. *Cell*, **94**, 481.

145. Li, H., Zhu, H., Chi-jie, X., and Yuan, J. (1998) Cleavage of BID by Caspase 9 mediates the mitochondrial damage in the Fas pathway of apoptosis. *Cell*, **94**, 491.

146. Perez, D. and White, E. (1998) E1B 19K inhibits Fas-mediated apoptosis through FADD-dependent sequestration of FLICE. *J. Cell. Biol.*, **141**, 1255.

147. Sarnow, P., Ho, Y. S., Williams, J., and Levine, A. J. (1982) Adenovirus E1b-58kd tumor antigen and SV40 large tumor antigen are physically associated with the same 54 kd cellular protein in transformed cells. *Cell*, **28**, 387.

148. Kao, C. C., Yew, P. R., and Berk, A. J. (1990) Domains required for in vitro association between the cellular p53 and the adenovirus 2 E1b 55K proteins. *Virology*, **179**, 806.

149. Yew, P. R. and Berk, A. J. (1992) Inhibition of p53 transactivation required for transformation by adenovirus early 1B protein. *Nature*, **357**, 82.

150. Bischoff, J. R., Kirn, D. H., Williams, A., Heise, C., Horn, S., Muna, M., et al. (1996) An adenovirus mutant that replicates selectively in p53-deficient human tumor cells. *Science*, **274**, 373.

151. Yew, P. R., Liu, X., and Berk, A. J. (1994) Adenovirus E1B oncoprotein tethers a transcriptional repression domain to p53. *Genes Dev.*, **8**, 190.

152. Martin, M. E. D. and Berk, A. J. (1998) Adenovirus E1B 55K represses p53 activation *in vitro*. *J. Virol.*, **72**, 3146.

153. Teodoro, J. G. and Branton, P. E. (1997) Regulation of p53-depenedent apoptosis, transcriptional repression, and cell transformation by phosphorylation of the 55-kilodalton E1B protein of human adenovirus type 5. *J. Virol.*, **71**, 3620.

154. Freyer, G. A., Katoh, Y., and Roberts, R. J. (1984) Characterization of the major mRNAs from adenovirus 2 early region 4 by cDNA cloning and sequencing. *Nucleic Acids Res.*, **12**, 3503.

155. Virtanen, A., Gilardi, P., Naslund, A., LeMoullec, J. M., Pettersson, U., and Perricaudet, M. (1984) mRNAs from human adenovirus 2 early region 4. *J. Virol.*, **81**, 822.

156. Halbert, D. N., Cutt, J. R., and Shenk, T. (1985) Adenovirus early region 4 encodes functions required for efficient DNA replication, late gene expression, and host cell shutoff. *J. Virol.*, **56**, 250.

157. Weinberg, D. H. and Ketner, G. (1983) A cell line that supports the growth of a defective early region 4 deletion mutant of human adenovirus type 2. *Proc. Natl. Acad. Sci. USA*, **80**, 5383.

158. Dobbelstein, M., Roth, J., Kimberly, W. T., Levine, A. J., and Shenk, T. (1997) Nuclear export of the E1b 55-kda and E4 34-kda adenoviral oncoproteins mediated by a rev-like signal sequence. *EMBO J.*, **16**, 4276.

159. Nordqvist, K., Ohman, K., and Akusjarvi, G. (1994) Human adenovirus encodes two proteins which have opposite effects on accumulation of alternatively spliced mRNAs. *Mol. Cell. Biol.*, **14**, 437.

160. Kanopka, A., Muhlemann, O., Petersen-Mahrt, S., Estmer, C., Ohrmalm, C., and Akusjarvi, G. (1998) Regulation of adenovirus alternative RNA splicing by dephosphorylation of SR proteins. *Nature*, 393, (6681), 185.

161. Nevels, M., Rubenwolf, S., Spruss, T., Wolf, H., and Dobner, T. (19970 The adenovirus E4orf6 protein can promote E1A/E1B-induced focus formation by interfering with p53 tumor suppressor function. *Prov. Natl. Acad. Sci. USA*, **94**, 1206.

162. Weiss, R. S., Lee, S. S., Prasad, B. V., and Javier, R. T. (1997) Human adenovirus early 4 region open reading frame 1 genes encode growth-transforming proteins that may be distantly related to dUTP pyrophosphatase enzymes. *J. Virol.*, **71**, 1857.

163. Javier, R. T. (1994) Adenovirus type 9 E4 open reading frame 1 encodes a transforming protein required for the production of mammary tumors in rats. *J. Virol.*, **68**, 3917.

164. Dobner, T., Horikoshi, N., Rubenwolf, S., and Shenk, T. (1996) Blockage by adenovirus E4orf6 of transcriptional activation by the p53 tumor suppressor. *Science*, **272**, 1470.

165. Steegenga, W. T., Riteco, N., Jochemsen, A. G., Fallaux, F. J., and Bos, J. L (1998) The large E1B protein together with the E4orf6 protein target p53 for active degradation in adenovirus infected cells. *Oncogene*, **16**, 349.

166. Lavoie, J. N., Nguyen, M., Marcellus, R. C., Branton, P. E., and Shore, G. C. (1998) E4orf4, a novel adenovirus death factor that induces p53-independent apoptosis by a pathway that is not inhibited by zVAD-fmk. *J. Cell Biol.*, **140**, 637.

167. Shtrichman, R. and Kleinberger, T. (1998) Adenovirus type 5 E4 open reading frame 4 protein induces apoptosis in transformed cells. *J. Virol.*, **72**, 2975.

168. Marcellus, R. C., Lavoie, J. N., Boivin, D., Shore, G. C., Ketner, G., and Branton, P. E. (1998) The early region 4 orf4 protein of human adenovirus type 5 induces p53-independent cell death by apoptosis. *J. Virol.*, **72**, 7144.

169. Kleinberger, T. and Shenk, T. (1993) Adenovirus E4orf4 protein binds to protein phosphatase 2A, and the complex down regulates E1A-enhanced junB transcription. *J. Virol.*, **67**, 7556.

170. Xia, Z., Dickens, M., Raingeaud, J., Davis, R. J., and Greenberg, M. E. (1995) Opposing effects of ERK and JNK-p38 MAP kinases on apoptosis. *Science*, 270, 1326.

171. Routes, J. M., Li, H., Bayley, S. T., and Ryan, S. K. D. J. (1996) Inhibition of IFN-stimulated gene expression and IFN induction of cytolytic resistance to natural killer cell lysis correlate with E1A-p300 binding. J. Immunol., **156**, 1055.

172. Tollefson, A. E., Ryerse, J. S., Scaria, A., Hermiston, T. W., and Wold, W. S. M. (1996) The E3- 11.6-kDa Adenovirus Death Protein (ADP) is required for efficient cell death: characerization of cells infected with and mutants. *Virology*, **220**, 152.

173. Tollefson, A. E., Scaria, A., Hermiston, T. W., Ryerse, J. S., Wold, L. J., and Wold, W. S. M. (1996) The adenovirus death protein (E3-11.6K) is required at very late stages of infection for efficient cell lysis and release of adenovirus from infected cells. *J. Virol.*, **70**, 2296.

174. Wold, W. S. M., Cladaras, C., Magie, S. C., and Yacoub, N. (1984) Mapping a new gene that encodes an 11,600-molecular-weight protein in the E3 transcription unit of adenovirus 2. *J. Virol.*, **52**, 307.

175. Scaria, A., Tollefson, A. E., Saha, S. K., and Wold, W. S. M. (1992) The E3-11.6K protein of adenovirus is an asn-glycosylated integral membrane protein that localizes to the nuclear membrane. *Virology*, **191**, 743.

176. Haussman, J., Ortmann, D., Witt, M., Veit, M., and Seidel, W. (1998) Adenovirus death protein, a transmembrane protein encoded in the E3 region, is palmitoylated at the cytoplasmic tail. *Virology*, **244**, 343.

177. Tollefson, A. E., Scaria, A., Saha, S. K., and Wold, W. S. M. (1992) The 11,600-M_W protein encoded by region E3 of adenovirus is expressed early but is greatly amplified at late stages of infection. *J. Virol.*, **66**, 3633.

Index